SOVIET MASER RESEARCH

MOLEKULYARNYE GENERATORY

МОЛЕКУЛЯРНЫЕ ГЕНЕРАТОРЫ

Transactions (Trudy) of the P.N. Lebedev Physics Institute, Volume XXI

SOVIET MASER RESEARCH

Edited by

Acad. D.V. Skobel'tsyn

Authorized translation from the Russian

CONSULTANTS BUREAU
NEW YORK
1964

ISBN 978-1-4684-0663-4 ISBN 978-1-4684-0661-0 (eBook)
DOI 10.1007/978-1-4684-0661-0

The original Russian text was published by the USSR Academy of Sciences Press in Moscow in 1963, as Volume XXI of the Transactions (Trudy) of the P. N. Lebedev Physics Institute.

Молекулярные генераторы

Труды физического института, том XXI

Library of Congress Catalog Card Number 64-16546

Softcover reprint of the hardcover 1st edition 1964

227 West 17th St., New York 11, N. Y.

CONTENTS

A THEORETICAL STUDY OF THE FREQUENCY STABILITY OF A MASER*

A. N. Oraevskii

INTRODUCTION

During the last decade, there have been great advances in the development of extremely stable molecular atomic standards of frequency (or of time). Such frequency standards can be used as absolute reference standards which do not require constant checking by astronomical observations. At present there have been developed two types of frequency standards: cesium standards, and frequency standards which use a maser with a beam of ammonia molecules [1-5].

The cesium frequency standard ("atomic clock") uses a line of the hyperfine structure of the cesium atom. This line is a discriminator with a large quality figure and is used to stabilize a quartz oscillator to a high degree of accuracy. If the quartz oscillator has a very high frequency stability it does not require continuous adjustment. In this case it is sufficient to check the frequency of the quartz oscillator from time to time with the spectrum line to introduce corrections for the secular deviation because of the aging of the quartz.

To obtain narrow spectral lines in atomic frequency standards, one uses the method of two spatially separated resonators [6]. The width of the spectral line obtained by this method with a beam of cesium atoms is about 100 cps at a frequency about 9000 Mc. The greatest progress in the perfecting of cesium frequency standards has been achieved by I. Essen. He has obtained an absolute frequency stability of the order of $\pm 1.5 \cdot 10^{-10}$ [1].**

In the molecular frequency standard ("molecular clock") one uses a maser operating on one of the inversion transitions of the ammonia molecule [2,5].

In the Physics Institute of the Academy of Sciences of the USSR a working prototype of a molecular clock based on a maser has been developed [2,3]. This apparatus employs a maser to secure frequency stabilization of a relatively poor quartz oscillator (quality of the quartz resonator $\sim 10^5$). The scheme of stabilization which has been developed guarantees an accuracy of the "locking" of the frequency of the quartz oscillator to that of the maser which is not worse than 10^{-10} [3].

The prospect of further increase of the frequency stability both in apparatus of the cesium-clock type and in masers is associated with the production of narrower spectral lines which still are sufficiently intense. In this connection Zacharias has considered the possibility of slowing down cesium atoms in the gravitational field of the earth [7]. This method, however, does not make it possible to obtain a beam of slow atoms of sufficient intensity.

Ramsey has considered a method for increasing the time of flight between two resonators by means of collisions between atoms of the working substance and a foreign body [8]. The collisions must be such that while they increase the time of flight of the atoms they produce practically no change of their internal state. A

* The literal translation of the term used in the Russian is "molecular generator." This will explain the frequent use in this paper of the word "generator" as a synonym or near-synonym for "maser."

** Very recently, communications have appeared announcing the achievement of frequency reproducibility of the order of $\pm 2 \cdot 10^{-11}$ by means of a cesium standard [66].

realization of this method through the development of a hydrogen maser in which the fundamental line width is of the order of 1 cps has been reported [9].

The development of very stable standards of frequency (or of time) which use an absorption line calls for the use of quartz oscillators with good frequency stability. At present there are quartz oscillators with a stability of 10^{-9} over a 24-hour period [10]. The placing of quartz oscillator circuits in liquid helium can greatly increase the stability of the quartz oscillators; in this case we can expect stabilities of their frequency of oscillation of 10^{-10} over a 24-hour period [11].

There are interesting possibilities in the use of systems excited by optical illumination to generate oscillations of high stability [12]. Recent papers [12, 13] have investigated the possibility of using for this purpose the hyperfine splitting of atoms of rubidium Rb^{87} in the ground state. To decrease the line width owing to collisions of the gas with the walls of the cavity, and also the Doppler broadening of the line, the gaseous rubidium is diluted with an inert gas at pressures up to 1 mm Hg. In such a system it has been possible to obtain a line width of ~100 cps. To secure self-excited operation it is indeed necessary to have a resonator quality of about 60,000, and this presents well-known technical difficulties.

The development of a frequency standard based on a maser is impossible without a detailed investigation of the way the frequency at which it oscillates depends on the various parameters of the apparatus and on the structure of the emission line of the molecule used to excite the oscillations of the maser.

Recently, a number of papers have appeared devoted to the study of the dependence of the frequency of a maser on various parameters [14-20]. The studies made in these papers, however, are far from exhausting all of the effects which influence the frequency of the oscillations of a maser. Furthermore, none of these papers has attempted to give any exact quantitative results for the behavior of the frequency of a maser as a function of such important characteristic quantities as the intensity of the molecular beam, the field strength in the sorting system, etc. The absence of any comparison between theoretically calculated values of the frequency of a maser with experimental results makes it impossible to answer the question as to whether all of the effects which most strongly influence the frequency of a maser have been taken into account in the construction of frequency standards.

The present paper is devoted to a theoretical treatment of a number of problems associated with the construction of a standard of frequency (or of time) based on a maser.

In the first section of this article we obtain equations of the maser which describe not only the monochromatic process, but an established process varying with the time in an arbitrary way. On the basis of these equations we consider the problem of the locking of the maser to an external force and study the monochromatic character of the oscillations of the maser; the effect of the waveguide system on the oscillations of the maser is taken into account.

In the second section we make an analysis of the effects which influence the frequency of the oscillations of a maser. On the basis of this analysis we give a calculation of the oscillation frequency of a maser as a function of various parameters. It is pointed out that because the frequency of the oscillations of a maser depends on their amplitude, there can be a shift of the frequency of oscillation of the maser when an external power penetrates the resonating cavity, in particular the power of an adjacent generator.

In the third section we analyze various methods for adjusting the frequency of a maser to that of a spectral line. We give a quantitative treatment of the possibility of tuning by modulation of the line width with an external magnetic field. A possibility is indicated for using two coupled resonators to tune the frequency of a maser to that of a spectral line.

In the fourth section we consider possibilities for the further increase of the frequency stability of masers. We propose a number of methods for obtaining "slow" molecules with an average speed much smaller than the thermal velocity at room temperature.

Section 1

ANALYSIS OF THE EQUATIONS OF A MASER

1. Derivation of the Equations of a Maser

The stationary self-oscillating process in a maser was first studied in [14, 21]. In these papers the conditions for self-excitation of a maser are obtained and the amplitude and frequency of the self-oscillations are calculated. The authors of [14, 21] also considered the question of the transmission band and the amplification coefficient of a molecular amplifier and showed that a molecular amplifier which uses a resonator is similar to a regenerative receiver, for which the transmission band decreases with increase of the amplification coefficient. In all of the calculations in [14, 21] it was assumed that the molecules have an exponential distribution of their times of flight τ through the resonator: $(1/\bar{\tau})e^{-\tau/\bar{\tau}}$. Subsequently the process of self-oscillation was also studied for the case of a beam of molecules with a fixed speed [15, 22]. In [23] the equations obtained in [21] were used to treat the effect of noise on the frequency of oscillation of a maser.

It must be pointed out that the processes studied in these papers are of a monochromatic or quasi-monochromatic nature, and the equations used in these papers are valid only for quasi-monochromatic processes.

In the present section we consider a number of questions which were not dealt with in the papers cited. Such problems as the calculation of the higher harmonics in a maser (calculation of the degree of monochromaticity of the oscillations of a maser) require a generalization of the equations obtained in [14, 15, 21, 22] to the case of oscillations of the maser which vary with the time in an arbitrary way. These paragraphs are devoted to the derivation of these equations.

We can construct the theory of a maser by regarding the active medium which excites the resonator as a medium with negative absorption, i.e., with a complex dielectric constant whose imaginary part is negative [21]. This sort of treatment, however, is valid only for monochromatic processes, since the concept of a dielectric constant applies only in this case. To derive the equations of a maser in a form not restricted to monochromatic oscillations it is most convenient to use the polarization of the active medium $N\mathbf{P}(\mathbf{x},t)$,* which the medium takes on under the action of electromagnetic radiation. The problem of the excitation of oscillations in a maser and of the amplification of a power supplied to a molecular (paramagnetic) amplifier reduces to the problem of the excitation of a resonator by an external current $\mathbf{J} = 4\pi N \frac{\partial \mathbf{P}}{\partial t}$. The polarization, in turn, depends on the field which excites the molecules in the resonator. Thus, the derivation of a closed system of equations for the maser consists of the solution of two problems: 1) the electrodynamic problem of the excitation of a resonator by an external current; 2) the quantum-mechanical problem of calculating the polarization $\mathbf{P}$ of the active medium under the action of an external electromagnetic field.

According to the theory of the excitation of resonators [24, 25], the variation with time of the field of a resonator belonging to a vibration of a particular type k is described by the equation

$$\frac{d^2\mathscr{E}_k}{dt^2} + \omega_k^2\mathscr{E}_k = -4\pi\frac{d^2}{dt^2}\int\limits_V N\,\mathbf{P}(\mathbf{x},t)\,\mathbf{E}_k(\mathbf{x})\,dv - \omega_k\oint \mathbf{S}_k\,d\mathbf{s}, \tag{1.1}$$

where $\mathscr{E}_k$ is the coefficient in the expansion of the field in terms of the eigenfunctions $\mathbf{E}_k(\mathbf{x})$ of the resonator, so that

$$\mathscr{E}(\mathbf{x},t) = \sum_k \mathscr{E}_k(t)\,\mathbf{E}_k(\mathbf{x}).$$

The integral $\oint \mathbf{S}_k\,ds$, taken over the surface of the resonator, represents the flux of power directed into the wall of the resonator and allows for the loss to the resonator wall for the given type of vibrations. This integral can be expressed in terms of the equivalent quality figure Q_k of the resonator [24, 25]:

*$\mathbf{P}(\mathbf{x},t)$ is the polarization of an individual molecule; N is the density of molecules in the active medium.

$$\omega_k \int S_k\, ds = \frac{\omega_k}{Q_k} \cdot \frac{d\mathcal{E}_k}{dt}. \tag{1.2}$$

Substituting Eq. (1.2) in Eq. (1.1), we obtain the final equation for the oscillations of a given type in a resonator under the influence of a polarization:

$$\frac{d^2\mathcal{E}_k}{dt^2} + \frac{\omega_k}{Q_k}\frac{d\mathcal{E}_k}{dt} + \omega_k^2\mathcal{E}_k = -4\pi N \frac{d^2}{dt^2}\int_V \mathbf{P}\mathbf{E}_k\, dv. \tag{1.3}$$

For a complete description of a maser it is necessary to have the equation which determines the dependence of $\mathbf{P}(\mathbf{x},t)$ on the field excited in the resonator. It can be seen from Eq. (1.3) that the oscillation of type k in the resonator is affected by the k-th "harmonic" of the polarization. Because of the nonlinear dependence of the polarization on the external field, the k-th harmonic of the polarization will depend not only on $\mathbf{E}_k$ but also on other types of oscillations. Since, however, the nonlinearity of the maser is small ($\sim 1/Q_k$), and also there is only a small excitation of those types of oscillation whose frequency differs decidedly from that of the molecular transition, the dependence of $\mathbf{P}(\mathbf{x},t)$ on other types of oscillations will be very slight. Therefore, in the rest of this section we shall not include this dependence. Therefore, according to Eq. (1.3), we have reduced the excitation of the resonator by the active medium to an analysis of the oscillations in an oscillating circuit with lumped constants.

To derive the equations which determine the coordinate and time dependences of the polarization $\mathbf{P}(\mathbf{x}, t)$ under the action of the field, we shall consider a quantum-mechanical system consisting of just two energy levels W_a and W_b. This assumption is entirely legitimate, since, for the ideal operation of a maser, it is necessary that the two working energy levels be sufficiently remote from the other energy levels so that these other levels will not have any effect on the frequency and shape of the emission line.

The stationary states W_a and W_b of the quantum-mechanical system we are considering are described by the wave functions $\psi_a e^{-iW_a/\hbar \cdot t}$ and $\psi_b e^{-iW_b/\hbar \cdot t}$, which are solutions of the equations

$$\hat{\mathcal{H}}_0\psi_a = W_a\psi_a, \quad \hat{\mathcal{H}}_0\psi_b = W_b\psi_b, \tag{1.4}$$

where $\mathcal{H}_0$ is the Hamiltonian of the noninteracting system.

When radiation $\mathcal{E}(\mathbf{x}, t)$ acts on the system it will no longer be in a stationary state, but will be described by a Ψ-function which depends on the time. It is natural to look for this Ψ-function in the form [26]

$$\Psi(\mathbf{x}, t) = a(\mathbf{x}, t)\psi_a + b(\mathbf{x}, t)\psi_b. \tag{1.5}$$

The Hamiltonian of a system which interacts with radiation with a dipole interaction is of the form

$$\hat{\mathcal{H}} = \hat{\mathcal{H}}_0 - \mu\mathcal{E}(\mathbf{x}, t), \tag{1.6}$$

so that $\Psi(\mathbf{x}, t)$ satisfies the equation

$$i\hbar\frac{\partial\psi}{\partial t} = [\hat{\mathcal{H}}_0 - \mu\mathcal{E}(\mathbf{x}, t)]\psi. \tag{1.7}$$

Using the fact that the functions ψ_a and ψ_b for the stationary states satisfy (1.4), we get the following equations to determine the coefficients $a(\mathbf{x}, t)$ and $b(\mathbf{x}, t)$:

$$\left.\begin{aligned} \frac{da}{dt} &= \frac{i}{\hbar}[-W_a a + \mu_{ab}\mathcal{E}b],\\ \frac{db}{dt} &= \frac{i}{\hbar}[-W_b b + \mu_{ab}^*\mathcal{E}a]; \end{aligned}\right\} \tag{1.8}$$

μ_{ab} is the matrix element of the dipole moment, which is defined by the equation

$$\mu_{ab} = \overset{*}{\mu}_{ba} = \int \psi_a \mu \psi_b dq. \tag{1.9}$$

The integration in Eq. (1.9) is taken over the configuration space of the molecule. It is also assumed that $\mu_{aa} = \mu_{bb} = 0$, which will always be true if the energy levels are not degenerate.

In the equations (1.8) the time derivatives are total derivatives, so that when the molecules are moving with speed v along the axis of the resonator we have

$$\frac{d}{dt} = \frac{\partial}{\partial t} + v\frac{\partial}{\partial z}.$$

To simplify our further equations, we shall hereafter suppose that $\mathcal{E}(\mathbf{x},t)$ does not depend on the coordinates. This is justified from a general physical point of view, since the nature of the processes in a maser as a self-oscillating system does not depend on the concrete configuration of the field.* This is also actually justified to some extent by the facts, since one usually employs in a maser an E_{010} wave in a cylindrical resonator, which does not depend on the coordinate z along the axis of the resonator, and the beam of molecules passes through near the axis of the resonator, where the dependence of the field on the transverse coordinate r is weak.

In this case, a, b, and **P** will not have any explicit dependence on the coordinate, so that the coordinate dependence is completely transferred to the dependence on the time the molecule has been in the resonator, $\tau = t - t_0$, where t_0 is the time at which the molecule enters the resonator. In fact, $z = v(t - t_0)$, so that

$$v\frac{\partial}{\partial z} = \frac{\partial}{\partial \tau},$$

and the right members of (1.8) will not have any explicit dependence on the coordinates. The polarization $\mathbf{P}(\tau, t)$ is determined by the relation

$$\mathbf{P}(\tau, t) = \int \Psi^* \mu \Psi \, dq. \tag{1.10}$$

Substituting for Ψ its expression in terms of ψ_a and ψ_b, we obtain an expression for $\mathbf{P}(\tau, t)$ in terms of the coefficients a and b:

$$\mathbf{P}(\tau, t) = \mu_{ab}\, ab^* + \mu_{ba}\, a^* b. \tag{1.11}$$

Using (1.8), we can show that $\mathbf{P}(\tau, t)$ satisfies the following equations:

$$\left.\begin{aligned} \frac{d^2\mathbf{P}}{dt^2} + \omega_{ab}^2 \mathbf{P} &= -2\,\omega_{ab}\cdot\mu_{ab}\frac{\mu_{ab}\mathcal{E}}{\hbar}\widetilde{R}, \\ \frac{d\widetilde{R}}{dt} &= \frac{2}{\hbar\omega_{ab}}\cdot\mathcal{E}\cdot\frac{d\mathbf{P}}{dt}, \end{aligned}\right\} \tag{1.12}$$

where

$$\left.\begin{aligned} \widetilde{R} &= |a|^2 - |b|^2, \\ \omega_{ab} &= \frac{1}{\hbar}(W_a - W_b). \end{aligned}\right\} \tag{1.13}$$

and

In (1.12) we have obtained the equations which describe the variation with time of the polarization $\mathbf{P}(\tau, t)$ of our system. These are not the final equations, however, since they do not take into account the following extremely important fact.

It is obvious that any particular particle cannot be for an arbitrarily long time a source of energy to sustain the oscillations in a maser or molecular amplifier. Some process is necessary which will restore (continuously or

*It must be noted here that if the length of the resonator is an even number of half wavelengths, it is very difficult to satisfy the condition for self-excitation of this type of oscillation.

periodically) some initial state in which the system existed before the operation started. In a maser which uses a molecular beam, this process is the removal from the resonator of the "spent" molecules and the entrance of new molecules which are in the upper energy level. For sealed masers [27] and paramagnetic amplifiers [28] this is a relaxation process. The result is that the complete ensemble of particles which takes part in the process of amplification or oscillation at a given time t consists of particles which have various times τ of interaction with the radiation field. To obtain the total polarization of the ensemble we must average the polarization of individual particles $\mathbf{P}(\tau, t)$ over the time τ. Since $\mathcal{E}$ does not depend on the time τ we can perform this averaging without solving (1.12). For this purpose we separate out from the total time derivative in (1.12) the derivative with respect to the time τ,

$$\frac{d}{dt} = \frac{\partial}{\partial \tau} + \frac{\partial}{\partial t}.$$

When we make this substitution (1.12) takes the form

$$\left.\begin{aligned} \left[\left(\frac{\partial}{\partial t} + \frac{\partial}{\partial \tau}\right)^2 + \omega_{ab}^2\right]\mathbf{P} &= -2\omega_{ab}\cdot\mu_{ab}\frac{\mu_{ba}\mathcal{E}}{\hbar}\tilde{R}, \\ \left(\frac{\partial}{\partial t} + \frac{\partial}{\partial \tau}\right)\tilde{R} &= 2\frac{1}{\hbar\omega_{ab}}\mathcal{E}\left(\frac{\partial}{\partial t} + \frac{\partial}{\partial \tau}\right)\mathbf{P}. \end{aligned}\right\} \qquad (1.14)$$

Multiplying all of the terms of (1.14) by $f(\tau)$ – the distribution function in the times τ – and integrating, we obtain equations which describe the time dependence of the average macroscopic polarization. If we assume that

$$f(\tau) = \frac{1}{\bar{\tau}}e^{-\tau/\bar{\tau}}, \qquad (1.15)$$

where $\bar{\tau}$ is the average time that a molecule spends in the resonator (or the relaxation time), the equations for the average values

$$\left.\begin{aligned} \mathbf{P}(t) &= \int_0^t \mathbf{P}(\tau, t)\cdot f(\tau)\,d\tau, \\ R(t) &= \int_0^t \tilde{R}(\tau, t)\cdot f(\tau)\,d\tau - R_0 \end{aligned}\right\} \qquad (1.16)$$

take the form

$$\left.\begin{aligned} \frac{d^2\mathbf{P}}{dt^2} + \frac{2}{\bar{\tau}}\frac{d\mathbf{P}}{dt} + \omega_{ab}^2\left(1 + \frac{1}{\omega_{ab}^2\bar{\tau}^2}\right)\mathbf{P} &= -2\omega_{ab}\cdot\mu_{ab}\frac{\mu_{ba}\mathcal{E}}{\hbar}(R + R_0), \\ \frac{dR}{dt} + \frac{1}{\bar{\tau}}R &= 2\frac{1}{\hbar\omega_{ab}}\mathcal{E}\left(\frac{dP}{dt} + \frac{1}{\bar{\tau}}\mathbf{P}\right). \end{aligned}\right\} \qquad (1.17)$$

In these equations,

$$R_0 = |a(\tau, t)|^2 - |b(\tau, t)|^2 \quad \text{for} \quad \tau = 0.$$

The quantity R_0 multiplied by the density N of molecules in the beam is the density of active molecules at the time that they enter the resonator.

Equations (1.17) together with Eq. (1.3) form a closed system of equations which describe oscillatory processes in the maser:*

---

*A similar system of equations has been obtained by a somewhat different method in [29]. For simplicity we shall hereafter assume $\mathbf{P} \parallel \mathbf{E}$, which is strictly true for an isotropic medium; $P = \int \mathbf{P}\mathbf{E}\,dv$, and the index k for the mode $\mathbf{E}_k$ is dropped, since we are considering only one type of oscillation.

$$\begin{aligned}
&\frac{d^2\mathcal{E}}{dt^2}+\frac{\omega_0}{Q_0}\cdot\frac{d\mathcal{E}}{dt}+\omega_0^2\mathcal{E}=-4\pi N\frac{d^2P}{dt^2},\\
&\frac{d^2P}{dt^2}+\frac{\omega_l}{Q_l}\cdot\frac{dP}{dt}+\omega_l^2P=-2\omega_{ab}\cdot\frac{|\mu_{ab}|^2}{\hbar}\mathcal{E}(R+R_0),\\
&\frac{dR}{dt}+\frac{1}{2}\frac{\omega_l}{Q_l}R=2\frac{1}{\hbar\omega_{ab}}\cdot\mathcal{E}\left(\frac{dP}{dt}+\frac{1}{2}\cdot\frac{\omega_l}{Q_l}P\right).
\end{aligned} \qquad (1.18)$$

In these equations Q_0 is the quality of the resonator, $Q_l = \frac{1}{2}\omega_l\bar{\tau}$,

$$\omega_l^2=\omega_{ab}^2\left(1+\frac{1}{\omega_{ab}^2\bar{\tau}^2}\right). \qquad (1.19)$$

The choice of a concrete form of the distribution of the molecules over the times τ imposes certain restrictions on (1.18). These equations cannot be applied in cases in which details of the initial shape of the emission line, which depends on the distribution of the molecules in times of flight $f(\tau)$, are important. For example, these equations do not describe the dependence of the frequency of oscillation of a maser on the intensity of the molecular beam, which dependence has its source in the difference between the true form of the function $f(\tau)$ and the exponential form (see Section 2). Finally, within the framework of our present idealization we cannot calculate effects which arise from the fact that a cavity resonator is a system with distributed parameters.

Equations (1.18) show that a maser is a system with "two and one-half" degrees of freedom. It can be seen from the equations for the quantity R that the nonlinearity of the maser is of an inertial nature. Apart from the higher harmonics, which are extremely small,

$$R=4\frac{Q_l}{\hbar\omega_l^2}\widetilde{\mathcal{E}P}, \qquad (1.20)$$

where the tilde denotes averaging over a time interval $\bar{\tau}$. For all processes whose frequency is larger than $1/\bar{\tau}$, R is constant up to terms of order Q_l^{-2}.

When we replace the third equation in (1.18) by the relation (1.20), we obtain the system of two equations

$$\begin{aligned}
&\frac{d^2\mathcal{E}}{dt^2}+\frac{\omega_0}{Q_0}\cdot\frac{\mathcal{E}d}{dt}+\omega_0^2\mathcal{E}=-4\pi N\frac{d^2P}{dt^2},\\
&\frac{d^2P}{dt^2}+\frac{\omega_l}{Q_l}\cdot\frac{dP}{dt}+\omega_l^2P=-2\frac{|\mu_{ab}|^2}{\hbar}\omega_{ab}\left(R_0+4\frac{Q_l}{\hbar\omega_{ab}^2}\widetilde{\mathcal{E}P}\right)\mathcal{E}.
\end{aligned} \qquad (1.21)$$

As is well known, a generator with an ideal inertial nonlinear element gives purely sinusoidal oscillations. Therefore, there will be practically no higher harmonics in a maser, as we shall hereafter confirm by a more exact calculation. The locking of a maser with an external force can occur only at a frequency close to that of the self-oscillation: there is no resonance of n-th type in a maser.

All of these assertions are correct without reservation as long as the condition

$$Q_l \gg 1$$

holds, as it practically always does with a great deal to spare.*

For convenience in our further work, we shall write (1.18) and (1.21) in dimensionless variables. We introduce the notation

---

*Strictly speaking, Eq. (1.21) is valid for cases in which the field (and the polarization) is of the form $A(t)\cos[\omega t+\varphi(t)]$, with

$$\frac{A'(t)}{A(t)}\sim\frac{\varphi'(t)}{\varphi(t)}\ll\frac{1}{\bar{\tau}}. \qquad (1.21a)$$

$$\frac{|\mu_{ab}|\,\tau}{\hbar}\,\mathscr{E} = x, \quad \frac{P}{\mu_{ab}} = v, \quad R = w,$$

$$\frac{\omega_0^2}{\omega^2} = 1 + 2\delta_1, \quad \frac{\omega_l^2}{\omega^2} = 1 + 2\delta_2, \quad \frac{\omega_0}{\omega}\frac{1}{Q_0} = 2h_1, \tag{1.22}$$

$$\frac{\omega_l}{\omega}\cdot\frac{1}{Q_l} = 2h_2, \quad 2k_1 = 4\pi N\,|\mu_{ab}|\,\hbar^{-1}\tau, \quad 2k_2 = 2h_2 R_0$$

and the dimensionless time t' = ωt, where ω is a quantity which has the dimensions of frequency and will be subsequently chosen with a value which is convenient for the solution of the problem. In this notation (1.18) and (1.21) take the following form:

$$\left.\begin{aligned}
\ddot{x} + 2h_1\dot{x} + x &= -2k_1\ddot{v} - 2\delta_1 x,\\
\ddot{v} + 2h_2\dot{v} + v &= -2k_2 x - 2\delta_2 v - 2h_2 xw,\\
\dot{w} + h_2 w &= 2h_2 x\,(\dot{v} + h_2 v)
\end{aligned}\right\} \tag{1.23}$$

and

$$\left.\begin{aligned}
\ddot{x} + 2h_1\dot{x} + x &= -k_1\ddot{v} - \delta_1 x,\\
\ddot{v} + 2h_2\dot{v} + v &= -k_2 x - 4h_2 x\cdot\widetilde{x\dot{v}} - \delta_2 v.
\end{aligned}\right\} \tag{1.24}$$

Let us estimate the orders of magnitude of the terms which appear in (1.23) and (1.24). For $\omega = \omega_0 \simeq \omega_l \simeq 10^{11}$ cps, $\bar{\tau} = 10^{-4}$ sec, N = 10^{12} cm^{-3}, and $Q_0 \simeq 10^3$, we have

$$2h_2 \simeq 10^{-7}, \quad 2h_1 \simeq 10^{-3}, \quad 2k_1 \simeq 10^{-2} - 10^{-3}, \quad 2k_2 \simeq 10^{-7}.$$

These estimates show that the nonlinear terms in the equations of a maser are small, so that a maser is a system of the Thomson type. A very effective method for the study of the equations which describe such systems is the so-called small-parameter method. For sufficiently precise adjustment of the frequency of the resonator to the frequency ω_l of the spectral line, (1.23) and (1.24) can be written in terms of a small parameter ε in the form

$$\left.\begin{aligned}
\ddot{x} + x &= \varepsilon\,(-2h_1\dot{x} - 2\delta_1 x - 2k_1\ddot{v}),\\
\ddot{v} + v &= \varepsilon\,(-2h_2\dot{v} - 2\delta_2 v - 2k_2 x - 2h_2 xw),\\
\dot{w} &= \varepsilon\,(-h_2 w + 2h_2 x\dot{v} + \varepsilon 2h_2^2 xv)
\end{aligned}\right\} \tag{1.25}$$

and

$$\left.\begin{aligned}
\ddot{x} + x &= \varepsilon\,(-2h_1\dot{x} - 2\delta_1 x - 2k_1\ddot{v}),\\
\ddot{v} + v &= \varepsilon\,(-2h_2\dot{v} - 2\delta_2 v - 2k_2 x + 4h_2^2 x\cdot\widetilde{x\dot{v}}).
\end{aligned}\right\} \tag{1.26}$$

One looks for the solution of these equations in the form of a formal series in powers of the dimensionless parameter ε, and then in the final solution sets ε equal to 1. The result is that we obtain series in the small quantities h_1, h_2, δ_1, δ_2, k_1, k_2.

2. The Self-Oscillation Process and the Higher Harmonics in a Maser

Here we shall derive the equation for the successive approximations which are obtained in the solution of (1.23) and (1.24) by the small-parameter method [30, 31]. At the same time, in the framework of this method we shall present results relating to the stationary self-oscillation process in a maser. Although these results are not original and are contained in a number of previous papers [14, 15, 21, 22], we have felt it necessary to present them here to preserve the continuity of the exposition, particularly since in what follows we shall have to use them repeatedly. We shall also give an estimate of the higher harmonics in a maser.

Equations (1.23) describe a system closely similar to that of two coupled conservative oscillators. Therefore, it is natural to assume the solution of the system in the form

$$x = R\cos(t+\varphi) + \varepsilon \sum_{k>1} [P^{(k)}\cos k(t+\varphi) + Q^{(k)}\sin k(t+\varphi)],$$
$$v = V\cos(t+\varphi) + U\sin(t+\varphi) + \varepsilon \sum_{k>1} [p^{(k)}\cos k(t+\varphi) + q^{(k)}\sin k(t+\varphi)], \tag{1.27}$$
$$w = W + \varepsilon \sum_{k>1} [S^{(k)}\cos k(t+\varphi) + T^{(k)}\sin k(t+\varphi)].$$

Here we have taken as the unit of time the reciprocal of the frequency of the radiation line, so that in (1.22) $\omega = \omega_7$ and $\delta_2 = 0$. The presence of the small parameter in (1.23) causes a slow change of the amplitudes R, $P^{(k)}$, $Q^{(k)}$,... and the phase φ. Therefore, following [30], we naturally introduce the small time $\theta = \varepsilon t$ and assume that R, $P^{(k)}$, $Q^{(k)}$,... are not only explicit functions of the parameter ε, but also depend on the small parameter through a slow variation of the time θ:

$$R = R(\varepsilon, \theta), \quad P^{(k)} = P^{(k)}(\varepsilon, \theta), \quad \ldots, \quad \varphi = \varphi(\varepsilon, \theta).$$

In this case,

$$\frac{d}{dt} = \frac{\partial}{\partial t} + \varepsilon \frac{\partial}{\partial \theta}. \tag{1.28}$$

We now substitute the expressions (1.27) in the equations (1.25) and make use of the relation (1.28). Equating the coefficients of $\sin k(t+\varphi)$ and $\cos k(t+\varphi)$ ($k = 0, 1, 2$) in the right and left members of the equations, we obtain the relations for the determination of the amplitudes R, $P^{(k)}$, $Q^{(k)}$, . . . and the phase φ. In doing so we must use the fact that the amplitudes and the phases themselves must be looked for in the form of power series in the small parameter ε:

$$\left.\begin{array}{l} R = R_0 + \varepsilon R_1 + \varepsilon^2 R_2 + \ldots \\ \cdots\cdots\cdots\cdots\cdots \\ \varphi = \varphi_0 + \varepsilon\varphi_1 + \varepsilon^2\varphi_2 + \ldots . \end{array}\right\} \tag{1.29}$$

For the determination of the quantities R_0, R_1, R_2, φ_0, φ_1, . . ., etc., we obtain a system of equations in successive approximations.

In our case, the first-approximation equations are of the form

$$\left.\begin{array}{l} -R_0\varphi_0' = -\delta R_0 + k_1 V_0, \\ -R_0' = h_1 R_0 + k_1 U_0, \end{array}\right\} \tag{1.30a}$$

$$\left.\begin{array}{l} -V_0\varphi_0' + U_0' = -h_2 U_0 - h_2 R_0 W_0 - k_2 R_0, \\ -U_0\varphi_0' - V_0' = h_2 V_0, \end{array}\right\} \tag{1.30b}$$

$$W_0' = -h_2 W_0 + h_2 R_0 U_0, \tag{1.30c}$$

where the prime denotes the derivative with respect to the "slow" time θ. These equations determine to first approximation the amplitude of the self-oscillations and the correction to the frequency of the oscillations of the maser given by the quantity φ_0'.

To calculate R_0 and φ_0' we must set $R_0' = 0$, $V_0' = U_0' = 0$ in (1.30). Solving the relations so obtained, we find

$$R_0^2 = \frac{k_1 k_2}{h_1 h_2} - 1 - \delta^2 \frac{1}{(h_1 + h_2)^2} \tag{1.31}$$

$$\varphi_0' = \delta \frac{h_2}{h_1 + h_2}. \tag{1.32}$$

In our notation the condition for self-excitation of the generator, $R_0^2 > 0$, takes the form

$$\frac{k_1 k_2}{h_1 h_2} > 1 + \delta^2 \frac{1}{(h_1 + h_2)^2} \tag{1.33}$$

If we substitute in Eqs. (1.32) and (1.33) the values of k_1, k_2, h_1, and h_2 from Eqs. (1.22), the condition for self-excitation takes the usual form [1-4]:

$$\eta = 4\pi \frac{|\mu_{ab}|^2}{\hbar} NQ\tau - (\omega_l - \omega_0)^2 Q_0^2 > 1, \tag{1.34}$$

and the frequency of the oscillations is bound to be given by

$$\omega = \omega_l \left(1 - \frac{\omega_l - \omega_0}{\omega_l} \cdot \frac{Q_0}{Q_l}\right). \tag{1.35}$$

The second-approximation equations for R_1, U_1, V_1, φ_1, . . . , etc. determine corrections to the amplitude and frequency of the self-oscillations. As can be seen from Eq. (1.35), in first approximation the frequency of the oscillations is exactly equal to that of the radiation line if $\omega_0 = \omega_l$.

Calculations show that this is true to arbitrary approximation if we neglect the higher harmonics.* Inclusion of the higher harmonics leads, however, to a shift of the oscillation frequency relative to the peak of the spectral line even for $\omega_0 = \omega_l$, and this shift depends on the amplitude of the self-oscillations.

To calculate this shift we write the equation in second approximation for R_1, V_1, U_1, W_1, φ_1 in the case $\delta = 0 (\omega_0 - \omega_l = 0)$:**

$$\left.\begin{aligned} -2R_0\varphi_1' = k_1V_1;\quad 2h_2U_1 &= -2h_2R_1W_0 - 2h_2R_0W_1 - k_2R_1 - 2h_2R_0S_0^{(2)};\\ -2R_1h_1 = k_1U_1;\quad 2h_2V_1 &= -2U_0\varphi_1' - 2h_2R_0T_0^{(2)};\\ W_1 &= -R_1U_0 + R_0U_1. \end{aligned}\right\} \tag{1.36}$$

Solving these equations we find that

$$\varphi_1' = -\frac{h_2}{h_1} k_1 T_0^{(2)}. \tag{1.37}$$

In our later calculations for the higher harmonics it will be shown that $T_0^{(2)} = \frac{1}{2} h_2 R_0 U_0$. Therefore, $\varphi_1' = -\frac{1}{4} h_2^2 R_0^2$. Thus, up to second order in ε with $\delta = 0$, the oscillation frequency of the generator is

$$\omega = \omega_l \left(1 - \frac{|\mu|^2 \mathscr{E}_0^2}{\hbar^2 \omega_l^2}\right). \tag{1.38}$$

Let us estimate the order of magnitude of the shift of the oscillation frequency from the peak of the spectral line. In the best case we can expect that the condition for self-excitation of the generator is satisfied by a factor 100, i.e., $\eta \simeq 100$. In this case, $R_0^2 \simeq 100$, and for $Q_l \simeq 10^7$ we have

$$\frac{R_0^2}{Q_l^2} = \frac{|\mu|^2 \mathscr{E}_0^2}{\hbar^2 \omega_l^2} \simeq 10^{-12}. \tag{1.39}$$

We see that the correction to the oscillation frequency caused by higher harmonics is extremely small and will be unimportant for the degrees of absolute stability obtained at present for the maser, which are of the order of 10^{-10}. To consider the problem of the existence of higher harmonics in the spectrum of the oscillations of a maser, it is necessary to write out the equations for the overtone amplitudes $P^{(k)}$, $Q^{(k)}$, $S^{(k)}$, $T^{(k)}$, $p^{(k)}$, and $q^{(k)}$.

It is clear from the form of (1.22) that the series for x and v in (1.27) will contain only odd harmonics and the series for w only even harmonics. Thus, in the expansions for x and v the first harmonic, and the largest one as to amplitude, will be the overtone with triple frequency, and in the expansion for w it will be the overtone with double frequency. Therefore, up to harmonics of still higher orders we have

*For a monochromatic stationary process of self-oscillation with neglect of the higher harmonics one can obtain an exact equation for the frequency of the self-oscillations [14].

**Equations (1.36) do not contain the amplitudes $P_0^{(k)}$, $Q_0^{(k)}$, $p_0^{(k)}$, $q_0^{(k)}$ of the higher harmonics of the field and the polarization, which it would seem must be present in the second-approximation equations. In what follows it will be shown that these quantities are simply equal to zero.

$$\left.\begin{aligned} x &= R\cos(t+\varphi) + \varepsilon\,[P^{(3)}\cos 3(t+\varphi) + Q^{(3)}\sin 3(t+\varphi)],\\ v &= V\cos(t+\varphi) + U\sin(t+\varphi) + \varepsilon\,[p^{(3)}\cos 3(t+\varphi) + q^{(3)}\sin 3(t+\varphi)],\\ w &= W + \varepsilon\,[S^{(2)}\cos 2(t+\varphi) + T^{(2)}\sin 2(t+\varphi)]. \end{aligned}\right\} \quad (1.40)$$

For simplicity we set $\delta = 0$, which means that φ_0' and V_0 are also equal to zero. In this case we get as the equation for the determination of $P^{(3)}$, $Q^{(3)}$, $p^{(3)}$, $q^{(3)}$, $S^{(2)}$, and $T^{(2)}$

$$\left.\begin{aligned} -4P^{(3)} + 3\varepsilon(1+h_1)Q^{(3)} &= \varepsilon 9k_1 p^{(3)},\\ -4Q^{(3)} - 3\varepsilon(1+h_1)P^{(3)} &= \varepsilon 9k_1 q^{(3)}, \end{aligned}\right\} \quad (1.41a)$$

$$\left.\begin{aligned} -4p^{(3)} + 3\varepsilon(1+h_2)q^{(3)} &= \varepsilon\,[k_2P^{(3)} + h_2(2P^{(3)}W + RS^{(2)})],\\ -4q^{(3)} - 3\varepsilon(1+h_2)p^{(3)} &= \varepsilon\,[k_2Q^{(3)} + h_2(2Q^{(3)}W + RT^{(2)})], \end{aligned}\right\} \quad (1.41b)$$

$$\left.\begin{aligned} 2T^{(2)} + \varepsilon h_2 S^{(2)} &= h_2 RU,\\ -2S^{(2)} + \varepsilon h_2 T^{(2)} &= h_2 RV. \end{aligned}\right\} \quad (1.41c)$$

As usual, we look for the values of the amplitudes in the form of series in ε:

$$\left.\begin{aligned} P^{(3)} &= P_0^{(3)} + \varepsilon P_1^{(3)} + \varepsilon^2 P_2^{(3)} + \dots\\ &\dots\dots\dots\dots\dots\dots\\ &\dots\dots\dots\dots\dots\dots\\ T^{(2)} &= T_0^{(2)} + \varepsilon T_1^{(2)} + \varepsilon^2 T_2^{(2)} + \dots. \end{aligned}\right\} \quad (1.42)$$

When we substitute Eqs. (1.42) in Eqs. (1.41) we can determine the coefficients in the series (1.42). In particular, we have from (1.41c)

$$T_0^{(2)} = \frac{h_2}{2} R_0 U_0,$$

which we have already used in the calculation of (1.38). For the other coefficients we get

$$P_0^{(3)} = Q_0^{(3)} = p_0^{(3)} = q_0^{(3)} = P_1^{(3)} = Q_1^{(3)} = Q_2^{(3)} = 0,$$

and

$$P_{(2)}^{(3)} = \frac{9}{32} k_1 h_2^2 R_0 U_0. \quad (1.43)$$

Thus we see that the harmonics in the maser appear only in third order in ε, which should guarantee that they are extremely small. In fact, the ratio of the power of the third harmonic to that of the first harmonic is

$$\left(\frac{P^{(3)}}{R_0}\right)^2 = \left[\frac{9}{16} h_1 h_2 R_0^2\right]^2. \quad (1.44)$$

Even for rather strong saturation ($R_0^2 \simeq 10$-30) the ratio (1.44) is extremely small:

$$\left(\frac{P^{(3)}}{R_0}\right)^2 \simeq 10^{-20}. \quad (1.45)$$

We see that in a maser the higher harmonics are strongly suppressed; a maser produces strictly monochromatic oscillations (to the accuracy of the noise line width).

3. Absence of the Effect of Pulling on the Oscillation Frequency of a Maser

Equations (1.24) show that a maser is similar to a generator with two coupled circuits. It could therefore be expected that in such a system there will occur pulling of the oscillation frequency such as occurs in an ordinary vacuum tube oscillator with two coupled oscillatory circuits [32].

For the phenomenon of pulling to exist it is necessary that the oscillation frequency of the maser be double-valued as a function of the lack of matching between the natural frequency of the resonator and the frequency of the molecular transition. Precisely this situation is observed in the ordinary two-circuit oscillator. The lack of uniqueness in the dependence of the frequency which is generated on the lack of matching of the frequencies of the two oscillatory circuits is directly connected with the existence of two resonance frequencies in the two-circuit system when the coupling is larger than the critical coupling. For the maser, however, the condition for self-excitation, Eq. (1.33),

$$\frac{k_1 k_2}{\delta_2^2 h_2^{-2} + 1} > h_1 h_2 \tag{1.46}$$

has precisely the meaning that the coupling must be larger than the critical coupling. Therefore, it would seem that as a system with two coupled circuits the maser must have two self-oscillation frequencies. Actually, however, as can be seen from Eq. (1.35), the oscillation frequency of the maser depends linearly on the frequency difference and is not a double-valued function.

To find the reason for this let us consider the action of an external force on a maser in which the condition for self-excitation is not quite satisfied. In this case (1.24) will take the form

$$\left.\begin{aligned} \ddot{x} + 2h_1\dot{x} + x &= -2k_1\ddot{v} - 2\delta_1 x + X\cos t, \\ \ddot{v} + 2h_2\dot{v} + v &= -2k_2 x - 4h_2 x \cdot \widetilde{x\dot{v}} - 2\delta_2 v. \end{aligned}\right\} \tag{1.47}$$

Solving them in the form

$$x = P\cos t + Q\sin t,$$
$$v = V\cos t + U\sin t$$

we get for the oscillation amplitude $R = (P^2 + Q^2)^{\frac{1}{2}}$ the expression

$$R = \frac{1}{2}\frac{X}{\sqrt{(\delta_1 + k^2\delta_2)^2 + (h_1 - k^2 h_2)^2}}. \tag{1.48}$$

Just as in ordinary systems, the amplitude R has two resonance maxima with respect to the frequency ω. The frequencies which correspond to the maxima are determined from the equation

$$\delta_1 + k^2\delta_2 = 0. \tag{1.49}$$

Here the role of the effective coupling coefficient is played by the expression

$$k^2 = \frac{k_1 k_2}{\delta_2^2 + h_2^2(1 + R^2)}, \tag{1.50}$$

For small values of R this goes over into expression (1.45). For sufficiently large R the effective coupling coefficient depends on the amplitude of the oscillations in the resonator.

Now let us assume that the condition for self-excitation is satisfied, so that self-oscillations occur in the maser. According to Eq. (1.31), the amplitude of the self-oscillations is determined from the equation

$$\frac{k_1 k_2}{\delta_2^2 h_2^{-2} + 1 + R^2} = h_1 h_2. \tag{1.51}$$

The meaning of condition (1.51) is, however, precisely that, in the case of self-oscillation, the effective coupling takes the critical value. Thus a maser always operates under conditions of critical coupling. This is the reason that the dependence of the oscillation frequency of a maser on the lack of exact matching is a single-valued one.

4. Locking of a Maser to an External Force

Experiments with masers frequently involve comparison of the oscillations of a maser with those of another source. The power from this source can penetrate into the maser and affect the self-oscillation process occurring in it. For a certain relation between the difference between the oscillation frequency of the maser and that of the external source on the one hand and the power acting on the maser on the other, a locking of the maser frequency to the external force acting on it can occur: the oscillations of the maser will occur in synchronism with the external signal [33].

We shall consider the action on a maser of a small external force whose frequency is close to that of the maser.* To do so we use (1.26), which for the case of a small external force $\varepsilon \cdot X \cos \omega t$ takes the form**

$$\left.\begin{aligned} \ddot{x} + x &= \varepsilon\,[2\,\delta_1 x - 2h_1\dot{x} - 2\,k_1\ddot{v} + X\cos t], \\ \ddot{v} + v &= \varepsilon\,[2\,\delta_2 v - 2h_2\dot{v} - 2k_2 x - 4h_2 x\,\widetilde{x v}]. \end{aligned}\right\} \tag{1.52}$$

Here we have introduced a dimensionless time by the substitution $t \to \omega t$, and

$$2\delta_1 = 1 - \frac{\omega_0^2}{\omega^2}, \quad 2\delta_2 = 1 - \frac{\omega_l^2}{\omega^2}.$$

As usual, we shall look for the solution of the equations in the form

$$x = P\cos t + Q\sin t,$$
$$v = V\cos t + U\sin t,$$

where P, Q, V, and U are slowly varying functions of the time. In the first approximation in ε we get the equations

$$\left.\begin{aligned} \frac{dP}{dt} &= \varepsilon\,[+\delta_1 Q - h_1 P - k_1 U] = \varepsilon\widetilde{P}, \\ \frac{dQ}{dt} &= \varepsilon\,[\delta_1 P - h_1 Q - k_1 V - X] = \varepsilon\widetilde{Q}, \\ \frac{dU}{dt} &= \varepsilon\,[-\delta_2 V - h_2 U - k_2 P - h_2 P(PU - QV)] = \varepsilon\widetilde{U}, \\ \frac{dV}{dt} &= \varepsilon\,[\delta_2 U - h_2 V - k_2 Q - h_2 Q(PU - QV)] = \varepsilon\widetilde{V}. \end{aligned}\right\} \tag{1.53}$$

The stationary process is determined from the relations

$$\frac{dP}{dt} = \frac{dQ}{dt} = \frac{dV}{dt} = \frac{dU}{dt} = 0. \tag{1.54}$$

This gives us four equations for the determination of P, Q, V, and U. The region of stability of this type of synchronization is calculated from the condition that the real parts of all of the roots of the equation

$$\begin{vmatrix} \frac{\partial\widetilde{P}}{\partial P} - \lambda & \frac{\partial\widetilde{P}}{\partial Q} & \frac{\partial\widetilde{P}}{\partial V} & \frac{\partial\widetilde{P}}{\partial U} \\ \frac{\partial\widetilde{Q}}{\partial P} & \frac{\partial\widetilde{Q}}{\partial Q} - \lambda & \frac{\partial\widetilde{Q}}{\partial V} & \frac{\partial\widetilde{Q}}{\partial U} \\ \frac{\partial\widetilde{V}}{\partial P} & \frac{\partial\widetilde{V}}{\partial Q} & \frac{\partial\widetilde{V}}{\partial V} - \lambda & \frac{\partial\widetilde{V}}{\partial U} \\ \frac{\partial\widetilde{U}}{\partial P} & \frac{\partial\widetilde{U}}{\partial Q} & \frac{\partial\widetilde{U}}{\partial V} & \frac{\partial\widetilde{U}}{\partial U} - \lambda \end{vmatrix} = 0 \tag{1.55}$$

*See also [34].

**We here suppose that the external power is so small that the oscillations produced will obey condition (1.21a).

are negative. These regions can be located according to the well-known method of Routh and Hurwitz, and for $\omega_0 \approx \omega_l$ we get the relation

$$\delta \leqslant \frac{1}{2} \frac{\Delta\omega_l}{\omega_l} \sqrt{\frac{P_{ext}}{P_0}}, \tag{1.56}$$

where $\Delta\omega_l$ is the original line width of the radiation of the molecular beam; P_{ext} is the power of the external signal; P_0 is the power of the maser acting as a self-oscillator; and $\delta = |\omega - \omega_l|/\omega_l$.

When condition (1.56) holds, sinusoidal oscillations with the frequency of the external force are established in the generator. If, on the other hand, condition (1.56) is not satisfied, the generator shows a polyharmonic behavior: there are self-oscillations, forced oscillations with the frequency of the external force, and interference terms which occur because of nonlinearity with the combination frequencies $m\omega_a \pm n\omega$, where ω_a is the frequency of the self-oscillations, ω is that of the external signal, and m and n are integers. These combination terms are weak, however, because of the inertial nature of the nonlinearity.

Equations (1.53) allow us to calculate the amplitude of the self-oscillations, which, in the case considered, will depend on the amplitude of the external force. Calculation from the equations gives for the amplitude R of the self-oscillations the formula

$$R = \left[R_0^2 - X_0^2 \frac{h_2^2}{\delta^2(\delta^2 + h_1^2)}\right]^{1/2}. \tag{1.57}$$

In this formula we have used the fact that in actual systems we always have $h_2 \ll h_1$; R_0 is the amplitude of the self-oscillation in the absence of the external signal, which is determined by Eq. (1.31).

This formula (1.57) for the amplitude of the self-oscillations of the maser which is under an external influence will be used in our later work (see Part 4 of Section 3).

5. The Influence of a Waveguide System on the Frequency of a Maser

Formula (1.35), which determines the oscillation frequency of a maser, was derived without taking into account the resonance properties of the waveguide system which, together with the power indicator, forms the load on the maser. There can, however, be a considerable effect on the oscillation frequency of a maser from the resonance properties of the waveguide system and changes in the load.

We can take these effects into account by considering the waveguide system of the maser and the crystal power detector as a resonance system with a certain equivalent quality figure. In this case, the maser can be treated in terms of two coupled resonant systems, one of which is excited by the molecular beam. To describe the oscillatory process, including effects of the waveguide system, we must adjoin to Eqs. (1.24) the equations of the waveguide system:

$$\left.\begin{aligned} \ddot{x} + 2h_1\dot{x} + x &= 2\delta_1 x - 2k_1\dot{v} + k_0\ddot{y}, \\ \ddot{y} + 2h_W\dot{y} + y &= 2\delta_W y + k_W\ddot{x}, \\ \ddot{v} + 2h_2\dot{v} + v &= 2\delta_2 v - k_2 x - 4h_2 x\cdot\widetilde{xv}, \end{aligned}\right\} \tag{1.58}$$

where y is the field intensity in the waveguide system, expressed in relative units, $2h_W = \frac{\omega_W}{\omega}\frac{1}{Q_W}$, ω_W is the resonance frequency of the waveguide system, Q_W is the quality of the waveguide system (including the detector), and k_0 and k_W are coupling coefficients. The solution of these equations for the stationary sinusoidal process gives the following expressions for the oscillation frequency for the maser:

$$\omega = \omega_l \cdot \left[1 + \frac{Q_{equ}}{Q_l} \cdot \left(\Delta_1 - \frac{k^2}{\Delta_2^2 + Q_W^{-2}} \cdot \Delta_2\right)\right]. \tag{1.59}$$

$$Q_{equ}^{-1} = Q_0^{-1} + k^2 Q_W, \quad k^2 = k_0 k_W, \quad \Delta_1 = \frac{\omega_0^2 - \omega_l^2}{\omega_l^2}, \quad \Delta_2 = \frac{\omega_W^2 - \omega_l^2}{\omega_l^2}.$$

It is convenient to express the coupling coefficient k^2 in terms of α, the ratio of the power entering the waveguide to the total power developed by the generator, and the "decoupling coefficient" $\beta = k_W/k_0$:

$$\frac{k^2}{\Delta_2^2 + Q_W^{-2}} = \frac{Q_W}{Q_0}\beta^{-1} \cdot \frac{\alpha}{1-\alpha}. \tag{1.60}$$

Substituting Eq. (1.60) in Eq. (1.59), we find that with effects of the waveguide system included, the oscillation frequency of the maser is given by the relation

$$\omega = \omega_l\left[1 + \frac{Q_{equ}}{Q_l}\left(\Delta_1 - \frac{Q_W}{Q_0}\beta^{-1}\cdot\frac{\alpha}{1-\alpha}\Delta_2\right)\right]. \tag{1.61}$$

For $Q_{equ} \approx 3000$, $\Delta_2 \approx 1/Q_W$, $\alpha \approx 0.1$, and no decoupling ($\beta = 1$) the added term brought in by the waveguide system is of the order 10^{-8}. This term can be compensated by a suitable tuning of the resonator. Small mismatches in the waveguide system, however, leading to variation of the coupling coefficient with the power α, can decidedly disturb the adjustment of the oscillation frequency of the maser to that of the spectral line. For example, a change of α by 10% shifts the oscillation frequency of the maser by $10^{-9}\,\omega_l$ cps. The influence of the waveguide system can be eliminated by increasing the decoupling coefficient β. For good operation of a maser it is desirable that β should be equal to 10-15, which corresponds to a power decoupling ~20 db. As a decoupling device one could use, for example, ferrite valves.

Section 2

DEPENDENCE OF OSCILLATION FREQUENCY OF A MASER ON VARIOUS PARAMETERS

1. Survey of Factors Influencing the Frequency of a Maser

It has been shown in a number of papers that masers have very high frequency stability – the relative change of the oscillation frequencies of two masers in 10 or 20 minutes does not exceed 10^{-11} [35, 20]. In spite of this high frequency stability of the oscillations of a maser, the frequency can undergo shifts over a comparatively wide range (~10^{-7}). Therefore, in the construction of molecular frequency standards it is necessary to choose a construction and a type of operation for the maser such that its oscillation frequency will differ as little as possible from that of the spectral line. The difference between the oscillation frequency of the maser and the frequency of the spectral line of the free molecule characterizes the absolute stability of the oscillation of the maser, which is given by the relation

$$\delta_{abs} = \left|\frac{\omega_l - \omega}{\omega_l}\right|,$$

where ω_l is the frequency of the spectral line of the free molecule and ω is the oscillation frequency of the maser.

As can be seen from Eq. (1.35), the oscillation frequency of a maser depends on three main quantities: (1) the frequency ω_l of the spectral line; (2) the natural frequency of the resonator ω_0, and (3) the width of the spectral line $\Delta\omega_l$. Let us consider the influence of these factors on the oscillation frequency of a maser.

In the simplest case, when Eq. (1.35) holds, the width of the spectral line depends only on the time of flight of molecules through the resonator. Actually, the width of the spectral line in a maser, as in other systems, is made up of the natural width, the Doppler width, and the broadening owing to the finite time of flight of molecules through the resonator. If we choose a type of oscillation for which the phase velocity of the electromagnetic wave in the direction of travel of the beam is infinite, the Doppler broadening will be determined by the angular aperture of the molecular beam [36] and can be made small by decreasing the aperture of the beam of molecules. The natural width of an emission line is also extremely small in the centimeter range of wavelengths so that with a suitable construction of a maser the line width is practically completely determined by the time of flight of the molecules through the resonator. In a maser an important part is played by the saturation effect, which contributes to the line width in a radio-spectroscopic observation of an emission line or an absorption line. If the distribution of times of flight of the molecules is according to the law $(1/\bar{\tau})e^{-\tau/\bar{\tau}}$, the saturation factor and the line width caused by the time of flight of molecules through the resonator make additive contributions to the total line width. In this case, the oscillation frequency of the maser does not depend on the saturation parameter. The actual distribution of velocity of the molecules in a molecular beam which has been through a sorting device differs from the function $(1/\bar{\tau})e^{-\tau/\bar{\tau}}$ and depends on the voltage on the sorting system. For the actual distribution the line does not have the Lorentz shape, and the widths caused by time of flight and by the saturation effect do not make additive contributions to the total line width [37, 38]. This causes the effective line width which appears in the expression for the oscillation frequency of the maser to depend on the saturation parameter, i.e., to depend in the final analysis on the intensity of the beam of active molecules entering the resonator. As has been noted above, the character of the time-of-flight distribution of the molecules is affected by the voltage on the sorting system, which means that the effective line width depends on this voltage.

In the derivation of (1.18) and the resulting formula (1.35) for the oscillation frequency of a maser, it was assumed that the molecular transition occurs only between two energy levels, so that the emission line consists of only one component of a definite frequency. In practice, however, it is hard to find a spectral line which consists of only one component; the presence of hyperfine interactions in the molecule leads to a splitting of the main transition into a number of hyperfine structure components, and with practically-attainable line widths many of these are not resolved. The line $J = 3$, $K = 3$ of the inversion spectrum of $N^{14}H_3$, which is usually used in masers, has a rich hyperfine structure caused by electric and magnetic interactions in the molecule [39]. The strongest of these – the electric quadrupole interaction of the N^{14} nucleus with the field of the molecule – is of order of magnitude 4 Mc [40]. The weakest interactions are those of magnetic type. They include the spin-spin interaction of the hydrogen and nitrogen nuclei, and the interaction of the spins of these nuclei with the magnetic field of the molecule caused by its rotation. They are of the order of 25-100 kc [40].

Owing to these interactions, the line of the inversion transition $J = 3$, $K = 3$ (and also other lines) is split into quite a number of components with respect to various quantum numbers $I_H I_N J F_1 F$,* which characterize the energy levels of the molecule when the hyperfine interactions are taken into account. The intervals between the individual components can considerably exceed the resolution threshold determined by the width of the spectral line. Therefore, strictly speaking, in order to characterize the transition, we must indicate in addition to the numbers J and K the changes of the numbers F_1 and F.

The component used in a maser is the most intense component of the inversion transition $J = 3$, $K = 3$, which is diagonal in the quantum numbers F_1 and F ($\Delta F_1 = 0$, $\Delta F = 0$).

If the constants of the quadrupole and magnetic couplings in ammonia were the same for the upper and lower inversion levels, the line $\Delta F_1 = 0$, $\Delta F = 0$ would consist of only one component. As has been shown in [40], however, the hyperfine interaction constants are different for the upper and lower inversion levels, which leads to a splitting of the line $J = 3$, $K = 3$, $\Delta F_1 = \Delta F = 0$ into twelve components corresponding to the twelve possible values of the quantum numbers F_1 and F for the level $J = 3$, $K = 3$ (Fig. 1b). At usual line widths these components are not resolved and appear in the emission from the molecules as a single spectral line. The peak position (frequency) of such a line depends on the ratios of the intensities of the individual components and on their frequencies. In the gaseous state the intensities of the components are determined by the thermal distribution of the molecules along the energy levels.

In the beam which has undergone sorting the ratios of the intensities of the components are changed, and the peak of the spectral line is shifted relative to that of the spectral line in the gas. This shift is not a constant quantity, but depends on the voltage on the sorting system, since the ratio of the intensities of the components varies with this voltage.** Moreover, the different components have different values of the matrix element of the dipole moment and show different degrees of saturation. The result is that the shift in question depends on the strength of the field in the resonator, i.e., in the final analysis the oscillation frequency in the maser depends on the intensity of the molecular beam.

---

*$\mathbf{F}_1 = \mathbf{I}_N + \mathbf{J}$, $\mathbf{F} = \mathbf{F}_1 + \mathbf{I}_H$, where $\mathbf{I}_N$ is the spin vector of the nitrogen nucleus, $\mathbf{I}_H$ is the vector of the total spin of the hydrogen nuclei, and $\mathbf{J}$ is the rotational angular momentum of the molecule. The vector diagram for the composition of the angular momenta in the $N^{14}H_3$ molecule is shown in Fig. 1a.

**The shift of the oscillation frequency of a maser relative to a frequency standard which uses the same absorption line of ammonia in the equilibrium gas was discovered in [18] and explained in [16]. The fact that this shift depends on the voltage was first pointed out in [41].

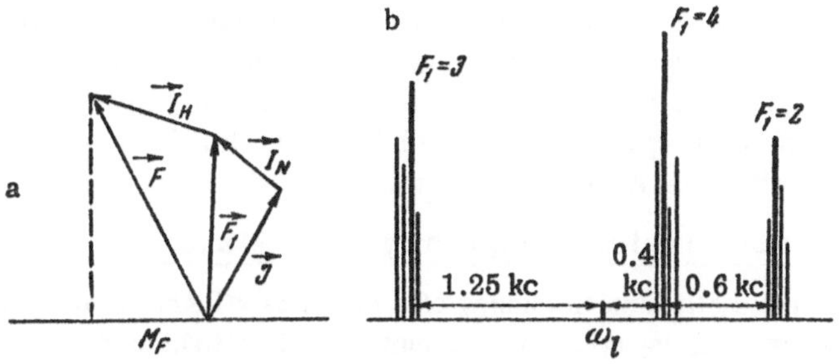

Fig. 1. Inversion line of $N^{14}H_3$, J = 3, K = 3, $\Delta F_1 = \Delta F = 0$. (a) Coupling scheme of angular momenta in the ammonia molecule; here J is the rotational angular momentum of the molecule, I_N is the spin of the nitrogen nucleus, and I_H is the total spin of the protons. (b) Structure of the inversion line of ammonia, J = 3, K = 3, $\Delta F_1 = \Delta F = 0$.

In the derivation of the equations of a maser in Section 1, the resonator was represented as an oscillating circuit with lumped parameters. This idealization does not take into account the propagation of the electromagnetic energy along the resonator. Actually, the molecules do not emit uniformly along the resonator, since the probability of emission increases as the time the molecule spends in the resonator increases. The result is that a wave occurs which transfers the energy along the resonator, and this causes a Doppler shift of the frequency of the spectral line even when one uses the mode E_{010} as the main type of oscillations in the resonator [15].

The magnitude and sign of this shift depend on the intensity of the beam of active molecules, since, for a small beam intensity, which means weak saturation, the emission of energy from the molecules occurs mainly as they are passing out of the resonator, and the flux of electromagnetic energy is directed against the motion of the molecules, which causes a decrease of the frequency. When the saturation is large, the molecules emit their energy in the first instants of their presence in the resonator, and owing to this the flux of energy is propagated in the direction of motion of the molecules, and this causes an increase of the oscillation frequency of the maser. This shift of the oscillation frequency of a maser can be compensated to a considerable degree if we send two oppositely moving identical beams through the resonator in a symmetrical fashion and take off the power from the maser exactly in the middle of the resonator [15].

Because of the presence of unresolved hyperfine-structure components and the nonuniformity of the radiation of the molecules along the resonator, the frequency of a maser can vary over a range $1\text{-}3 \cdot 10^{-9}$ (see Parts 2 and 3 of this section). The frequency of the resonator has a still stronger effect: retuning of the resonator frequency within the limits of its transmission band changes the oscillation frequency of a maser by an amount equal to the width of the spectral line ($\sim 10^{-7}$). We can point out a number of other effects which influence the oscillation frequency of a maser.

First, there is the effect of nearby resolved lines. Although such lines do not take a direct part in the process of generation, they change the dielectric permeability of the medium and thus influence the effective frequency of the resonator. The inversion lines of ammonia which are nearest to the lines J = 3, K = 2 and J = 3, K = 3, which are commonly used, are separated from them by 100 Mc or more [42], so that their effect on the frequency of a maser is negligible. There can be a stronger influence from neighboring lines caused by the hyperstructure. Such lines are primarily the magnetic components of the hyperfine structure $\Delta F_1 = 0$, $\Delta F = \pm 1$. They are situated symmetrically relative to the main line $\Delta F_1 = \Delta F = 0$, and for the inversion transition J = 3, K = 3 they are separated from the main line by 50 kc [40]. The effect of nearby lines is considered in Part 5 below. Under ordinary circumstances it is evidently not much larger than 10^{-11}.

Second, there is the frequency shift of the spectral line owing to the second-order Doppler effect; when the speed of the molecules is $v = 5 \cdot 10^4$ cm/sec, this amounts to $v^2/c^2 \approx 3 \cdot 10^{-12}$. Finally, there are corrections to the oscillation frequency of a maser caused by the finite time of passage of molecules through the resonator [see Eq. (1.19)] and by the higher harmonics in the maser [Eq. (1.38)]. These corrections also do not exceed 10^{-12}.

In summarizing this discussion we can say that to an accuracy of the order of 10^{-11} to 10^{-12} the oscillation frequency of a maser depends on: 1) the natural frequency of the resonator; 2) the shift of the peak of the spectral line because of the presence of unresolved hyperfine-structure components; 3) lack of uniformity of the emission of the molecules along the length of the resonator, which leads to a Doppler shift of the oscillation frequency of the maser; 4) the effective width of the spectral line, which changes with changing intensity of the molecular beam and with changes of the voltage on the sorting system.

2. Calculation of the Oscillation Frequency of a Maser

Here we shall calculate the oscillation frequency of a maser as a function of the various parameters of the generator. In this calculation we shall not take into account the effect indicated as point 3 in the list above and which we shall henceforth refer to as "effect III," since we shall suppose that it has been eliminated by using two symmetrical beams moving in opposite directions. The other influences on the oscillation frequency of a maser cannot be eliminated by any improvements in the construction of the generator,* and they must be taken into account in the calculation.

This problem cannot be solved by means of the system of equations (1.18), since these equations do not take into account the features of the time-of-flight distribution of the molecules which have been pointed out in Part 1 of this section, and are valid only for a one-component emission line. Equation (1.3) is of a more general character, and when we make a suitable calculation of the polarization **P** we can use this equation to calculate the dependence of the oscillation frequency of a maser on the various parameters. Therefore, in calculating the oscillation frequency in a maser we shall use the equation

$$\frac{d^2\mathscr{E}}{dt^2} + \frac{\omega_0}{Q}\cdot\frac{d\mathscr{E}}{dt} + \omega_0^2\mathscr{E} = -4\pi\frac{d^2P}{dt^2}. \tag{2.1}$$

Since, in the present case, we are interested in the frequency of stationary harmonic self-oscillations in a maser, we can represent $\mathscr{E}(t)$ and P(t) in the form

$$\mathscr{E}(t) = \mathscr{E}_0 e^{-i\omega t}, \quad P(t) = P_0 e^{-i\omega t}, \tag{2.2}$$

where $\mathscr{E}_0$ is the amplitude and ω is the frequency of the stationary oscillation.

The quantity $\mathscr{E}_0$ can always be taken to be real. The polarization P_0 describes the properties of the molecular beam and is in general complex:

$$P_0 = P_0' + iP_0''. \tag{2.3}$$

Substituting Eq. (2.2) in Eq. (2.1), we get the following equations for the determination of the amplitude and frequency:

$$\left.\begin{aligned}(\omega_0^2 - \omega^2)\mathscr{E}_0 &= 4\pi N\omega^2 P_0',\\ -\frac{\omega\omega_0}{Q}\cdot\mathscr{E}_0 &= 4\pi N\omega^2 P_0''\end{aligned}\right\} \tag{2.4}$$

or, upon dividing the first equation by the second,

$$\frac{P_0'}{P_0''} = \frac{\omega_0^2 - \omega_l^2}{\omega\omega_0}Q \simeq 2Q\frac{\omega_0 - \omega}{\omega}. \tag{2.5}$$

If the line consists of several independent components, then

$$P_0 = \sum_k P_{0k}, \tag{2.6}$$

where P_{0k} is the polarization of the molecular beam caused by the presence of the k-th component in the emission

*The effects of hyperfine-structure components could be eliminated only by choosing a different transition, or a different substance having no hyperfine splitting of the energy levels.

line. Thus, the calculation of the polarization P_0 when there are several independent components in the line reduces to the calculation for one component.

Equations (1.12) are quite suitable for this calculation, since they are not restricted to an averaging of the polarization over a particular distribution of the molecules as to their times of flight τ. If we neglect the higher components, which make only a small contribution which we are neglecting in this calculation, Eqs. (1.12) can be easily solved for the monochromatic process. The polarization $P_{0k}(\tau)$ calculated by solving these equations is to be averaged over the time of flight τ, after which one calculates the oscillation frequency of the maser by solving (2.5).

The calculation of the polarization $P_{0k}(\tau)$ is carried out in Appendix I. From this calculation we obtain the relation

$$\left.\begin{aligned} P'_{0k}(\tau) &= \frac{\omega_k-\omega}{\gamma_k^2}\cdot\frac{|\mu_k|^2}{\hbar}\left(1-\frac{\sin\gamma_k\tau}{\gamma_k\tau}\right)\cdot\mathscr{E}_0,\\ P''_{0k}(\tau) &= -\frac{|\mu_k|^2}{\hbar}\cdot\frac{1}{\gamma_k}\cdot\frac{1-\cos\gamma_k\tau}{\gamma_k\tau}\cdot\mathscr{E}_0, \end{aligned}\right\} \qquad (2.7)$$

so that

$$\left.\begin{aligned} P'_0 &= \mathscr{E}_0\sum_k\left\{\frac{\omega_k-\omega}{\gamma_k^2}\cdot\frac{|\mu_k|^2}{\hbar}\int\left(1-\frac{\sin\gamma_k\tau}{\gamma_k\tau}\right)\cdot f_k(\tau)\,d\tau\right\},\\ P''_0 &= \mathscr{E}_0\sum_k\left\{-\frac{|\mu_k|^2}{\hbar}\cdot\frac{1}{\gamma_k^2}\cdot\int\frac{1-\cos\gamma_k\tau}{\gamma_k\tau}\cdot f_k(\tau)\,d\tau\right\}. \end{aligned}\right\} \qquad (2.8)$$

For convenience we represent the frequency of the k-th component of the line in the form

$$\omega_k = \omega_l + \eta_k,$$

where ω_l is the frequency of the transition in the molecule in the absence of hyperfine splitting of the energy levels.

We shall assume that $|\eta_k/\omega_l| \ll 1$ and $|(\omega_0-\omega_l)/\omega_l| \ll 1$, and look for the solution of Eq. (2.5) close to ω_l. The result of the calculation is

$$\omega = \omega_l\left[1-\frac{\omega_l-\omega_0}{\omega_l}\cdot\frac{Q}{Q_l}+\Delta\right]. \qquad (2.9)$$

In this formula,

$$Q_l = \omega_l\frac{\sum_k J_k^s}{\sum_k \gamma_k J_k^c};\quad \Delta = \frac{1}{\omega_l}\,\frac{\sum_k \eta_k J_k^s}{\sum_k J_k^s}; \qquad (2.10)$$

$$\left.\begin{aligned} J_k^s &= \int\left(1-\frac{\sin\gamma_k\tau}{\gamma_k\tau}\right)\cdot f_k(\tau)\,d\tau;\\ J_k^c &= \int\frac{1-\cos\gamma_k\tau}{\gamma_k\tau}\cdot f_k(\tau)\,d\tau. \end{aligned}\right\} \qquad (2.11)$$

The further calculation of the oscillation frequency of the maser involves finding the time-of-flight distribution of the molecules for each hyperfine-structure component. This distribution is closely connected with the properties of the sorting of molecules which are in one or another energy state. Therefore, we shall carry out the further calculation for the example of the inversion transition line J = 3, K = 3, $\Delta F_1 = \Delta F = 0$ of $N^{14}H_3$, which is commonly used to excite oscillations in masers.

As has already been pointed out, this line consists of twelve components, owing to the electric quadrupole and magnetic dipole interactions of the nuclei with each other and with the field of the molecule. Because of the weakness of the magnetic interaction we would need to take it into account only in calculating to accuracies of 10^{-11} and better. Therefore, in our calculation we shall neglect the magnetic hyperfine structure, although, generally speaking, there would be no particular difficulty in including it.

When the magnetic hyperfine structure is neglected the quantum state of a molecule can be sufficiently characterized by giving the numbers I_N, J, K, F_1, and M_{F_1}, whose meaning is clear from Fig. 1a, and the transition used in the maser will contain three quadrupole components corresponding to the three values of F_1: 2, 3, and 4 (Fig. 1b). The index k in (2.10) and (2.11) should be replaced by $F_1M_{F_1}$,* where M_{F_1} is the projection of the angular momentum F_1 on the direction of the electric field in the resonator.

The time-of-flight distribution $f_k(\tau)$ of the molecules depends on the energy of interaction of a molecule with the field of the sorting system. The interaction energy is determined by the projection of the total angular momentum F_1 of the molecule on the direction of the field in the sorting system, which we denote by M'_{F_1}. The direction of the field in the sorting system may not be the same as the direction of the field in the resonator, so that in the general case $M'_{F_1} \neq M_{F_1}$.

We can, however, express the time-of-flight distribution $f_{F_1M_{F_1}}(\tau)$ of the molecules in terms of distributions $f_{F_1M'_{F_1}}(\tau)$ by using the equation

$$f_{F_1M_{F_1}}(\tau) = \sum_{M'_{F_1}=-F_1}^{M'_{F_1}=F_1} a^{F_1}_{M_{F_1}M'_{F_1}} f_{F_1M'_{F_1}}(\tau), \tag{2.12}$$

where $a^{F_1}_{M_{F_1}M'_{F_1}}$ is the probability that a molecule has the value M_{F_1} for the projection of its total angular momentum along the direction of the field in the resonator, if the projection of its total angular momentum along the direction of the field in the sorting system is M'_{F_1}.

The probabilities $a^{F_1}_{MM'}$ are equal to squares of matrix elements for the transformation of the wave function under rotation of the coordinate systems, and can be determined from the Wigner D function [43]:

$$a^{F_1}_{MM'} = |D^{F_1}_{MM'}(\alpha, \beta, \gamma)|^2, \tag{2.13}$$

where α, β, and γ are the angles of the rotation of the coordinate system.

If one uses for the excitation of the maser the oscillation mode E_{010} in a cylindrical resonator, and uses a quadrupole condenser to sort the molecules, then the electric field in the sorting system is perpendicular to that in the resonator. This means that $\beta = \pi/2$, and

$$a^{F_1}_{MM'} = \overline{\left|D^{F_1}_{MM'}\left(\alpha, \frac{\pi}{2}, \gamma\right)\right|^2}, \tag{2.14}$$

where the bar indicates averaging over the angles α and γ. The values of the coefficients given by the formula (2.14) for $F_1 = 2, 3$, and 4 (the line J = 3, K = 3 of $N^{14}H_3$) are given in Table I.

The time-of-flight distribution of the molecules depends in an important way on the manner of their emergence from the source of the molecular beam. If the diameter d of the opening of the grid in the source of the molecular beam is much larger than the thickness a of the foil from which the grid is made, the flow of the gas out of the source is a molecular (Knudsen) flow.

In the grids ordinarily used, $a \simeq d$, so that, at sufficiently high pressures, the nature of the flow is intermediate between Knudsen flow and viscous flow [44-46]. Evidently, however, it is still closer to molecular flow, and in deriving the time-of-flight distribution $f_{F_1M'_{F_1}}(\tau)$ of the molecules we shall use the Maxwell velocity distribution of the molecules, which is proportional to the expression

$$N_{JK} \cdot e^{-mv^2/2kT} \cdot v^3 \sin\vartheta \cdot \cos\vartheta, \tag{2.15}$$

*We omit the indices I_N, J, K since they are the same for all components of the lines in question.

TABLE I. Values of Coefficients $a^{F}_{M_F M'_F}$

F_1	M_F	M'_F = 0	1	2	3	4
4	0	36/256	0	40/256	0	70/256
	1	0	36/256	8/256	28/256	56/255
	2	40/256	8/256	16/256	56/256	28/256
	3	0	28/256	56/256	36/256	8/256
	4	70/256	56/256	28/256	8/256	1/256
3	0	0	48/256	0	80/256	—
	1	48/256	4/256	40/256	60/256	
	2	0	40/256	64/256	24/256	
	3	80/256	60/256	24/256	4/256	
2	0	64/256	0	96/256	—	—
	1	0	64/256	64/256		
	2	96/256	64/256	16/256		

where m is the mass of the molecule, T is the temperature of the source of the molecular beam, and k is Boltzmann's constant.

Furthermore, we shall assume that the molecules are distributed over the hyperfine-structure components according to equal probabilities, since the energy of the quadrupole interaction is much smaller than kT. Molecules with the distribution given by Eq. (2.15) come to the entrance of the sorting system. At the exit of this system we shall have a distribution which differs from Eq. (2.15). We shall assume that the sorting system is so long that during the passage through it molecules that are in the lower inversion level are practically completely removed from the molecular beam, so that at the exit from the sorting system we have a beam of molecules which are only in the upper inversion level. In practice, this is always true even for small voltages on the sorting system. For example, if a voltage of about 3 kV is applied to a sorting system which consists of a quadrupole condenser, then in a length 10-15 cm of the quadrupole condenser practically all molecules which are in the lower inversion level are removed from the molecular beam. It is not true, however, that all of the molecules which are in the upper inversion level will remain in the molecular beam.

The passage into the resonator will be completely successful for all of the "upper" molecules which go into the sorting system at angles ϑ such that

$$\tan \vartheta \leqslant \frac{R}{L}, \tag{2.16}$$

where R is the radius of the sorting system and L is its length.

For the case in which the angle is such that $\tan \vartheta > R/L$, then only those molecules get into the resonator for which the relation

$$\frac{mv^2}{2} \cdot \sin \vartheta \leqslant W_{F_1 M'_{F_1}} \tag{2.17}$$

holds, where $W_{F_1 M'_{F_1}}$ is the maximum energy of interaction of molecules which are in the level F_1, M'_{F_1} with the field of the sorting system.

If we introduce the time of flight of a molecule through the resonator, $\tau = l/(v \cos \vartheta)$, where l is the length of the resonator, then when we use Eq. (2.16) and Eq. (2.17) we get the following distribution of active molecules as to their times of flight through the resonator:

$$f_{F_1 M_{F_1}}(\tau) = AN_{JK}\frac{1}{\tau^3}\begin{cases} e^{-\bar{\tau}^2/\tau^2}\left(1 - e^{-\frac{R^2}{L^2}\cdot\frac{\bar{\tau}^2}{\tau_M^2}}\right) & \tau_M \leqslant \tau < \infty, \\ e^{-\bar{\tau}^2/\tau^2}\left(e^{-\frac{R^2}{L^2}\cdot\frac{\bar{\tau}^2}{\tau_M^2}} - e^{-\frac{R^2}{L^2}\cdot\frac{\bar{\tau}^2}{\tau^2}}\right) & 0 < \tau < \tau_M, \end{cases} \tag{2.18}$$

where

$$\bar{\tau}^2 = \frac{ml^2}{2kT}, \quad \tau_M = \frac{R}{L}\sqrt{\frac{m}{2W_{F_1 M'_{F_1}}}}\cdot l. \tag{2.19}$$

The maximum energy of interaction of the ammonia molecules with the field of the sorting system depends on the voltage U on the sorting system. This interaction energy is much larger than the energy of quadrupole coupling and, whereas when they emerge from the sorting system the state of the molecules is determined by the quantum numbers $I_N J K F_1 M'_{F_1}$, in the sorting system itself the presence of the strong field "breaks" the coupling of the vectors $\mathbf{I}_N$ and $\mathbf{J}$, so that the conserved quantum numbers will be $I_N M'_{I_N}$; JKM'_J, where M'_{I_N} and M'_J are the projections of the spin of the nitrogen nucleus and of the rotational angular momentum of the molecule along the direction of the electric field in the sorting system.

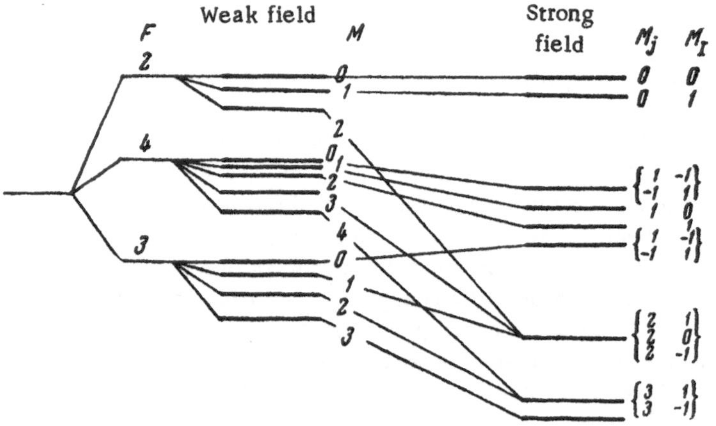

Fig. 2. Diagram of the correspondence between the quantum numbers $I_N M_{I_N}$; JM_J and $F_1 M'_{F_1}$ for the inversion line J = 3, K = 3 of $N^{14}H_3$ as the molecule is taken from a weak electric field into a strong field [47].

In the case of a strong interaction which is much larger than the quadrupole coupling, the energy of interaction of an ammonia molecule with the external electric field can be calculated according to [42] and is given by

$$W = \frac{\hbar\omega_{\text{inv}}}{2}\left[\left(1 + 4\mu_0^2\frac{K^2 M_J'^2}{J^2(J+1)^2}\cdot\frac{E_m^2}{\hbar^2\omega_{\text{inv}}^2}\right)^{1/2} - 1\right]; \tag{2.20}$$

ω_{inv} is the frequency of the inversion transition without including the hyperfine structure ($\omega_l \simeq \omega_{\text{inv}}$); E_m is the maximum field strength in the sorting system, which depends on U in a way fixed by the construction of the sorting system. For a quadrupole condenser $E_m = U/R$, where R is the radius of the entrance aperture in the quadrupole condenser.

We see that W depends on the quantum number M'_J. The state with the quantum numbers $I_N M'_{I_N}$; JKM'_J is uniquely defined by giving the quantum numbers $I_N JKF_1 M'_{F_1}$. A diagram of the relation between the quantum

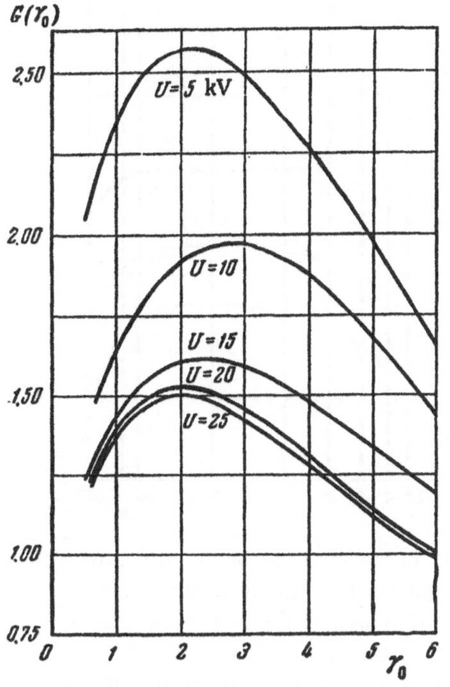

Fig. 3. Curves for the function $G(U, \gamma_0)$.

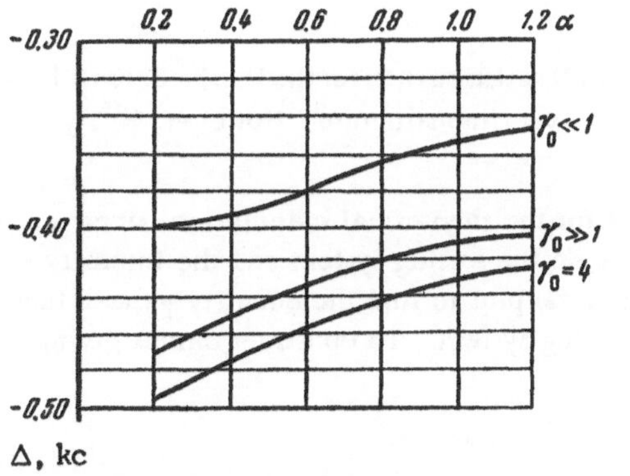

Fig. 4. The shift of the peak of the spectral line of $N^{14}H_3$, J = 3, K = 3, $\Delta F_1 = \Delta F$ = 0, owing to the presence of hyperfine structure, as a function of the saturation γ_0 and the voltage on the sorting system

$$\left(\alpha = 4\mu_0^2 \frac{K^2 M_I^2}{J^2(J+1)^2} \cdot \frac{U^2}{R^2} \cdot \frac{1}{h^2\omega_{\text{inv}}^2}\right).$$

numbers $I_N M'_{I_N} J K M_J$ and $I_N J K F_1 M'_J$ is shown in Fig. 2 [47]. By referring to this diagram of the correspondence between quantum numbers we can determine what maximum energy $W_{F_1 M'_{F_1}}$ is involved in the distribution of times of flight of the molecules which contribute to any particular component of the emission line. The distribution $f_{F_1 M'_{F_1}}(\tau)$ is then completely determined, and this makes it possible to calculate the oscillation frequency of the maser as a function of the voltage of the sorting system and the saturation parameter γ, which is a function of the number of active molecules entering the resonator.

For the calculations it is convenient to write (2.9) in the form

$$\omega = \omega_l \left[1 - \frac{\omega_l - \omega_0}{\omega_l} \cdot \frac{Q_0}{Q_l} G(U, \gamma_0) + \Delta(U, \gamma_0)\right], \quad (2.21)$$

where*

$$\frac{1}{Q_l} = \frac{2}{\omega_l \overline{\tau}}; \quad \gamma_0 = \frac{\mu_0 \mathcal{E}_0}{\hbar} \cdot \frac{K}{\sqrt{J(J+1)}} \overline{\tau};$$

$$G = \frac{\sum \gamma_{F_1 M_{F_1}} J^c_{F_1 M_{F_1}}}{\sum J^s_{F_1 M_{F_1}}}; \quad \Delta = \frac{1}{\omega_l} \cdot \frac{\sum \eta_{F_1} J^s_{F_1 M_{F_1}}}{\sum J^s_{F_1 M_{F_1}}}. \quad (2.22)$$

The quantities G and Δ have been calculated as functions of the voltage U and the saturation γ_0 with an electronic computing machine for the case of a quadrupole condenser with the ratio L/R = 30 and of the E_{010} wave in the cylindrical resonator. The results of the calculation are shown in Figs. 3 and 4.

As can be seen from Eq. (2.21), the function $G(U, \gamma_0)$ determines the variation of the oscillation frequency of a maser, as it depends on the tuning of the natural frequency ω_0 of the resonator to the frequency ω_l of the spectral line. The presence of hyperfine structure gives an additive correction to the frequency which does not depend on the tuning of the resonator. As can be seen from the curves, the change of the frequency caused by the presence of hyperfine structure when the saturation parameter is varied is a quantity of the order of $2 \cdot 10^{-9}$; when the voltage is changed from 4 to 25 kV, the change of the frequency is about $3\text{-}5 \cdot 10^{-9}$. Therefore, to secure high absolute stability of the oscillation frequency of a maser ($\sim 10^{-10}$ and higher), it is necessary to use emission lines in which there is no hyperfine structure.

A line of this sort in $N^{14}H_3$ is the inversion transition line J = 3, K = 2.** A disadvantage of this line is its relatively weak

*It is convenient to split off from the quantity $\gamma_{F_1 M_{F_1}}$ a factor γ_0 which does not depend on the quantum numbers $F_1 M_{F_1}$; μ_0 is the dipole moment of the nonrotating molecule, so that

$$\mu_{F_1 M_{F_1}} = \mu_0 \cdot \frac{K M_F}{J(J+1)} \cdot \frac{F_1(F_1+1) + J(J+1) - I_N(I_N+1)}{2F_1(F_1+1)}.$$

**For this line there is no quadrupole hyperfine structure, but there is a magnetic hyperfine splitting of the line

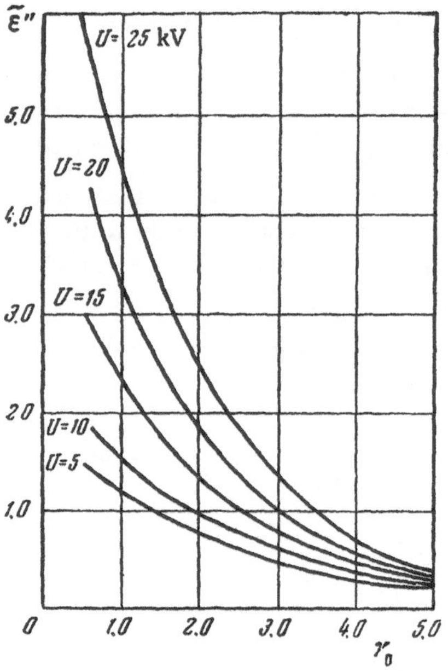

Fig. 5. Curves of the imaginary part of the complex polarizability $\tilde{\varepsilon}''$ of the molecular beam, calculated according to Eq. (2.25).

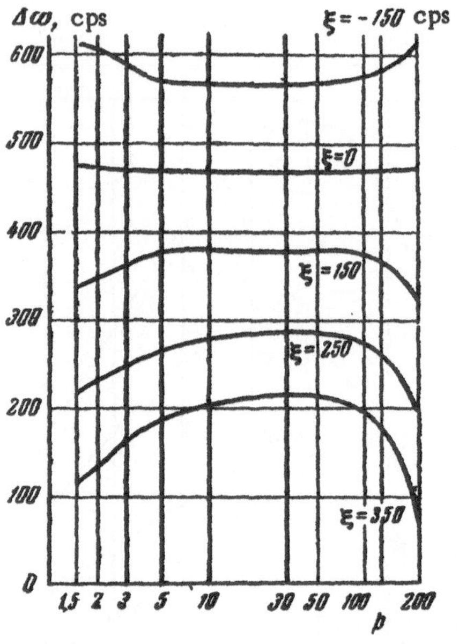

Fig. 6. Dependence of the oscillation frequency of a maser on the pressure in the source of the molecular beam, for the line J = 3, K = 3 of $N^{14}H_3$ when effect III is compensated. Pressure p is given in relative units on a logarithmic scale. $\xi = \frac{Q}{Q_l}(\nu_p - \nu_l)$

intensity (for the line J = 3, K = 2 the condition for self-oscillation is less well satisfied than for the line J = 3, K = 3 by a factor of 10-12). In $N^{15}H_3$ the line J = K = 3 is an intense line without hyperfine structure. However, $N^{15}H_3$ is very expensive.

The use of the saturation parameter as a variable is convenient for the theoretical calculation, since γ_0 appears naturally in (2.21) and (2.22). On the other hand, the voltage on the sorting system and the intensity of the molecular beam are independent variables. Therefore, it would be helpful to find the quantity γ_0 as a function of the intensity of the molecular beam and the voltage on the sorting system. To construct curves giving the quantity γ_0 we use Eq. (2.4):

$$\varepsilon'' = \frac{1}{Q_0}, \tag{2.23}$$

where $\varepsilon'' = 4\pi P_0''/\mathcal{E}_0$ is the imaginary part of the complex polarizability of the molecular beam and is given by the relation

$$\varepsilon'' = 4\pi AN_{JK} \frac{|\mu_0|^2}{\hbar} \tau^2 \frac{1}{\gamma_0^2} \sum_{F_1 M_{F_1} M'_{F_1}} \gamma_{F_1 M_{F_1}} a^{F_1}_{M_{F_1} M'_{F_1}} J^c_{F_1 M_{F_1}}. \tag{2.24}$$

Here $AN_{JK} = g_{JK} \frac{N_0}{\Sigma_{st}} e^{-E_{JK}/kT}$, where N_0 is the total number of molecules in the beam, Σ_{st} is the partition function for a given molecule, and g_{JK} is the statistical weight of the state.

---

[40]. This splitting is small, however, and the frequency shift owing to the magnetic hyperfine structure is of an order of magnitude not greater than 10^{-11}. A calculation of the hyperfine structure of the line J = 3, K = 2 has been made in [48].

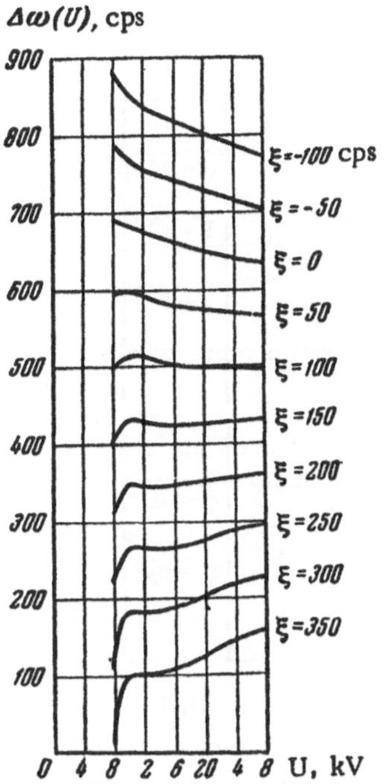

Fig. 7. Dependence of oscillation frequency of a maser on the voltage on the sorting system, for the line J = 3, K = 3 of $N^{14}H_3$ when effect III is compensated. (Curves constructed on the basis of calculation.)

If we take the factor $4\pi AN_{JK}\frac{|\mu_0|^2}{\hbar}\bar{\tau}^2$ over into the left member then for a given construction of the sorting system the right member of Eq. (2.23) is a universal function of U and γ_0:

$$\tilde{\varepsilon}'' = \frac{1}{\gamma_0^2} \sum_{F_1 M_{F_1} M'_{F_1}} \gamma_{F_1 M_{F_1}} a^{F_1}_{M_{F_1} M'_{F_1}} J^c_{F_1 M_{F_1}}. \qquad (2.25)$$

The dependence on the number of molecules, the quality figure of the resonator, etc., will be completely determined by the right member of this equation.

The quantity $\tilde{\varepsilon}''$ has also been calculated with an electronic computing machine and is shown in Fig. 5. If we fix the voltage on the sorting system, the intensity of the molecular beam, and the quality figure of the resonator and know the quantity $\bar{\tau}$, we can use Fig. 5 to find γ_0. The value of γ_0 so found can be used to calculate the oscillation frequency of the maser and its power for prescribed values of the beam intensity and the voltage on the sorting system.

Thus, in order to calculate the oscillation frequency of a maser as a function of the pressure, it is necessary to know the flux of molecules from the source of the molecular beam as a function of the pressure in the source. Unfortunately, it is difficult to calculate the flux of molecules from the source of the molecular beam, since the flow of the gas out of the source differs from viscous flow and also from molecular flow, being of an intermediate character [44-46], which has not been studied theoretically. Moreover, when the pressure in the source of the molecular beam is rather large (~2-4 mm Hg) the vacuum in the bulb of the maser becomes very poor, and this damages the molecular beam and decreases the power of the oscillations of the maser [17, 44, 45, 49]. When the pressure is made still larger the oscillations cease altogether. Thus, at sufficiently high pressures, the amplitude of the oscillations of the maser (and thus also the quantity γ_0) is not determined uniquely by the flux of molecules from the source, but depends on the condition of the vacuum in the bulb of the maser, and this still further complicates the dependence on the pressure in the source of the molecular beam. Therefore, in calculating the oscillation frequency, it is simplest to use an experimental curve of the dependence of the power of the maser on the parameters of the apparatus (in particular, the pressure in the source of the molecular beam).

In calculating the oscillation frequency of a maser working with the line J = 3, K = 3 of $N^{14}H_3$, we have used the results of measurements of the power of the maser as a function of the pressure in the source of the molecular beam [50]. It is true that these measurements are of a comparative character and enable us to determine only the relative change of γ_0 with a change of the pressure, but this is enough to calculate the nature of the variation of the oscillation frequency of the maser with changes of the pressure in the source of the molecular beam. The results of the calculation are shown in Fig. 6.

The calculation of γ_0 as a function of the voltage for a fixed pressure in the source of the molecular beam is not particularly difficult and can be made by using Fig. 5. We have in this way made a calculation of the oscillation frequency of a maser (line J = 3, K = 3 of $N^{14}H_3$) as a function of the voltage on the sorting system. The results of the calculation are shown in Fig. 7.

By using the dependence of the function G on the pressure and the voltage we can make a calculation of the oscillation frequency of a maser working on the line J = 3, K = 2 of $N^{14}H_3$. As has already been pointed out, the inversion line of $N^{14}H_3$, J = 3, K = 2 differs from the line J = 3, K = 3 by the absence of quadrupole hyperfine structure in the first approximation of perturbation theory. The presence of a quadrupole coupling in the second approximation of perturbation theory and of a magnetic hyperfine structure can be neglected, since the inclusion

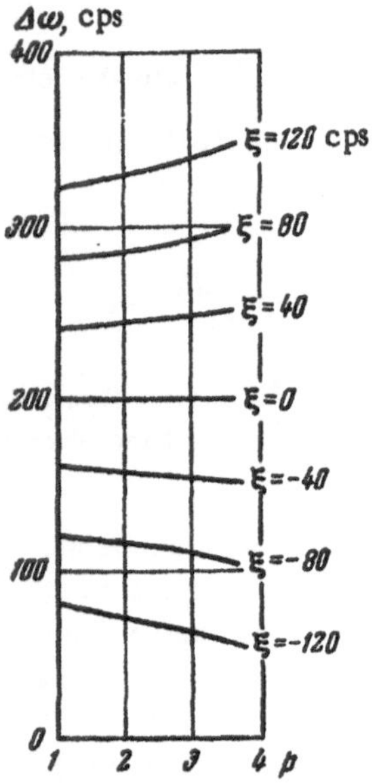

Fig. 8. Dependence of oscillation frequency of a maser on the pressure in the source of the molecular beam for the line $J = 3$, $K = 2$ of $N^{14}H_3$, with effect III compensated. Pressure is given in relative units (the curves are drawn on the basis of calculations).

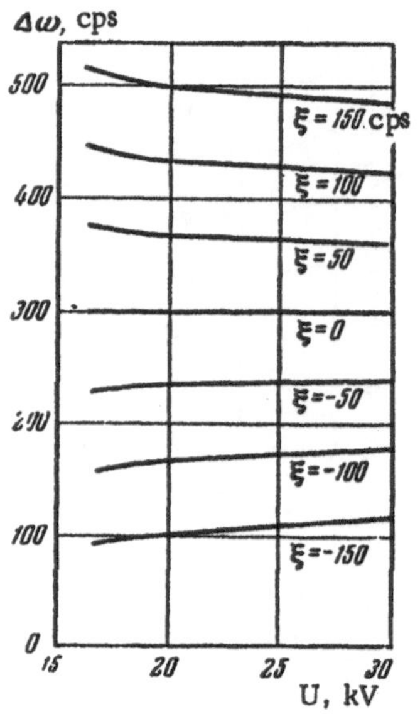

Fig. 9. Dependence of oscillation frequency of a maser on the voltage on the sorting system for the line $J = 3$, $K = 2$ of $N^{14}H_3$, with effect III compensated (the curves are drawn on the basis of calculations).

of these effects introduces changes in the oscillation frequency of a maser of the order of only 10^{-11}. Therefore, in calculating the oscillation frequency of a maser working on the line $J = 3$, $K = 2$ of $N^{14}H_3$, we are to set $\Delta = 0$ in Eq. (2.21).

The function $G(U, \gamma_0)$ calculated for the line $J = 3$, $K = 3$ of $N^{14}H_3$ will be correct also for the line $J = 3$, $K = 2$ if, in Fig. 3, we reduce the scale on the axis of γ_0 by a factor 1.5, since the matrix element of the dipole moment for an inversion transition of ammonia is proportional to K. Thus, if we know γ_0 as a function of the pressure for the line $J = 3$, $K = 2$, we can calculate the oscillation frequency of the maser. Unfortunately, we do not know of any papers in which there were measurements of the amplitude of the oscillation of a maser working on the line $J = 3$, $K = 2$. Therefore, we have used the γ_0 obtained for the line $J = 3$, $K = 3$ [50] and have allowed for the facts that: 1) the statistical weight of the inversion levels $J = 3$, $K = 2$ of ammonia is only half of that for the levels $J = 3$, $K = 3$; 2) the matrix element of the dipole moment for the inversion transition $J = 3$, $K = 2$ is smaller by a factor of 1.5 than that for the transition $J = 3$, $K = 3$, which means that there is less effective sorting of molecules which are in the level $J = 3$, $K = 2$.

On this basis we have calculated the oscillation frequency of a maser working on the line $J = 3$, $K = 2$ of $N^{14}H_3$ as a function of the pressure in the source of the molecular beam and of the voltage on the sorting system. The results of the calculation are shown in Figs. 8 and 9.

3. Comparison of the Theoretical and Experimental Characteristics of a Maser

For a comparison of our calculations with experimental data we have used the results of [50, 16, 51]. In [50] the oscillation frequency of a maser working with the line $J = 3$, $K = 3$ with one counter beam was studied as a function of the pressure in the source of the molecular beam and of the voltage on the sorting system. In [16] a detailed study was made of the dependence of the frequency of a maser ($J = 3$, $K = 3$) on the voltage on the sorting system. Finally, in [51], the dependence of the oscillation frequency on the pressure and voltage are given for a maser working on the line $J = 3$, $K = 2$. Furthermore, in [51], the dependence of the pressure was found both

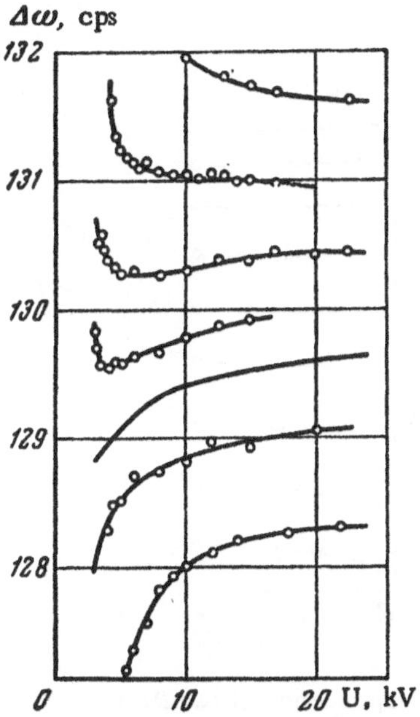

Fig. 10. Dependence of oscillation frequency of a maser on the voltage on the sorting system, for the line J = 3, K = 3 (experimental curves from [16]).

with effect III compensated (two identical beams in opposite directions), and without compensation of this effect (single beam), and this permits a more detailed comparison of the calculated and experimental characteristics. Figures 10-12 (see also Figs. 13 and 18 of [6, 52]) present curves taken from the papers we have cited.

Comparing the theoretically calculated curves for the line J = 3, K = 3 with the experimental curves, we can note a rather good qualitative agreement between them. The difference in the behavior of the frequency as a function of the pressure, which is most noticeable for small differences between ω_l and ω_0, is explained by the influence on the oscillation frequency of the maser of effect III, which was not compensated in the determination of these experimental characteristics and which has a considerable effect on the oscillation frequency of a maser for small mistunings.

In the theoretical plot (see Fig. 7) of the variation of the oscillation frequency of a maser with the voltage there is a steep change of the frequency at small voltages, which is less pronounced in the experimental plot (see Fig. 18 of [52]). To explain this we note that the cause of a steep change of the frequency is a rapid change of the effective line width at small voltages owing to the finite ratio R/L of the radius of the sorting system to its length: for finite R/L the relative number of slow molecules in the beam increases with increase of the voltage. If, on the other hand, L/R is infinite, then a change of the voltage affects only the number of active molecules but does not change the ratio between the "slow" and " fast" molecules.

It is now easy to explain the apparent discrepancy between the theoretically calculated (see Fig. 7) and experimental (see Fig. 18 of [52]) variations of the frequency with changes of the voltage. It was assumed in the calculation that L/R = 30; in the generator used in [50], the value was L/R = 50. Therefore, the dependence of the frequency on the voltage measured in [50] should be somewhat flatter, and the sharp change of the oscillation frequency with change of the voltage should appear for smaller values of U than is the case in the calculation. In [50], however, the range of voltages below 8-10 kV was not studied. In [16], where the sorting system had approximately the same length-to-radius ratio as in [50], the author measured a range of voltages down to 4-5 kV (see Fig. 10). As can be seen from this figure, there is in this case a sharp change of the oscillation frequency of the maser at small values of U.

Let us turn to the experimental data for a generator working on the line J = 3, K = 2 (see Figs. 11 and 12). In the measurements for the curves shown in Fig. 11a, a generator with two oppositely moving beams was used, so that effect III was compensated. Comparing these characteristics with those calculated theoretically for a generator using the same line, we see that the theoretical and experimental characteristics are of precisely the same type.

For a quantitative comparison of the calculated results with experiment, both for the generator using the line J = 3, K = 2 and for that using the line J = 3, K = 3, one needs absolute measurements of the power of the maser, which are hard to make with a sufficient degree of accuracy, so that no such measurements were made in the experimental papers in question. For just this reason we could not give the scale of abscissas in absolute units in our curves from the theoretical calculations (see Figs. 6 and 9). It must be noted, however, that by changing the scale on the axis of pressures one can secure good agreement of the calculated curves (see Fig. 8) with the experimental curves (see Fig. 11a) for the line J = 3, K = 2.*

We note that there is no choice of the scale on the pressure axis which will bring the experimental curves

*The securing of absolute coincidence between the theoretical curves and the experimental curves for the line J = 3, K = 3 has no meaning, since the theoretical characteristics have been calculated for a generator in which effect III is compensated and in the experiment this effect was not compensated.

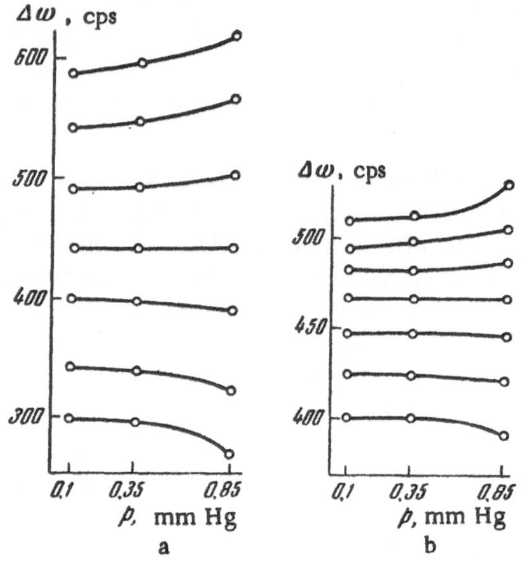

Fig. 11. Dependence of oscillation frequency of a maser on the pressure in the source of the molecular beam, for the line J = 3, K = 2 of $N^{14}H_3$ (experimental curves from [51]). (a) With effect III compensated;(b) for a generator with a single molecular beam.

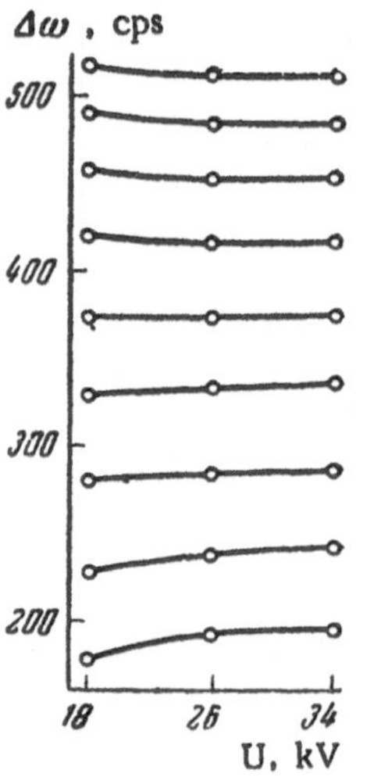

Fig. 12. Dependence of oscillation frequency of a maser on the voltage on the sorting system for the line J = 3, K = 2 of $N^{14}H_3$ with effect III compensated (experimental curves from [51]).

for the line J = 3, K = 2 which were taken without compensation of effect III (see Fig.11b) into agreement with the theoretical curves calculated for the case of compensation of effect III. This is a rather convincing quantitative proof of the correctness of our ideas on which the calculation was based. Thus, we can state that as the result of our calculation a formula has been obtained which describes to good accuracy the dependence of the oscillation frequency of a maser on the various parameters.

4. The Effect of an External Force on the Oscillation Frequency of a Maser

In Part 4 of Section 1 we have considered the action of an external force on an excited maser and have calculated the region of locking of the oscillations of the maser to the external force. The condition for locking to occur is of the form

$$\delta \leqslant \frac{1}{2} \Delta\omega_l \sqrt{\frac{P_{ext}}{P_0}}, \tag{2.26}$$

where δ is the mistuning between the peak of the spectral line and the frequency of the applied force, $\Delta\omega_l$ is the original width of the spectral line, P_{ext} is the power of the external force, and P_0 is the power of the maser in self-oscillatory operation.

If condition (2.26) is not satisfied, a biharmonic condition will exist in the generator – there will be forced vibrations with the frequency of the external force and also self-oscillations. The amplitude R of the self-oscillations depends on the applied force X, so that [cf. Eq. (1.57)]

$$R^2 = R_0^2 - \frac{X^2}{\delta_2} \cdot \left(\frac{\Delta\omega_l}{\Delta\omega_0}\right)^2, \tag{2.27}$$

where $\Delta\omega_0$ is the transmission band of the resonator and R_0 is the amplitude of the self-oscillations without the action of the external force. It can be seen from Eq. (2.27) that for small mistuning even a small external force can cause a large change in the amplitude of the self-oscillations.

It has been shown in the present section that the difference between the actual shape of the line and the Lorentz shape, which it has when the distribution of times of flight through the resonator for the molecules has the form $(1/\bar{\tau})e^{-\tau/\bar{\tau}}$, leads to a considerable dependence of the oscillation frequency of a maser on the amplitude (γ_0). Therefore, in actual apparatus there will not only be locking of the maser when condition (2.26) is satisfied, but also a change of the oscillation frequency of the maser under the action of the external force, even if (2.26), the condition for locking, is not satisfied.

Experimental studies of the dependence of the frequency of a maser on various parameters, tests of the frequency stability, etc., are usually made by comparing the oscillations of two masers. It is then possible for the power of one maser to penetrate into the other. If the frequencies of the two masers are nearly equal, then even a weak interaction can change the frequencies of both masers in comparison with the frequencies of noninteracting masers. This peculiar mutual pulling of the frequencies of two masers must be taken into account in order to obtain reliable experimental results. It is best to eliminate the mutual influence of the masers by using decoupling devices.

5. Influence of Nearby Lines on the Oscillation Frequency of a Maser and the Effect of Nonadiabatic Conditions in the Focusing of the Molecules

Here we shall estimate the effect of nearby lines on the oscillation frequency of a maser. To calculate the frequency of the maser we can use Eq. (2.5), taking P to be the sum

$$P = P_{\text{main}} + \sum_i P_i, \tag{2.28}$$

where P_{main} is the polarization of the beam caused by the main transition and $\sum_i P_i$ is the sum of the polarizations caused by the presence of "extraneous" nearby lines. Assuming that the distance from the main line to the nearest adjacent line is much larger than the line width, we can take $\sum_i P_i \ll P_{\text{main}}$. In this case (2.5), the equation for determining the oscillation frequency of a maser, can be put in the following form:

$$\frac{P'_{\text{main}}}{P''_{\text{main}}} = 2Q\frac{\omega_0 - \omega}{\omega} - \sum_i \frac{P'_i}{P_{\text{main}}}. \tag{2.29}$$

As usual, we shall look for the oscillation frequency of the maser in the form

$$\omega = \omega_l(1 + \delta). \tag{2.30}$$

Substituting Eq. (2.30) in Eq. (2.29) and neglecting terms δ^n with $n > 1$, we find that

$$\delta \equiv \frac{\omega - \omega_l}{\omega_l} = \frac{1}{Q_l}\sum_i \frac{P'_i}{P''_{\text{main}}} + \frac{Q}{Q_l}\cdot\frac{\omega_0 - \omega_l}{\omega_l}G. \tag{2.31}$$

For $\omega_0 = \omega_l$ we have

$$\frac{\omega - \omega_l}{\omega_l} = \frac{1}{Q_l}\cdot\sum_i \frac{P'_i}{P''_{\text{main}}}. \tag{2.32}$$

For the calculation of the right member of Eq. (2.32) we assume for simplicity that the distribution of the times of flight of molecules through the resonator is given by the law $(1/\bar{\tau})e^{-\tau/\bar{\tau}}$ and that the generator is one which uses the inversion transition line $J = 3$, $K = 3$, $\Delta F_1 = \Delta F = 0$ of $N^{14}H_3$. This line has eighteen symmetrically located magnetic satellites caused by transitions $\Delta F_1 = 0$, $\Delta F = \pm 1$. The distance from the main transition to these side components is about 50 kc. The intensities of the transitions $\Delta F_1 = 0$, $\Delta F = \pm 1$ are smaller than that of the main transition by approximately a factor 100. If, besides this, we use the fact that the distance from the main line to these satellites is several tens of times greater than the line width of the emission, i.e., the transitions $\Delta F_1 = 0$, $\Delta F = \pm 1$ are nonresonance transitions, the condition $\sum_i P_i \ll P_{\text{main}}$ is here satisfied quite well.

Therefore, in the calculation we can assume that the transitions $\Delta F_1 = 0$, $\Delta F = \pm 1$ have practically no effect on the occupation numbers of the energy levels, and that the distribution of the molecules over the energy levels is completely determined by the main transitions $\Delta F_1 = \Delta F = 0$.

In this case, a simple calculation using (2.8) shows that

$$\frac{|\omega-\omega_l|}{\omega_l}=\frac{1}{Q}\,\frac{\sum\limits_{F_1,F,M_F}\left|\mu_{F\to F+1}^{F,M_F}\right|^2\cdot\left|N_{F,M_F}^{F_1}-N_{F+1,M_F}^{F_1}\right|\cdot\left|(\omega_l-\omega_{F\to F+1}^{F})\tau\right|^{-1}}{\sum\limits_{F_1,F,M_F}\left|\mu_{F\to F}^{F,M_F}\right|^2 N_{F,M_F}^{F_1}\left(1+\left|\gamma_{F,M_F}^{F_1}\right|^2\right)^{-1}}, \tag{2.33}$$

where $\mu_{F\to F\pm1}^{F,M_F}$ is the matrix element of the dipole moment for transitions $\Delta F_1 = 0$, $\Delta F = \pm 1$; $\mu_{F\to F}^{F,M_F}$ is the matrix element of the dipole moment for transitions $\Delta F_1 = \Delta F = 0$; $N_{F,M_F}^{F_1}$ is the number of molecules in the state F_1, F, M_F; and $\omega_{F\to F\pm1}^{F_1}$ is the frequency of the transition $F, F_1 \to F \pm 1, F_1$.

In Eq. (2.33) we have used the fact that the frequencies of the side lines $\Delta F_1 = 0$, $\Delta F = \pm 1$ are symmetrically located relative to the main frequency $\omega_l = \omega_{F\to F}^{F_1}$, so that

$$\omega_l=\frac{1}{2}\left(\omega_{F\to F+1}^{F_1}+\omega_{F\to F-1}^{F_1}\right). \tag{2.34}$$

Because of the fact that kT is much larger than the energy of the hyperfine interaction, in the state of thermodynamic equilibrium we have for any two states F_1FM_F, $F_1F'M_F'$ the relation

$$N_{F,M_F}^{F_1}=N_{F',M_F'}^{F_1}. \tag{2.35}$$

For adiabatic sorting,condition (2.35) is violated, and the difference of the fractional populations of two states is of the same order of magnitude as the average population. In this case, the mistuning given by formula (2.33) can be quite sizable: for $|\omega_l-\omega_{F\to F+1}^{F_1}|\simeq 50$ kc, $\bar{\tau} = 2 \cdot 10^{-4}$ sec, $Q_l = 10^7$, and $\gamma^2 = 5\text{-}10$,

$$\left|\frac{\omega_l-\omega}{\omega_l}\right|\simeq 10^{-8}-5\cdot 10^{-9}. \tag{2.36}$$

As has been pointed out in [40], however, when the sorting of the molecules occurs in a quadrupole condenser, the condition for adiabaticity

$$\omega\bar{\tau}\ll 1 \tag{2.37}$$

is not satisfied for frequencies of the order of 50 kc.* Because of this, on the emergence of the molecules from the quadrupole condenser there will be intense "pure magnetic" transitions between the levels of the magnetic hyperfine structure, which will equalize the populations of the levels.

In [40] the intensities of the magnetic transitions $\Delta F_1 = 0$, $\Delta F = \pm 1$ were observed to be equal to a high degree of accuracy. Therefore, in the usual apparatus, the frequency shift caused by the presence of nearby lines evidently does not exceed 10^{-10}. For the sublevels caused by the quadrupole hyperfine structure, separated by distances of the order of 1.5-2 Mc, the adiabaticity condition (2.37) is usually satisfied, and the quadrupole hyperfine structure levels characterized by quantum numbers $F_1M_{F_1}$ contain different numbers of molecules after the sorting. It is true that this does not lead to any appreciable influence of the allowed components of the quadrupole hyperfine structure on the oscillation frequency of the maser, since these levels are rather far from ω_l (1.5 Mc), but it does cause the influence of unresolved quadrupole components which has been noted above (effect II, see Part 1 of this section).

*In order of magnitude the time is a/v, where a is the distance in which the electric field changes from zero to its maximum value at the entrance to the sorting system (or at the exit), and v is the average velocity of the molecules along the axis of the sorting system.

Complete equalization of the populations of the quadrupole sublevels after the sorting would eliminate effect II. This equalization can be accomplished by using a radio-frequency field (~1.5 Mc) to induce pure quadrupole transitions in the beam of sorted molecules before they enter the resonator.

Section 3

METHODS FOR TUNING THE OSCILLATION FREQUENCY OF A MASER TO THE FREQUENCY OF A SPECTRAL LINE

1. General Remarks on the Tuning of the Oscillation Frequency of a Maser to the Frequency of a Spectral Line

As was shown in the preceding section, the oscillation frequency of a maser can differ from the frequency of the spectral line, and the difference between the oscillation frequency of the maser and the frequency of the spectral line varies with changes of the parameters of the apparatus. Therefore, for a frequency standard, it is necessary to secure the maximum possible agreement between the oscillation frequency of the maser and the frequency of the spectral line. Equation (2.21) shows that the smaller the mistuning between the resonator frequency and the peak of the spectral line, the closer the agreement between the oscillation frequency of the maser and the frequency of the spectral line. When the mistuning is zero, the influence of such factors as a change of the intensity of the molecular beam or the voltage on the sorting system is very much smaller, and for a single spectral line is given only by effect III (nonuniform emission by the molecules along the length of the resonator).

In practice it is hard to determine the natural frequency of a loaded resonator to a sufficient degree of accuracy because of the distortion of the amplitude – phase characteristics of the resonator by the waveguides which are connected to it [see Eq. (1.61)]. When the resonator quality is $\sim 4\text{-}8 \cdot 10^3$ and that of the spectral line is $\sim 10^7$, it is hard to secure a relative coincidence of the oscillation frequency of the maser with the frequency of the spectral line to an accuracy greater than 10^{-8}. A more effective method of tuning the frequency of the resonator to that of the spectral line is to use the dependence of the oscillation frequency of the maser on the various parameters.

From Eq. (2.21),

$$\omega = \omega_l \left[1 - \frac{Q}{Q_l} \cdot \frac{\omega_l - \omega_0}{\omega_l} G(U, \gamma) + \Delta \right], \tag{3.1}$$

it follows that for $\omega_0 \neq \omega_l$ the oscillation frequency ω will change if we change (1) $G(U, \gamma)$ by modulating the intensity of the molecular beam or the voltage on the sorting system, or (2) the quality (width) Q_l of the spectral line. Usually, by observing the difference frequency of two masers (the one being tuned and an auxiliary device) while one readjusts the frequency of the resonator, one can secure a decrease of the dependence of the difference frequency on the voltage of the sorting system or the pressure in the source of the molecular beam. In generators which use the line J = 3, K = 3 of $N^{14}H_3$, it has been possible in this way to secure an absolute stability of the order of 10^{-9} [42, 46, 48-50, 53, 54].

In the most general case the procedure of tuning the frequency of a maser to that of a spectral line by changing the parameters of the maser consists in using readjustment of the resonator frequency to achieve a minimum change of the oscillation frequency with changes of a parameter of the maser over a fixed range. The frequency obtained in this way is regarded as the standard frequency. Because of effects II and III, however, it may not be the same as ω_l. Moreover, the frequency obtained by varying one of the parameters will depend on the other parameters, and for different values of these other parameters one will in general obtain different oscillation frequencies of the maser [49, 53, 54].

If, however, one uses in a maser two symmetrical oppositely moving molecular beams and a line without hyperfine structure, so that effects II and III are absent, then, according to Eq. (2.21), the frequency of the maser is given by the formula

$$\omega = \omega_l \left[1 - \frac{Q}{Q_l} \cdot \frac{\omega_l - \omega_0}{\omega_l} \cdot G(U, \gamma_0) \right]. \tag{3.2}$$

In this case, one can use a precise tuning to secure coincidence of the oscillation frequency of the maser with the frequency ω_l of the spectral line, and this coincidence will not depend on any of the parameters. When one knows the dependence of the oscillation frequency of a maser on the various parameters one can determine the possible value of the absolute frequency stability of the generator.

Let the oscillation frequency depend on the mistuning and on a certain parameter which can be varied over the range* $x_{min} < x < x_{max}$. Let us denote by $\Omega(\xi)$ the maximum change of the frequency when x is varied over the range (x_{min}, x_{max}) with fixed ξ, i.e.,

$$\Omega(\xi) = \max\{|\omega(\xi, x_1) - \omega(\xi, x_2)|\}, \quad x_{min} < x_{1,2} < x_{max}. \tag{3.3}$$

Also, let ξ_0 be the value of the mistuning for which

$$\Omega(\xi_0) = \min \Omega(\xi). \tag{3.4}$$

Then the maximum obtainable absolute frequency stability δ_m of the maser which can be obtained by tuning with changes of the parameter x is given by

$$\delta_m = \frac{\Omega(\xi_0)}{\omega_l}. \tag{3.5}$$

This value δ_m, however, can be attained only when one uses exact measurements of the difference frequency of two masers. If there is an inaccuracy α in the measurement, then the true value of the stability is $\delta > \delta_m$. We adopt the definition that $\Delta\xi = |\xi_0 - \xi|$ is the change of the mistuning for which

$$|\Omega(\xi_0 + \Delta\xi) - \Omega(\xi_0)| \simeq \alpha. \tag{3.6}$$

The absolute stability at $\alpha = 0$ is determined by the maximum value of the change of frequency over the ranges $(\xi_0 - \Delta\xi, \xi_0 + \Delta\xi)$ and (x_{min}, x_{max}):

$$\left.\begin{array}{c} \delta = \max|\omega(\xi_1, x_1) - \omega(\xi_2, x_2)|, \\ \xi_0 - \Delta\xi < \xi_{1,2} < \xi_0 + \Delta\xi, \\ x_{min} < x_{1,2} < x_{max}. \end{array}\right\} \tag{3.7}$$

The interval $\Delta\xi$, which depends on α, is determined from Eq. (3.6). If $\Delta\xi$ is small and the function $\Omega(\xi)$ is continuous, we can write an explicit expression for $\Delta\xi$ by expanding the left member of Eq. (3.6) in a series:

$$\Delta\xi = \left[\frac{d\Omega}{d\xi}\right]^{-1}_{\xi=\xi_0} \cdot \alpha. \tag{3.8}$$

These arguments enable us to estimate the absolute stability from the characteristics of the maser.

As can be seen from Fig. 10, for a maser with one beam working on the inversion line J = 3, K = 3 of $N^{14}H_3$, we have $\Omega(\xi)/\omega_l \approx 10^{-9}$, so that the maximum absolute frequency stability of such a device is $\sim 10^{-9}$.

Elimination of effect III by the use of two oppositely moving equal molecular beams makes it possible to obtain a greater absolute stability. For such a generator (Fig. 6), $\delta_m = 5\text{-}7 \cdot 10^{-10}$. The oscillation frequency of a maser which uses a line without hyperfine structure and operates with two identical and oppositely moving beams is described by (3.2). For such a device $\Omega(\xi_0) = 0$ for $\xi_0 = 0$, so that the absolute stability is determined by nothing but the accuracy of the frequency measurements.

If we take the parameter x to be the pressure p, then (3.8), the relation for the frequency determined by formula (3.2), can be rewritten in the form

$$\frac{|\omega_0 - \omega_l|}{\omega_l} \cdot \frac{Q}{Q_l} \cdot \max|G(p_1) - G(p_2)| = \alpha, \quad p_{min} < p_{1,2} < p_{max}. \tag{3.9}$$

*One can choose the range of variation of the parameter which is most suitable for the tuning by using the characteristics of the maser.

From this we have

$$\frac{\omega_0 - \omega_l}{\omega_l} \cdot \frac{Q}{Q_l} = \frac{\alpha}{\max |G(p_1) - G(p_2)|}, \tag{3.10}$$

and the absolute stability of the generator will be given by

$$\delta = \frac{\alpha G_m}{\max |G(p_1) - G(p_2)|}, \tag{3.11}$$

where G_m is the maximum value of G(p) in the range

$$p_{min} < p < p_{max}.$$

It is not hard to find from Figs. 8 and 11a the values of the quantities which are involved in relation (3.11). Substitution of these values into Eq. (3.11) shows that to tune the oscillation frequency of the maser to the frequency of the spectral line J = 3, K = 2 ($\nu_l = \omega_l / 2\pi \approx 2.2 \cdot 10^{10}$ cps) with accuracy $\sim 10^{-10}$ by changing the pressure it is necessary to measure the difference frequency with an accuracy down to ~1 cps. As can be seen from Figs. 9 and 12, tuning by varying the voltage requires a greater accuracy in the frequency measurements (~0.2-0.3 cps).

2. The Method of Modulating the Line Width by an External Magnetic Field

One can also carry out the tuning of the frequency of a maser by changing the width of the spectral line through the action of an external magnetic field [15, 17]. The application of an external magnetic field splits the spectral line into several components, and owing to this there is an increase in the effective width of the line. By modulating the strength of the magnetic field we change Q_l and along with it we change the oscillation frequency of the maser if $\omega_0 \neq \omega_l$. Only an exact tuning of the resonator frequency to the peak of the spectral line will remove the change of the oscillation frequency.

We shall calculate the effect of the magnetic field on the width of the spectral line and give an estimate of the maximum value of the magnetic field strength which can be used without causing an appreciable shift of the peak of the spectral line.

For the use of the method of modulating the line width by an external magnetic field it is necessary that the direction of the magnetic field producing the modulation be perpendicular to the electric field in the resonator which is produced by the induced transition between the molecular energy levels.* In this case, the selection rules allow transitions $\Delta M = \pm 1$, and we get radiation of several different frequencies. If, on the other hand, the magnetic field producing the Zeeman modulation is parallel to the electric field of the oscillations in the resonator, then the selection rules give $\Delta M = 0$, and we get radiation of only one frequency from the molecules. These facts are illustrated in Fig. 13 for the case in which the magnetic field splits each inversion level into two sublevels, corresponding to a total angular momentum of the molecule of $\frac{1}{2}$.

For simplicity we shall base our further calculation on this case. When the external magnetic field is applied, two spectral lines are produced caused by the two parameters of the energy levels (Fig. 13a); transitions between levels of different pairs are forbidden and therefore the polarizations corresponding to the two spectral lines are additive:

$$\left.\begin{aligned} P' &= P'_{\omega+\delta} + P'_{\omega-\delta}, \\ P'' &= P''_{\omega+\delta} + P''_{\omega-\delta}. \end{aligned}\right\} \tag{3.12}$$

To calculate the oscillation frequency of the maser we use our already known relation (2.5):

$$\frac{P'}{P''} = 2Q\frac{\omega_0 - \omega}{\omega}. \tag{3.13}$$

*For a cylindrical resonator with the waves of type E_{010} excited, the direction of the electric field is along the axis of the resonator. Consequently, the direction of the magnetic field must be perpendicular to the axis of the resonator.

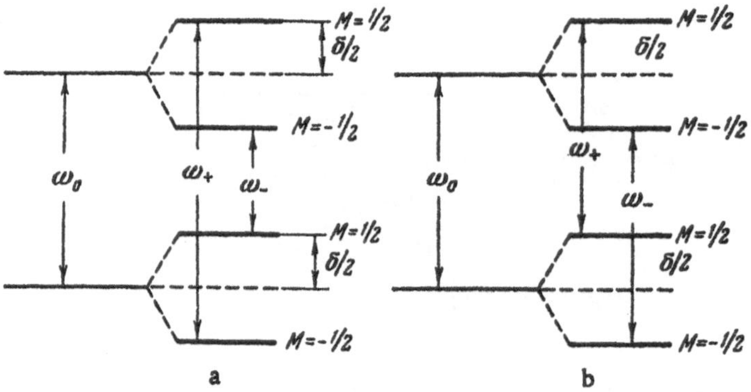

Fig. 13. Schematic representation of the energy levels of the molecule. a) $M = \pm 1$; $\omega_+ = \omega_0 + \delta$; $\omega_- = \omega_0 - \delta$. b) $M = 0$, $\omega_+ = \omega_-$. The thin and broken lines are the levels without the magnetic field; the heavy lines are the levels with the effect of the magnetic field included.

We can calculate the value of $P_{\omega \pm \delta}$ by using relations (2.7) and the velocity distribution $(1/\bar{\tau})e^{-\tau/\bar{\tau}}$. Substituting the values of P' and P" so obtained in Eq. (3.13), we calculate the oscillation frequency of the maser and find for it the expression

$$\omega = \omega_l \left[1 - \frac{\omega_l - \omega_0}{\omega_l} \frac{Q_0}{Q_l} \cdot \frac{1 + \gamma^2 + \Delta^2}{1 + \gamma^2 - \Delta^2} \right], \tag{3.14}$$

where γ is the saturation parameter, equal to $(\mu \mathcal{E}_0/\hbar)\bar{\tau}$, and $\Delta = \delta/\omega_l \bar{\tau}$.

It can be seen from a comparison of Eqs. (3.11) and (3.14) that, when the magnetic field is applied, the effective line width is changed and becomes equal to

$$\Delta\omega_{\text{eff}} = \Delta\omega_{l0} \frac{1 + \gamma^2 + \Delta^2}{1 + \gamma^2 - \Delta^2}. \tag{3.15}$$

From the condition $\gamma^2 > 0$ for the existence of the process of generation, it follows that the quantity Δ^2 must satisfy the relation

$$\Delta^2 < \eta^{-1}, \tag{3.16}$$

where η is given by (1.34). Formulas (3.14) and (3.15) are valid only provided that, in the absence of a magnetic field, the molecular transition used in the generator consists of only a single line. This is actually not the case. The magnetic hyperfine structure and the difference between the coupling constants of the hyperfine interactions for the upper and lower inversion levels (see Part 1 of Section 2) make the line for the transition $J = 3$, $K = 2$, $\Delta F_1 = \Delta F = 0$ consist of six unresolved components. Therefore, the peak of the total spectral line will shift when the external magnetic field is applied, and the larger the value of Δ the larger will be this shift.

To estimate the shift of the peak of the spectral line when the external magnetic field is applied, we assume that the line of the transition consists of two unresolved components with frequencies ω_1 and ω_2. When there is modulation by an external field, each of these components is doubled: $\omega_1 \to \begin{matrix} \omega_1 + \delta_1 \\ \omega_1 - \delta_1 \end{matrix}$ and $\omega_2 \to \begin{matrix} \omega_2 + \delta_2 \\ \omega_2 - \delta_2 \end{matrix}$. The real part of the polarizability, $\varkappa' = P'/\mathcal{E}_0$, is the sum of four terms:

$$\varkappa' = \sum_{i=1}^{2} \varkappa' (\omega_i \pm \delta_i). \tag{3.17}$$

The peak of the spectral line is determined by the point on the frequency scale where $\varkappa'$ is zero. In the absence of an external magnetic field $\delta_i = 0$, and when there are two unresolved components, the peak of the line is determined from the equation

$$\sum_{i=1}^{2} \varkappa'(\omega_i) = 0. \tag{3.18}$$

For $\delta_i \neq 0$, the peak of the line is shifted, and its position can be found from the relation

$$\sum_{i=1}^{2} \varkappa'(\omega_i \pm \delta_i) = 0. \tag{3.19}$$

We must find the quantity $\omega_l - \omega_l'$, where ω_l is the zero of Eq. (3.18) and ω_l' is the zero of Eq. (3.19). Assuming, as before, the Lorentz shape for $\varkappa'$ [which corresponds to the time-of-flight distribution $(1/\bar{\tau})e^{-\tau/\bar{\tau}}$], we get for this difference

$$\omega_l - \omega_l' = 3\left\{\sum_{i=1}^{2} \frac{I_i \Delta_i^2 (\omega_i - \omega_l) \cdot \omega_l^{-1}}{1+\gamma_i^2} \cdot \left[\sum_{i=1}^{2} \frac{I_i}{1+\gamma_i^2}\right]^{-1}\right\}. \tag{3.20}$$

For the oscillation frequency of the maser we find the expression

$$\omega = \omega_l \left[1 - \frac{\omega_l - \omega_0}{\omega_l} \cdot \frac{Q_0}{Q_l} \left(1 + 2\frac{\sum_{i=1}^{2} I_i \Delta_i^2}{\sum_{i=1}^{2} \frac{I_i}{1+\gamma_i^2}}\right) + 3\,\frac{\sum_{i=1}^{2} \frac{I_i \Delta_i^2 (\omega_i - \omega_l)/\omega_l}{1+\gamma_i^2}}{\sum_{i=1}^{2} \frac{I_i}{1+\gamma_i^2}}\right]. \tag{3.21}$$

Here I_1 and I_2 are the relative intensities of the components of the line; ω_1 and ω_2 are the frequencies of these components;

$$\omega_l = \frac{I_1\omega_1 + I_2\omega_2}{I_1 + I_2}; \quad \Delta_1 = g_1\beta H\bar{\tau}; \quad \Delta_2 = g_2\beta H\bar{\tau};$$

g_1 and g_2 are the Zeeman splitting factors corresponding to the first and second components of the line; and β is the nuclear magneton. Since the frequency of the tuned generator must not depend on the external magnetic field, it is necessary that

$$2\,\frac{\omega_l - \omega_0}{\omega_l} \cdot \frac{Q_0}{Q_l} \cdot \frac{\sum_i \frac{I_i \Delta_i^2}{1+\gamma_i^2}}{\sum_i \frac{I_i}{1+\gamma_i^2}} - 3\,\frac{\sum_i \frac{I_i \Delta_i^2 (\omega_i - \omega_l)}{(1+\gamma_i^2)\,\omega_l}}{\sum_i \frac{I_i}{1+\gamma_i^2}} = 0,$$

from which we have

$$\frac{\omega_l - \omega_0}{\omega_l} \cdot \frac{Q_0}{Q_l} = 1.5\,\frac{\sum_i \frac{I_i \Delta_i^2 (\omega_i - \omega_l)}{(1+\gamma_i^2)\,\omega_l}}{\sum_i \frac{I_i \Delta_i^2}{1+\gamma_i^2}}.$$

Thus, the frequency of the tuned generator will be given by

$$\omega = \omega_l \left[1 - 1.5\,\frac{\sum_i \frac{I_i g_i^2 (\omega_i - \omega_l)}{(1+\gamma_i^2)\,\omega_l}}{\sum_i \frac{I_i g_i^2}{1+\gamma_i^2}}\right]. \tag{3.22}$$

It can be seen from Eq. (3.22) that the frequency of a generator tuned with modulation by a magnetic field does not coincide with the peak of the spectral line given by formula (3.18), but depends on the intensities and frequencies of the individual components of the line.

Therefore, the frequency of a generator tuned by varying the pressure will be different from that of a generator tuned by modulation with a magnetic field. This frequency difference is of the order of 50-70 cps.*

The formulas we have found are valid for the Lorentz line shape. Deviation of the actual line shape from the Lorentz shape leads us to a dependence of ω on the pressure in the source of the molecular beam and on the voltage on the sorting system, if $\omega_l - \omega_0 \neq 0$. Therefore, the frequency of a maser which is obtained by means of magnetic tuning will, in general, depend on the parameters of the device if $\omega_1 \neq \omega_2$, i.e., if the line consists of more than one component.

The inversion line J = 3, K = 2, $\Delta F = 0$, $\Delta\alpha = 0$ of $N^{14}H_3$ consists** of six transitions: $\frac{9}{2} \to \frac{9}{2}$, $\frac{7}{2}- \to \frac{7}{2}-$, $\frac{7}{2}+ \to \frac{7}{2}+$, $\frac{5}{2}- \to \frac{5}{2}-$, $\frac{5}{2}+ \to \frac{5}{2}+$, $\frac{3}{2}+ \to \frac{3}{2}+$. The frequencies of these transitions are somewhat different because of differences in the interaction constants for molecules which are in the upper and lower energy levels [10, 55], so that the line J = 3, K = 2, $\Delta F = \Delta\alpha = 0$ consists of six components. This difference of the magnetic interaction constants has not been experimentally measured because it is so small.

An order-of-magnitude estimate shows that the six components of the line J = 3, K = 2, $\Delta F = \Delta\alpha = 0$ of $N^{14}H_3$ are distributed over a range of some tens of cycles per second. Therefore, according to Eq. (3.22), the value of ω obtained in an experiment which used magnetic tuning differs from ω_l by some tens of cycles per second.

A rigorous calculation of the oscillation frequency of a maser for an arbitrary number of energy levels, each of which splits up into 2F + 1 components in a magnetic field, is extremely complicated. The complication arises because of the selection rules for the case of a magnetic field applied perpendicular to the electric field in the resonator: $\Delta M = \pm 1$. In this case the total polarizability of the molecule is not equal to the sum of the polarizabilities caused by the separate components of the line. In the case of weak saturation, however, there is additivity of the polarizabilities contributed by the separate components, and a calculation of the frequency of a maser on the assumption that γ does not exceed 0.1-0.2 leads to the formula

$$\omega = \omega_l \left[1 - \frac{\omega_l - \omega_0}{\omega_l} \cdot \frac{Q_0}{Q_l} \left(1 + 2 \frac{\sum_i I_i \Delta_i^2}{\sum_i I_i} \right) + 3 \frac{\sum_i I_i \eta_i \Delta_i^2}{\sum_i I_i} \right].$$

Here I_i is the intensity of the i-th component of the line, summed over all Zeeman transitions; for the lines J = 3, K = 2, $\Delta F = \Delta\alpha = 0$ of $N^{14}H_3$ we have $i \equiv (F, \alpha)$; ω_l is the mean frequency of the spectral line in the absence of the magnetic field,

$$\omega_l = \frac{\sum_i I_i \omega_i}{\sum_i I_i},$$

*As has already been pointed out, the presence of several hyperfine structure components in the ammonia line $\Delta F_1 = \Delta F = 0$ is due to differences between the hyperfine structure constants for the upper and lower inversion levels of ammonia. This difference Δq has been determined for the quadrupole constant in [40]; no measurements have been made of the corresponding difference Δm for the magnetic constants. We shall estimate this difference by assuming that $\Delta m / \Delta q$ is equal to the ratio of the constants in question.

**Since for the line J = 3, K = 2 the number F_1 is not a conserved quantum number, to characterize the energy states with $F = \frac{7}{2}$ and $F = \frac{5}{2}$, one introduces a supplementary index α which takes two values (+) and (−) (see [48]).

TABLE II. g-Factors for the Line J = 3, K = 2 of $N^{14}H_3$

F	α	$g_{F\alpha}$	F	α	$g_{F\alpha}$
9/2		1.0	5/2	+	0.8
7/2	+	1.2	5/2	—	—0.1
7/2	—	0.7	3/2		—0.3

and $\eta_i = (\omega'_i - \omega_l)/\omega_l$, so that $\sum_i I_i\eta_i = 0$; $\Delta_{i\equiv F,\alpha} = g_{F\alpha}\beta H\bar{\tau}$, where $g_{F\alpha}$ is the factor for the Zeeman (spectroscopic) splitting.

The Hamiltonian for the interaction of the ammonia molecule with the external magnetic field H is of the form

$$\mathcal{H} = g_{JK}\beta \mathbf{J}\mathbf{H} + g_{I_N}\beta \mathbf{I}_N\mathbf{H} + g_{I_N}\beta \mathbf{I}_H\mathbf{H} \equiv g_{F\alpha}\beta \mathbf{F}\mathbf{H},$$

so that

$$g_{F\alpha} = g_{JK}\langle F, \alpha | J | F, \alpha\rangle + g_{I_N}\langle F, \alpha | I_N | F, \alpha\rangle + g_{I_H}\langle F, \alpha | I_H | F, \alpha\rangle.$$

Since g_{JK}, g_{I_N} and g_{I_H} are known and have the respective values [46] $0.560 - 0.076 \cdot K^2J^{-1}(J+1)^{-1}$; 0.403; and 5.59, there is no difficulty in calculating $g_{F\alpha}$. The results of the calculation for the line J = 3, K = 2 of $N^{14}H_3$ are given in Table II.

As was shown in Part 1 of this section, tuning by means of the pressure leads to coincidence of the oscillation frequency of the maser to an accuracy of 10^{-10}. This makes it possible to determine the difference of the frequencies of a maser when it is tuned by means of modulating the pressure in the source of the molecular beam and when it is tuned by using an external magnetic field, and to calculate the difference of the magnetic interaction constants for molecules in the upper and lower inversion levels. To do this it is necessary to express the frequencies of the components in terms of this difference of the magnetic constants.

For lines of $N^{14}H_3$ with K not a multiple of 3, the magnetic constants which determine the hyperfine interaction are 3 in number. Therefore, strictly speaking, it is necessary to have the results of tunings for 3 lines. But all the inversion lines of $N^{14}H_3$ except J = 3, K = 2 have quadrupole splittings in the first order of perturbation theory which are much larger than the magnetic splitting, and the effect of the magnetic satellites on the tuning will be "smeared out" by the stronger influence of the quadrupole structure of the line.* In our opinion, however, a natural assumption is that the change δa_i of the magnetic interaction constant on going from one inversion level to the other is proportional to the value of the constant a_i itself, so that the ratio does not depend on the index i:

$$\frac{\delta a_i}{a_i} = \Delta_m = \text{const.} \tag{3.23}$$

In this case, the frequencies of the components in the line J = 3, K = 2, $\Delta F = \Delta\alpha = 0$ can be expressed in terms of the energy of the hyperfine splitting, namely:

$$\omega_{F\alpha} = \omega_l + E_{F\alpha} \cdot \Delta_m, \tag{3.24}$$

where $E_{F\alpha}$ is the energy of the hyperfine splitting of the level $F\alpha$ when one does not take into account the difference of the interaction constants for the upper and lower inversion levels.

When we use Eq. (3.24) we get for the difference of the frequencies of a generator tuned by means of

*The molecule $N^{15}H_3$ does not have a quadrupole hyperfine structure, so that when this molecule is used the measurement of the magnetic interaction constants can be made more exactly. A calculation of the hyperfine structure and an investigation of this molecule will be presented in another article.

changes of pressure and by means of an external field the formula

$$\delta = 1.5\Delta_m \cdot \left\{ \frac{\sum\limits_{F\alpha} I_{F\alpha} E_{F\alpha} \Delta^2_{F\alpha}}{\sum\limits_{F\alpha} I_{F\alpha}\Delta^2_{F\alpha}} - \frac{\sum\limits_{F\alpha} I_{F\alpha} E_{F\alpha}}{\sum\limits_{F\alpha} I_{F\alpha}} \right\}.$$

The values of $E_{F\alpha}$ have been calculated in [48], and

$$I_{F\alpha} = \sum_{M_F M'_F} |\langle KJF\alpha M_F | \hat{d}_x | KJF\alpha M'_F \rangle|^2,$$

$$M'_F = M_F \pm 1,$$

where $\langle KJF\alpha M_F | \hat{d}_x | KJF\alpha M'_F \rangle$ are matrix elements of the component of the electric dipole moment of the molecule, which can be calculated in accordance with [56]. From such a calculation we find that

$$\delta = 2.3 \cdot 10^4 \cdot \Delta_{m}, \text{ cps.}$$

By determining δ experimentally we can determine Δ_m, and along with it the difference of the magnetic constants for the upper and lower inversion levels, according to Eq. (3.23).

By using Eq. (3.11) we can estimate the possibilities of the method of magnetic modulation. Here the role of the function G is played by the expression in the large parentheses in Eq. (3.21). For $\omega / Q_l \simeq 1$ kc and $\gamma \simeq 2$ in order to tune a generator to accuracy of $\sim 10^{-10}$ it is necessary to measure the difference frequency of two generators to an accuracy of 1-0.5 cps. Here, however, because of the previously noted dependence of the frequency of the generator tuned by modulating the magnetic field on the parameters of the apparatus, we cannot speak of an absolute tuning (stability), but only of reproducibility of the frequency for a given construction of the generator and a particular set of conditions of operation.

3. The Method of Two Coupled Resonators

It was pointed out in the preceding section that the largest influence on the oscillation frequency of a maser is that of the natural frequency of the resonator.

In [57] two coupled resonators were used to decrease the variation of frequency of a maser because of thermal drift of the natural frequency of the resonator. In this part we consider the possibility of using two coupled resonators to increase the absolute frequency stability of a maser.

The oscillation frequency of a maser with two coupled resonators and a molecular beam sent through one of them is given by Eq. (1.59), if in this equation we replace Q_W and ω_W by Q_2 and ω_2 – the quality figure and the natural frequency of the resonator without the molecular beam–

$$\omega = \omega_l \left[1 - \frac{Q_{\text{equ}}}{Q_l} \cdot \frac{\Delta_1 \Delta_2 + \Delta_1 \cdot Q_2^{-2} - k^2 \Delta_2}{\Delta_2^2 + Q_2^{-2}} \right]. \tag{3.25}$$

In this formula, $\Delta_1 = \omega_l - \omega_1$, $\Delta_2 = \omega_l - \omega_2$. For our further argument it is convenient to rewrite Eq. (3.25) in the form

$$\omega = \omega_l \left[1 - \frac{1}{2Q_l} \cdot \frac{\omega_l - \omega_{-I}}{\Delta\omega} \cdot \frac{\omega_l - \omega_0}{\Delta\omega} \cdot \frac{\omega_l - \omega_I}{\Delta\omega} \right]. \tag{3.26}$$

In writing the formula (3.26) we have assumed that $Q_1 \simeq Q_2 = Q_0$; $\Delta\omega = \omega_l / Q_0$ is the transmission band of an individual resonator; and ω_{-I}, ω_0, and ω_I are the roots of the equation

$$y(\omega) = \Delta_1 \Delta_2 + \Delta_1 Q_2^{-2} - k^2 \Delta_2 = 0. \tag{3.27}$$

The structure of Eq. (3.26) is very similar to that of the formula for the oscillation frequency of a maser with one resonator, Eq. (1.35), if we write it in terms of the transmission band of the resonator:

$$\omega = \omega_l\left[1 - \frac{1}{Q_l}\cdot\frac{\omega_l - \omega_0}{\Delta\omega}\right]. \qquad (3.28)$$

The difference between (3.26) and (3.28) is that, in the former equation, there are three factors of the form $(\omega_l - \omega_k)/\Delta\omega$. Therefore, if ω_{-l}, ω_0, and ω_l are sufficiently nearly equal, then for a given accuracy of agreement between ω_0 and ω_l the oscillation frequency of a maser with two coupled resonators will coincide with ω_l to a higher accuracy than that of a generator with one resonator.

For example, assuming $Q_l \simeq 10^7$ and $(\omega_{-l} - \omega_l) \leq 0.1\,\Delta\omega$, to obtain an absolute stability of about 10^{-10} it is necessary that ω_0 agree with ω_l to an accuracy of 10% of the transmission band of the resonator. In the case of a single resonator, in order to obtain this same absolute stability it is necessary to tune the natural frequency ω_0 of the resonator to the frequency ω_l of the spectral line to an accuracy of 0.1% of the transmission band of the resonator.

The tuning of the oscillation frequency of a maser to the frequency of the spectral line by means of two resonators can be thought of in the following way.

1. One finds the roots of (3.27).
2. One matches these roots by varying the parameters of the resonators (their natural frequencies ω_1 and ω_2 and the coupling coefficient k^2).
3. One secures agreement between the line frequency ω_l and the roots of (3.27).

It is known from the theory of coupled circuits [58] that the roots of (3.27) agree exactly with the resonance frequencies of two coupled circuits, which are determined from the condition that the reactance of the coupled system be equal to zero. Experimentally it is hard to follow the pure dependence of the reactance of two coupled circuits on the frequency of the applied signal, but one can easily observe on an oscilloscope the amplitude of the oscillation for both the first and the second resonators. Therefore, the question arises as to the relation between the resonance frequencies [or, what is the same thing, the roots of (3.27)] of a system of two coupled circuits and the characteristic points of the amplitude characteristics of the signals in the first and second circuits (resonators). These characteristic points are the extreme values of the amplitude as a function of the frequency of the applied signal.

As has been shown in [58], the extreme values of the amplitude E_1 in the first circuit, on which the signal acts directly, occur for values of the frequency ω of the applied signal which satisfy the relations

$$\omega - \omega_{00} = 0,\ \ \omega - \omega_{00} = \pm\,\omega_{00}\,\{[k^4 + 2k^2(Q_2^{-2} + Q_1^{-1}Q_2^{-1})]^{1/2} - Q_2^{-2}\}^{1/2}. \qquad (3.29)$$

In the second or "passive" circuit the extreme values of the amplitude E_2 occur at the points

$$\omega - \omega_{00} = 0,\ \ \omega - \omega_{00} = \pm\,\omega_{00}\left\{k^2 - \frac{1}{2}(Q_1^{-2} + Q_2^{-2})\right\}^{1/2} \qquad (3.30)$$

and coincide with each other for

$$k = \frac{1}{\sqrt{2}}(Q_1^{-2} + Q_2^{-2})^{1/2}. \qquad (3.31)$$

In (3.29) and (3.30) it is assumed that the partial frequencies of the two circuits are equal:

$$\omega_1 = \omega_2 = \omega_{00}. \qquad (3.32)$$

When relation (3.32) holds, (3.27) can be solved easily, and we get the following values for ω_{-l}, ω_0, and ω_l:

$$\omega_{-l} = \omega_{00} - (k^2 - Q_2^{-2})^{1/2},\ \ \omega_0 = \omega_{00},\ \ \omega_l = \omega_{00} + (k^2 - Q_2^{-2})^{1/2}. \qquad (3.33)$$

From a comparison of formulas (3.29) and (3.33) it can be seen that the required roots of (3.27) are most nearly equal to the maxima of the amplitude of oscillation in the second, passive circuit. For absolutely identical circuits ($\omega_1 = \omega_2$ and $Q_1 = Q_2$), the roots of (3.27) coincide exactly with the extremal points of the amplitude of the

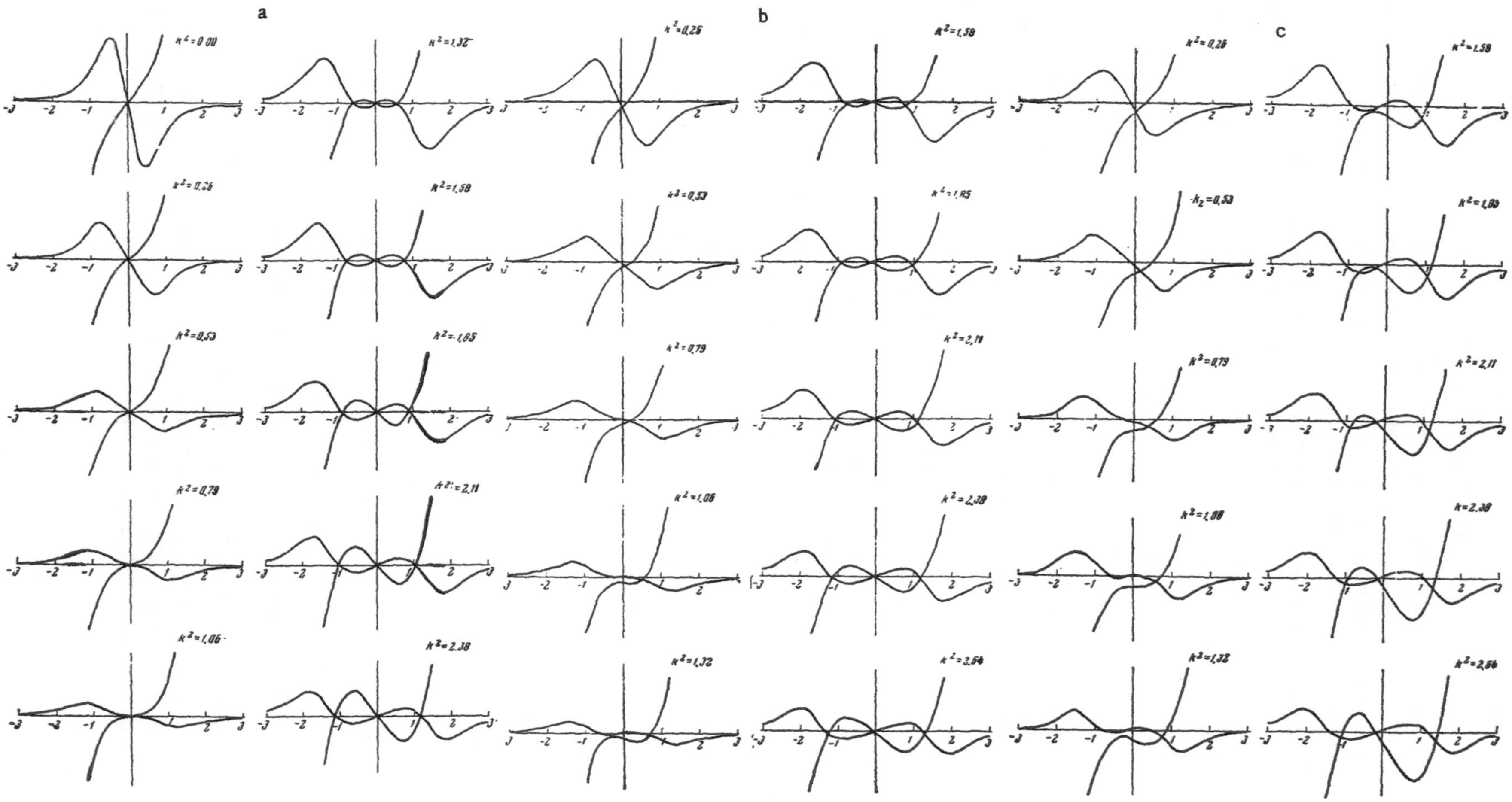

Fig. 14. Plots of the derivative of the amplitude in the second resonator and of the quantity $y(\omega) = \Delta^3 + \delta\Delta^2 + \Delta(1 - k^2) + \delta$ as functions of the frequency of the applied signal for various values of the coupling coefficient k. The abscissa is the quantity $(\omega_2 - \omega)/\Delta\omega = \Delta$; curves a are for $\delta = 0$; curves b are for $\delta = 0.025$; curves c are for $\delta = 0.5$.

second circuit. If $Q_1 \neq Q_2$, then there is a difference between the roots of (3.27) and the points where the amplitude E_2 is a maximum, which amounts to a quantity $\Delta\omega(\delta Q/Q)^{\frac{1}{2}}$, where $\delta Q = |Q_2 - Q_1| << Q_0$.

Figure 14 shows the derivative of the amplitude in the second circuit as a function of the frequency of the applied signal for various values of the coupling coefficient k and the mistuning $\delta = \omega_2 - \omega_1$. For comparison we show a plot of the function $y(\omega)$ in the left member of (3.27). It can be seen from Fig. 14 that agreement between the roots of (3.27) and the zeros of the derivative (or the extreme values of the amplitude) is possible only when $\delta = 0$. Furthermore, one can choose the coupling constant k so that all three roots come together at a single point.

It follows from this that it is necessary to have a visual criterion for the equality of the partial frequencies in the tuning of the resonators. Such a visual criterion exists. In fact, it can be shown that if, when one observes the signals in the first and second circuits, it happens simultaneously that the derivative of the amplitude in the first circuit forms four symmetrical "loops" (Fig. 15b) and the derivative of the amplitude in the second circuit forms two symmetrical "loops" with a small flattening in the middle (Fig. 15a), this means that $\delta = 0$ and $k = Q_0^{-1}$. Then the roots of (3.27) are equal to each other and coincide with the zeros (more exactly, with the triple zero) of the derivative of the amplitude in the second circuit.

If the partial frequencies are unequal, or if the condition for critical coupling $k = Q_0^{-1}$ is violated, there is a loss of symmetry in the pattern shown in Fig. 15.

The accuracy of this method of tuning depends on the difference of the qualities of the resonators δQ. For $\delta Q/Q \approx 0.1$-0.5 the absolute stability which can be achieved is of the order of 10^{-9} to $5 \cdot 10^{-10}$.

For this method of tuning to be realized it is necessary that the coupling with the waveguide system be reduced by the use of directional couplers, since the waveguides have a large influence. For example, with a coupling which decreases the quality of the resonators by a factor of 1.5, the zeros of the derivative of the amplitude in the second circuit can be altered because of the effect of the waveguides by 30 to 50% of the transmission band of the resonator.

The method of tuning the oscillation frequency of a maser by the use of two coupled resonators makes it unnecessary to have thermostatic control of the resonators, as has been pointed out in [58]. This property of two coupled resonators is graphically demonstrated by Fig. 15. In fact, the change of $y(\omega)$ on a change of the frequency of the external action can be interpreted as a change of the oscillation frequency of the maser with a synchronous change of the partial frequency of the resonators ($\omega_1 - \omega_2 = \delta = \text{const}$).

There is a region $\Delta\Omega$ in which the dependence of the curves of $y(\omega)$ on the frequency is weak, and since the change of temperature is the same in the two coupled resonators, this means that there is only a weak dependence of the oscillation frequency of the maser on the temperature if ω is in the range $\Delta\Omega$.

A shortcoming of the use of two resonators is that one must have a larger number of active molecules for the excitation of the maser than for a maser with one resonator, since, for $k = Q_0^{-1}$, the equivalent quality of the system is only half the value for a single resonator.

Section 4

THE ENHANCEMENT OF THE FREQUENCY STABILITY OF MASERS BY THE USE OF BEAMS OF SLOW MOLECULES

1. Remarks on the Question of Increasing the Length of the Resonator

As has already been pointed out, enhancement of the frequency stability of a maser depends on the production of narrow spectral lines. If we have been successful in eliminating the effects on the oscillation frequency of a maser from such factors as the Doppler broadening of the line and the frequency shifts owing to unresolved hyperfine structure and nonuniformity of the emission of the molecules along the length of the resonator (effect III), then the main factor which determines the frequency stability of the maser will be the time of flight of the molecules through the resonator. Under such conditions the frequency stability of the oscillation of the maser increases with increase of the time of flight. The time of flight can be increased in two ways: by increasing the length of the resonator, and by decreasing the average speed of the molecules in the beam.

An increase of the length of the resonator requires a decrease of the aperture of the molecular beam. In fact, for any prescribed aperture ϑ_0 there is an optimal length l_{opt} of the resonator for which the line width caused by the time of flight through the resonator is equal to the Doppler width of the line caused by the fact that the beam is not precisely parallel.

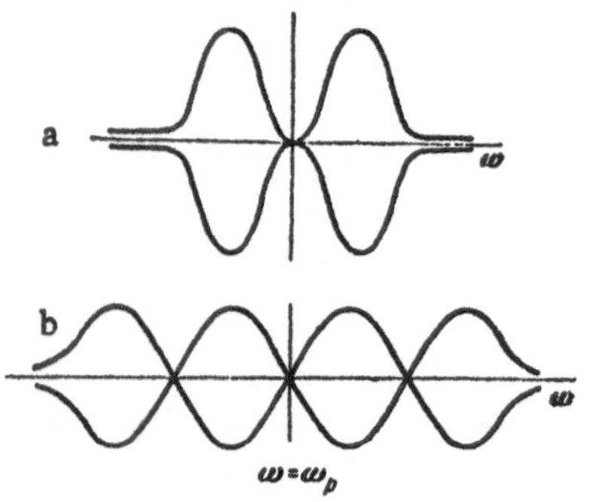

Fig. 15. Oscillogram of the derivative of the amplitude of the oscillations for critical coupling and δ = 0. (a) In the second or passive resonator; (b) in the first resonator.

Let us make an estimate of the dependence of l_{opt} on the quantity ϑ_0, the aperture of the molecular beam. We shall assume, for the sake of simplicity, that the velocity distribution of the molecules when they enter the resonator is of the form

$$f(v, \vartheta) \sim e^{-\beta v^2} v^3 \sin\vartheta \cos\vartheta, \qquad (4.1)$$
$$0 \leqslant \vartheta \leqslant \vartheta_0.$$

Using this distribution we can easily calculate the line width caused by the Doppler effect because of the transverse component of the velocity:

$$\frac{\Delta\omega_{\text{Dop}}}{\omega_l} = \frac{1}{c} \cdot \frac{\int v \sin\vartheta f(v, \vartheta)\, dv\, d\vartheta}{\int f(v, \vartheta)\, dv\, d\vartheta} = \frac{1}{2}\sqrt{\frac{\pi}{\beta c^2}} \cdot \vartheta_0. \qquad (4.2)$$

The line width caused by the finite time of flight of the molecule through the resonator is given by

$$\frac{\Delta\omega_l}{\omega_l} = \frac{1}{\omega_l l} \cdot \frac{\int v \cos\vartheta f(v, \vartheta)\, dv\, d\vartheta}{\int f(v, \vartheta)\, dv\, d\vartheta} = \frac{3}{2}\sqrt{\frac{\pi}{\beta}} \cdot \frac{1}{\omega_l l}. \qquad (4.3)$$

Equating expressions (4.2) and (4.3), we find the optimal length of the resonator l_{opt}:

$$l_{opt} \simeq 3\frac{\lambda}{\vartheta_0}, \qquad (4.4)$$

where

$$\lambda = \frac{c}{\omega_l}.$$

For the ammonia line J = 3, K = 3, we have $\lambda \approx 0.2$ cm, and for $\vartheta_0 \approx 0.1$ this gives for l_{opt} a value of about 6-10 cm.

The aperture $\vartheta_0 \approx 0.1$ corresponds to about 6°. The usual devices have this sort of angle of capture of the beam. To decrease $\Delta\omega_l$ by increasing l by, for example, a factor of 10, it is necessary to decrease ϑ_0 by a factor of 10. This leads to a decrease of the number of molecules by a factor of 100, since the number of molecules is proportional to ϑ_0^2 (or more exactly, to $\sin^2\vartheta_0$).

The condition for self-excitation, which is proportional to $\bar{\tau}^2$ or l^2 [see Eq. (1.33)], remains unchanged. The power of the oscillations is decreased by a factor of 100. One can increase the power of the oscillations in the maser by using a large number of narrowly directed beams. Therefore, it seems at first glance that the use of narrowly directed beams together with an increase of the length of the resonator opens up attractive prospects for enhancing the stability of the oscillation of a maser. Actually, there are great difficulties in the way of the use of narrowly directed beams of molecules to increase the frequency stability of a maser.

As has been shown in [15], the shift ξ of the oscillation frequency of a maser caused by nonuniformity of the emission from the molecules along the length of the resonator is proportional to $\frac{1}{Q_l} \cdot \frac{l^2}{\lambda^2}$. Since $Q_l \sim l$, we have $\xi \sim l_{opt}$. According to the estimates made in [15], for the length of resonator usually employed ($l \approx 10$ cm) we have $\xi \approx 2 \cdot 10^{-9}$. To compensate this shift to an accuracy of 10^{-10} we must have two oppositely moving beams which are kept symmetrical to an accuracy of 10%. If we wish to increase the frequency stability of the maser by increasing its length, for example to bring the stability to 10^{-11} (increasing the length by a factor of 10), the degree of symmetry of the beams which must be secured to compensate the shift ξ becomes 0.1% (!), because if the length is ten times the usual length we have $\xi \approx 10^{-8}$. It is extremely difficult in practice to produce two oppositely moving beams whose intensities are kept equal to an accuracy of 0.1%.

We note also that even if there is ideal compensation of the shift of the oscillation frequency caused by nonuniformity of the emission along the length of the resonator, there is a limit on the length of the resonator such that an increase of the length beyond this limit will give no gain in the stability even for arbitrarily small apertures of the molecular beam.

Indeed, in spite of the fact that two ideally symmetrical beams in opposite directions eliminate the frequency shift, the electric field is still nonuniform along the axis, although it is symmetric relative to the center of the resonator. The nonuniformity of the field leads to a Doppler broadening of the line which is of the order of magnitude of the frequency shift with one beam. Since this broadening increases with the length of the resonator, an increase of the length is profitable only up to the point at which this Doppler broadening of the line becomes equal to the broadening of the line because of the time of flight. This gives us the relation

$$\frac{1}{Q_t} \cdot \frac{1}{Q_l} \cdot \frac{l^2}{\lambda^2} = \frac{1}{Q_l},$$

where Q_t is a dimensionless quantity which has the meaning of a quality figure and is defined in [15]. From this we have

$$l_{opt} = \lambda \sqrt{Q_t}. \tag{4.5}$$

It was assumed in [15] that $Q_t = 10^4$; for $\lambda = 1.25$ cm this gives

$$l_{opt} \simeq 10^2 \text{ cm}. \tag{4.6}$$

The existence of a limit on the length of the resonator and the difficulty associated with the production of two beams of intensities which are equal to a high degree of accuracy make the method of enhancing the frequency stability of a maser by increasing the length of the resonator a not very promising one. Because of this, we have reason to consider possibilities for enhancing the frequency stability of a maser which are associated with the use of beams of "slow molecules," i.e., molecules whose average velocity is much smaller than the thermal velocity at ordinary (room) temperatures.

Possible methods for producing the beam of slow molecules are: (1) Removal of the molecules with large speeds from an ordinary molecular beam, so as to produce a sort of "cutoff" of the velocity distribution of the molecules on the high-speed side; (2) slowing down of the molecules; (3) lowering the effective temperature of the molecular beam.

We shall now consider these methods for producing a beam of slow molecules and shall analyze the possibilities for further enhancement of the frequency stability of a maser by the use of slow molecules.

2. Production of a Beam of Slow Molecules by Reflecting Molecules from a Potential Barrier

A cutting-off of a molecular beam on the high-speed side of the distribution can be obtained by reflecting the beam of ammonia molecules from a potential barrier. As the potential barrier one can use a nonuniform electric field. As is well known [14], in a nonuniform field the force on molecules which are in the upper inversion level is in the direction of decreasing absolute value of the field strength, and the force on molecules which are in the lower inversion level is in the direction of increasing field strength. If a molecule which is in the upper inversion level encounters a potential barrier consisting of an electric field, it is reflected from the barrier; on the other hand, a molecule which is in the lower inversion level passes through the barrier. Thus, the molecules which are reflected are active molecules which can be used to excite a resonator.

The only "upper" molecules which are reflected from the barrier, however, are those whose velocities satisfy the relation

$$\frac{1}{2} m (\mathbf{v} \operatorname{grad} W(\mathbf{r}_0))^2 \leqslant W(\mathbf{r}_0) \cdot |\operatorname{grad} W(\mathbf{r}_0)|^2, \tag{4.7}$$

where $\mathbf{r}_0$ is the point where the potential energy $W(\mathbf{r})$ of a molecule in the electric field has its maximum value.

This relation determines the maximum speed of a molecule in the reflected beam. If $W(\mathbf{r}_0)$ is sufficiently small, the molecules in the beam will have small speeds. As for the main features of the construction of such a system, it is desirable to choose a field such that the equipotentials

$$W(\mathbf{r}) = \text{const} \tag{4.8}$$

are focusing surfaces. Then when we place the source of the molecular beam at the focus we obtain a reflected beam with a small aperture Ω, although the aperture Ω_0 of the incident beam is rather large. The aperture of the reflected beam can be determined by using Liouville's theorem on the constancy of the phase volume in conservative systems. In our case,

$$S \cdot \Omega = \text{const}, \tag{4.9}$$

where S is the cross-sectional area of the beam and Ω is the solid angle which defines the aperture of the beam.

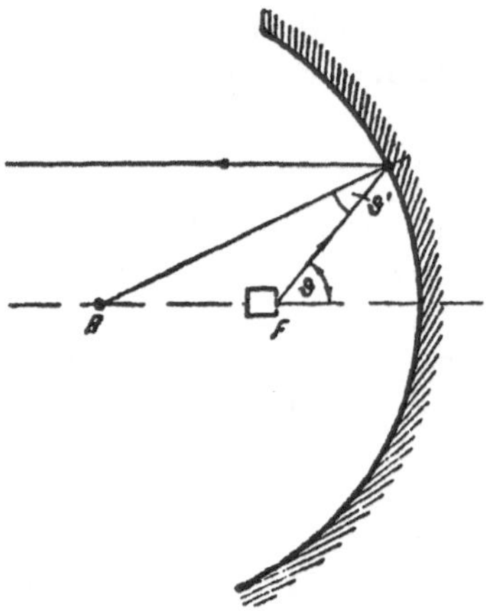

Fig. 16. Diagram for the calculation of the number of active molecules in the case of reflection from an ideal "mirror." R is the center of the sphere at whose surface reflection occurs, and F is the focus of the "mirror." The source of the molecular beam is placed at the focus. The focal distance is half of the radius of the spherical surface of the mirror.

We now calculate the number of active molecules for the case of reflection from an ideal spherical mirror. The number of molecules per second emerging from a source of area s and with speed in the range (v, v + dv) is

$$dN = ns\left(\frac{m}{2\pi kT}\right)^{3/2} e^{-\beta v^2} v^3\, dv \sin\vartheta \cos\vartheta\, d\vartheta\, d\varphi. \tag{4.10}$$

Approximately half of the molecules (with accuracy up to $\hbar\omega_{inv}/kT \approx 0.001$) are in the upper inversion level. To get the number of molecules reflected from the mirror we must integrate the quantity dN/2 over the angles and speeds which satisfy relation (4.7). If ϑ is the angle at which a molecule is emitted relative to the axis of the system (Fig. 16) and ϑ' is the angle between the radius to the point of reflection by the spherical mirror and the path of the molecule coming from the source, then relation (4.7) takes the form

$$0 \leqslant v^2 \leqslant 2\frac{W_m}{m}\cos^{-2}\vartheta \tag{4.11a}$$

and we must also have

$$0 \leqslant \vartheta \leqslant \vartheta_m. \tag{4.11b}$$

Integration of dN/2 over the ranges indicated gives us the number of active molecules reflected from the mirror. When we use the fact that $W_m/kT \ll 1$, we find that this number of molecules is

$$N_a = \frac{1}{8} n\bar{v}s\left(\frac{W_m}{kT}\right)^2 F(\vartheta_m), \tag{4.12}$$

where

$$F(\vartheta_m) = \int_0^{\vartheta_m} \frac{\cos\vartheta \sin\vartheta}{\cos^4\vartheta'}\, d\vartheta. \tag{4.13}$$

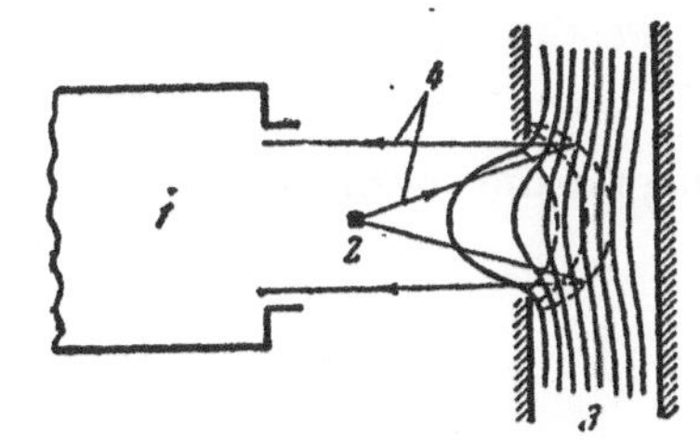

Fig. 17. Schematic diagram of a maser working with a beam of slow molecules reflected from a "mirror." (1) Resonator; (2) source of molecular beam; (3) sorting system, consisting of the potential barrier formed by a plate condenser with a round hole in one of the plates; solid lines are the equipotentials of the electrostatic field; broken lines are the levels of equal potential energy of the interaction of molecules with the field; (4) path of a molecule.

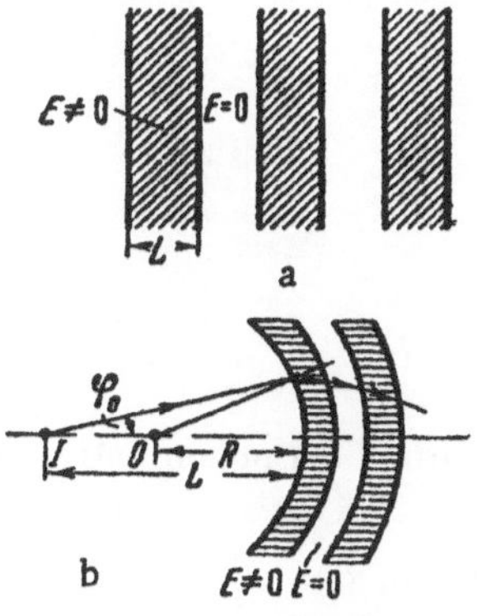

Fig. 18. Schematic diagram of sections for slowing down molecules. (a) Decelerating sections in the form of plane-parallel plates (the sections in which the electric field is different from zero are shaded); (b) decelerating sections in the form of spherical surfaces with a common center O. For focusing it is necessary that the distance L from the first surface to the source of molecules be larger than the radius R of this surface.

The ratio of the number of active molecules N_a to the total flux of molecules coming from the source $N_0 = \frac{1}{4} n\bar{v}s$ is

$$\frac{N_a}{N_0} = \frac{1}{2}\left(\frac{W_m}{kT}\right)^2 F(\vartheta_m). \tag{4.14}$$

For $\vartheta_m \simeq 45°$ we have $F(\vartheta_m) \simeq 1$. Therefore, when we lower the speed by a factor of 10 we lose about a factor of 10^4 in the number of molecules as compared with the total flux. Since, in an ordinary maser, only about $\frac{1}{100}$ of the total flux of molecules is used, if the volume of the resonator is the same, the condition for self-excitation remains unchanged, since the coefficient of self-excitation is proportional to the square of the time of flight through the resonator.

A generator of this type is most conveniently realized with a beam of molecules of deuterated ammonia, since, in this case, the input aperture into the resonator can be made sufficiently large [59]. The focusing potential barrier can be provided by the field between two plates with a hole in one of them (Fig. 17). The shape of these plates is to be chosen from a calculation of the best focusing so as to combine the largest possible original aperture of the molecular beam with a small angular spread of the flux of molecules entering the resonator.

3. Slowing Down of a Beam of Molecules by a Nonuniform Electric Field

Potential barriers can be used to slow the molecules down, as well as to produce a cutoff in their speed. If a molecule which is in the upper inversion level comes into an electric field with which it has the interaction energy W, its speed is reduced from the value v_0 to a value v such that the total energy is conserved:

$$\frac{mv_0^2}{2} = W + \frac{mv^2}{2}. \tag{4.15}$$

We get the maximum possible number of molecules by slowing down the molecules which have a speed lying close to the most probable speed, which is of the order of magnitude $(kT/m)^{1/2}$. The electric fields which can be achieved, however, are such that $W_m \simeq 0.01$ kT. Therefore, the speed of a molecule cannot be lowered much by one act of slowing down. We can, however, use a multiple slowing down by a system consisting of a succession of sections having an electric field and spaces with zero field (Fig. 18a). As it enters the field a molecule which is in the upper inversion level is slowed down. When it comes out of the field it is accelerated again to its original speed. If, however, it is transferred by radiation to the lower level while it passes through the field, then it is again slowed down as it emerges. It is then to be returned to the upper level by radiation, etc.

The probability of transition of a molecule from one level (1) to the other (2) when the frequency of the external alternating field is tuned to resonance with that of the molecular transition is given by the relation [26]

$$w_{1\to 2} = \sin^2\frac{\gamma t}{2}; \quad \gamma = \frac{|\mu_{1\to 2}|}{\hbar}\cdot\mathcal{E}, \tag{4.16}$$

where $\mu_{1\to 2}$ is the matrix element of the dipole moment of the molecule for the transition 1 → 2, $\mathcal{E}$ is the amplitude of the alternating field, and t = L/v, where v is the speed of the molecule.

In order for the molecule to be sure to go from one level to the other during the time of flight, we must have the condition

$$\frac{\gamma L}{v} = \pi(2m+1), \quad m \text{ is an integer.} \tag{4.17}$$

Since the beam of molecules is nonmonochromatic in speed, condition (4.17) cannot be satisfied for all of the molecules at once, and therefore it is possible for molecules to be lost during the process of slowing down. The total number of molecules at the output after passing through n sections will be given by the formula

$$N_{\text{dec}} = N_0 \frac{\int\limits_{v_0}^{\infty} F(v)\,dv \prod\limits_{k=1} \sin^2 \frac{\gamma_k L_k}{2v_k}}{\int\limits_{v_0}^{\infty} F(v)\,dv}. \tag{4.18}$$

F(v) is the velocity distribution function of the molecules,

$$v_k = \left(v_0^2 - \frac{2W}{m}k\right)^{1/2}, \quad v_0 = \left(\frac{2W}{m}n\right)^{1/2},$$

and N_0 is the total number of molecules in the given solid angle.

Expression (4.18) has its largest value if condition (4.19) is satisfied for the most probable speed v_m:

$$\frac{\gamma_k L_k}{\sqrt{v_m^2 - \frac{2W}{m}k}} = \pi. \tag{4.19}$$

For the Maxwell velocity distribution

$$v_m = \left(\frac{2}{3}\beta\right)^{-1/2}. \tag{4.20}$$

The integration of expression (4.18) for the Maxwell velocity distribution can be carried out numerically. The results of a calculation made with an electronic computing machine are as follows:

Number of decelerating sections, n	75	80	85	90	95
Number of molecules slowed down, N_{dec}/N_0	0.068	0.064	0.058	0.051	0.043

A decelerating system constructed with uniform fields between plane-parallel plates cannot be used, since, in this case, the deceleration decreases only the component of the velocity which is perpendicular to the plane of the plate, and thus defocuses the molecular beam.

One can, however, arrange decelerating sections in the form of spherical mirrors so that the system will focus the beam, as can be seen from Fig. 18b.

In this system the original angle of emergence φ_0 will be changed as the deceleration proceeds and be decreased to $\pi - \varphi_0$. When the angle of emergence has become equal to $\pi - \varphi_0$ one can identically repeat the decelerating system if the amount of slowing down which has been achieved is still insufficient.

As can be seen from the calculation of N_{dec}, the slowing down of molecules is also associated with the loss of a large proportion of the particles. If we assume that we can use molecules emerging within an angle of ~10°, then N_0 amounts to ~1% of the total number of molecules and the ratio of the number of decelerated molecules to the total number is $\sim 5 \cdot 10^{-4}$.

4. Lowering of the Temperature of a Molecular Beam by Collisions Between the Ammonia Molecules and a Cold Inert Gas

It is impossible to lower the temperature of the molecular beam by cooling the source; a small decrease of the temperature of the source does not give any appreciable result, since the speed of the molecules decreases only as $T^{\frac{1}{2}}$. A large decrease of the temperature of the source (to a few degrees on the absolute scale) leads to freezing out of the working substance (ammonia).

One can, however, lower the temperature of the ammonia by means of collisions between the ammonia molecules and a cold gas. If the ammonia does not interact with the cold gas in any other way than through collisions, the kinetic energy of the ammonia molecules will gradually be lowered to the energy of the cold gas. Helium at a temperature of the order of a few degrees Kelvin can be used as the cold gas.

The conditions for such a system to function efficiently are as follows: (1) The time for establishing the equilibrium of the mixture of helium and ammonia must be less than the time between two successive collisions of an ammonia molecule with another ammonia molecule or with the walls of the vessel. (2) There must be adequate heat exchange so that the temperature of the helium is not appreciably increased.

The first condition leads to the relation

$$\frac{n_A \sigma_{A\to A}}{m_A^{1/2}} \ll \frac{n_H \sigma_{A\to H}}{\sqrt{2}\, \mu_{AH}^{1/2}}, \tag{4.21}$$

where $\sigma_{A\to A}$ is the cross section for ammonia–ammonia collisions; $\sigma_{A\to H}$ is that for ammonia–helium collisions; n_A and n_H are, respectively, the density of ammonia and helium molecules; m_A is the mass of the ammonia molecule; and μ_{AH} is the reduced mass of the system of an ammonia molecule and a helium molecule.

The average temperature of the mixture can be calculated from the thermal balance. If the source of ammonia molecules emits N_A molecules per second which are at the temperature T_A, and N_H is the number of molecules of helium which collide with the wall of the vessel in one second, then

$$N_A T_A + N_H T_H = (N_A + N_H)\, T, \tag{4.22}$$

where T is the temperature of the mixture and T_H is the temperature of the walls of the vessel (and of the cold helium).

From this equation we have

$$T = T_H \frac{N_H}{N_A + N_H} + T_A \frac{N_A}{N_A + N_H}. \tag{4.23}$$

For $N_A/N_H \simeq 0.01$, which assures that condition (4.21) is satisfied, and $T_A \simeq 100$°K,

$$T = T_H + 1\text{°K}, \tag{4.24}$$

which is quite satisfactory.

An important parameter which characterizes the slowing down of the molecules is the average distance traveled by a molecule while it is being slowed down from the energy kT_A to an energy ~kT. This distance $\sqrt{\overline{R_T^2}}$ is connected with the mean free path l of the molecule being slowed down by the relation

$$\sqrt{\overline{R_T^2}} = \left(2 \frac{l}{\xi} \ln \frac{T_A}{T}\right)^{1/2}, \tag{4.25}$$

where

$$\xi = 1 + \frac{\alpha}{1-\alpha} \ln \alpha, \quad \alpha = \left(\frac{m_A - m_H}{m_A + m_H}\right)^2. \tag{4.26}$$

One can obtain formula (4.25) by using the "age theory" of the slowing down of neutrons in nuclear reactors [60]. In the case considered here,

$$\sqrt{\overline{R_T^2}} \simeq 0.1 \text{ cm}. \tag{4.27}$$

One important point must be noted. In thermodynamic equilibrium at a temperature 4-5°K, the ammonia molecules are practically all in the ground state $J = 0$, $K = 0$. The ground state cannot be used for generating radiation, since for $J = 0$, $K = 0$ there is no inversion doublet. The situation is saved by the fact that the time for establishing the equilibrium state of the rotational energy is much larger than that for establishing equilibrium of the translational energies, because of the fact that transitions with $\Delta K \neq 0$ are strongly forbidden even in collisions between molecules.

There are two main difficulties in the way of the realization of this method of slowing down ammonia molecules by collisions with cold helium. The first is connected with the admission of the ammonia molecules into a reservoir of cold helium. The second is that of producing a high vacuum when a beam of helium molecules is present. To obtain a high vacuum it is now necessary to use adsorbing materials (for example, carbon) cooled with liquid helium.

5. Analysis of the Effectiveness of the Methods Considered in Parts 2 and 3

The selection of slow molecules from the beam by removal of the molecules with larger speeds, and also slowing down of the beam by an external field, destroys the smoothness of the velocity distribution, and this leads to a change of the shape of the emission line. It was shown in Section 2 that the formula for the oscillation frequency of a maser involves not the average time $\bar{\tau}$ which the molecule spends in the resonator, but a certain effective quantity $\bar{\tau}/G$ which depends strongly on the line shape. A change of the line shape can increase the value of $\bar{\tau}/G$, and in spite of the increase of $\bar{\tau}$ the gain in frequency stability will be small. We shall explain this with an example.

In the case of a single-speed beam the oscillation frequency of the maser is given by the relation [15]

$$\omega = \omega_l \left[1 - \frac{Q_0}{Q_l} \cdot \frac{\omega_l - \omega_0}{\omega_l} G(\gamma\xi_0)\right], \tag{4.28}$$

where $Q_l = \omega_l l / 2v_0$, where v_0 is the speed of the molecules in the beam, $\xi_0 = l / v_0$, γ is the saturation parameter, and

$$G(\gamma\xi_0) = \frac{1 - \cos\gamma\xi_0}{1 - \sin\gamma\xi_0/\gamma\xi_0}. \tag{4.29}$$

The function $G(\gamma\xi_0)$ in formula (4.28) for the case of a single-speed beam differs decidedly from the corresponding function for an actual distribution and can vary from 0 to 3. It reaches its maximum value of 3 for $\gamma\xi_0 \to 0$. Under ordinary conditions $G(\gamma\xi_0) \approx 1$.

Let us now assume that we have increased Q_l by a factor of 10 by decreasing v_0. As we have seen earlier, the production of slow molecules involves the loss of a large number of particles. This in turn can lead to a decided decrease of $\gamma\xi_0$, with the result that $G(\gamma\xi_0)$ increases (for example, from unity to a quantity of the order of 3), and then the effective quality of the line will increase not by a factor of 10, but only by a factor of 3-3.5.

Although this example is an idealized case, nevertheless it shows that we must obtain answers to two questions:

1. Under what conditions will a decrease of the average speed of the molecules lead to a linear increase of the effective quality of the line with increase of $\bar{\tau}$?
2. What actual gain in frequency stability of the oscillation of a maser can be obtained by the methods proposed above (Parts 2 and 3)?

We shall now try to answer these questions.

Condition (1.34) for self-excitation of a maser can be written in the form

$$\eta = \lambda Q\bar{\tau} > 1, \tag{4.30}$$

where

$$\lambda = 4\pi |\mu|^2 \hbar^{-1} N_a,$$

and

$$\bar{\tau} = \int \tau F(\tau)\, d\tau. \tag{4.31}$$

For a beam of molecules with a cutoff on the side of high speeds (see Part 2),

$$\left.\begin{aligned} F(\tau) &= A\frac{d\tau}{\tau^6}, \\ \tau_{min} &\leqslant \tau < \infty, \end{aligned}\right\} \tag{4.32}$$

where A is a normalization constant equal to $\left[\int\limits_{\tau_{min}}^{\infty} F(\tau)\, d\tau\right]^{-1}$, and $\tau_{min} = l/v_{max}$.

In fact, the beam of molecules emerging from the source has a Maxwellian distribution of speeds,

$$e^{-\beta v^2} v^3, \quad 0 \leqslant v < \infty.$$

The "cutoff" molecular beam has the same distribution except that $0 \le v \le v_{max}$. If we wish to decrease the average speed in the beam by an order of magnitude in comparison with the thermal speed, then $\beta v_{max}^2 \simeq 0.01$, and we can replace $e^{-\beta v^2}$ by unity in the entire range $(0, v_{max})$. Then, introducing the variable $\tau = l/v$, we get the distribution (4.32).

When the distribution (4.32) is used, Eq. (2.4), which describes the stationary harmonic process in the maser, can be rewritten in the form

$$\lambda (\omega_l - \omega)\, \tau_{min}^2 J_s(z) = 2(\omega_0 - \omega), \tag{4.33a}$$

$$\lambda \tau_{min} J_c(z) = \frac{1}{Q}, \tag{4.33b}$$

where $z = \gamma \tau_{min}$, and

$$\left.\begin{aligned} J_s(z) &= 5z^3 \int\limits_z^{\infty} \left(1 - \frac{\sin\xi}{\xi}\right) \frac{d\xi}{\xi^6}, \\ J_c(z) &= 5z^4 \int\limits_z^{\infty} \frac{1-\cos\xi}{\xi} \frac{d\xi}{\xi^6}. \end{aligned}\right\} \tag{4.34}$$

The mistuning of the oscillation frequency of the maser relative to the frequency ω_l of the molecular transition which characterizes the absolute stability of the oscillation of the generator is seen from Eq. (4.33a) to depend only on z and τ_{min}:

$$\frac{\omega_l - \omega}{\omega_l} = 2Q\, \frac{\omega_l - \omega_0}{\omega_l} \cdot \frac{1}{\omega_l \tau_{min}} \cdot \frac{J_c(z)}{J_s(z)}. \tag{4.35}$$

Therefore, if we increase τ_{min} (and also $\bar{\tau}$, which in this case is equal to $\frac{5}{4}\tau_{min}$), keeping the value $z = \gamma\tau_{min}$, we obtain a linear increase of the absolute stability of the maser with the quantity τ_{min}. As can be seen from relation (4.33b), which can be written in the form

$$J_c(z) = \left(\frac{4}{5} \lambda Q \bar{\tau}\right)^{-1}, \tag{4.36}$$

the constancy of the value of z is equivalent to constancy of the self-excitation coefficient of the generator $\lambda Q \bar{\tau} = \frac{5}{4} \lambda Q \tau_{min}$.

It is interesting to give an estimate of the quantity J_c/J_s for large changes of z. A simple calculation shows that

$$0.83 \leqslant \frac{J_c}{J_s} \leqslant 2.25, \quad \text{if} \quad 0 \leqslant z < \infty. \tag{4.37}$$

Thus, when there are large changes of the self-excitation coefficients, the quantity J_c/J_s does not change much, and we have a linear increase of the absolute stability with increase of τ_{min} over practically the entire range of variation of the self-excitation coefficient. At the same time, inequality (4.37) shows that it is somewhat more advantageous to work with large values of the self-excitation coefficient.

When the molecules are slowed down to low speeds (Part 3) we can regard the distribution of the molecules as uniform over a range of speeds $\Delta v = v_0$, i.e.,

$$F(v) = \text{const} \quad \text{for} \quad 0 \leqslant v \leqslant v_0.$$

In this case also it can be shown that with a constant self-excitation coefficient the stability of the maser increases in proportion to $\bar{\tau}$.

This conclusion is justified, however, only when we neglect the natural width of the spectrum in the maser, which is caused by fluctuations. As has been shown in [23], at a temperature of the order of 100° Kelvin, the largest contribution to the natural width $\Delta\omega$ of the spectrum comes from the thermal fluctuations of the electromagnetic field in the resonator. According to [23], the width of the spectrum caused by these fluctuations is given by

$$\Delta\omega = \alpha \frac{kT_r}{P_0} \cdot \frac{1}{\bar{\tau}^2}, \tag{4.38}$$

where P_0 is the power of the generator; α is a dimensionless coefficient which depends on the line shape and is of the order of magnitude of unity; T_r is the temperature of the resonator.

According to Eq. (1.31), the power of a maser can be written in the form

$$P_0 = N_a \frac{\hbar\omega_l}{2} \cdot \frac{\eta - 1}{\eta}, \tag{4.39}$$

where N_a is the flux of active molecules through the resonator. From Eq. (4.14) we have

$$N_a = N_0 \left(\frac{\bar{\tau}_0^2}{\tau_{min}^2} \right)^2, \tag{4.40}$$

where

$$\bar{\tau}_0^2 = \frac{ml^2}{2kT}.$$

Comparing Eqs. (4.38), (4.39), and (4.40), we see that the fluctuation width $\Delta\omega$ increases quadratically with increase of τ_{min} for constant η:

$$\Delta\omega = 2\alpha \frac{kT_r}{\hbar\omega_l} \cdot \frac{\eta}{\eta - 1} \cdot \frac{1}{N_0 \bar{\tau}_0^4} \tau_{min}^2. \tag{4.41}$$

It is clear from this that there is no advantage in increasing τ_{min} to values larger than

$$\tau_{opt} \simeq (\alpha kT_r)^{-1/4} \cdot \left(Q_0 \frac{\omega_l - \omega_0}{\omega_l} \cdot \frac{\eta - 1}{\eta} \hbar N_0 \bar{\tau}_0^4 \right)^{1/4}. \tag{4.42}$$

It follows from Eq. (4.42), however, that when τ_{min} is decreased by a factor of 10, the thermal noise at room temperature (~300°K) will no longer be of any importance.

The argument we have given shows that the use of slow molecules in a maser opens up new possibilities for increasing the frequency stability of the oscillations. In spite of the large decrease of the number of active molecules in beams of slow molecules, the margin of self-excitation achieved in present generators ($\eta \sim 30$) and the possibility of increasing the quality of the resonators makes it quite feasible to develop generators which operate with slow molecules. This can lead to an increase of the stability of such generators by a factor of ~10 to 100 as compared with present generators.

SUMMARY

1. Equations of a maser are obtained which describe a stationary process with an arbitrary variation with time; the condition for locking of a maser to an external force is obtained, and it is demonstrated theoretically that the oscillations of a maser are highly monochromatic.

2. The frequency of the oscillations of a maser is calculated as a function of various parameters; the results of the calculation are in good agreement with the experimental data, which justifies the assertion that the ideas on which the calculation is based are correct; on the basis of the calculated characteristics it is shown that with a maser with two oppositely moving molecular beams and operated with the line J = 3, K = 3 of $N^{14}H_3$ it is possible to obtain a stability of $5\text{-}7 \cdot 10^{-10}$; to obtain a higher stability ($\sim 10^{-10}$) it is necessary to use a line without quadrupole hyperfine structure (for example, the line J = 3, K = 2 of $N^{14}H_3$).

3. An analysis is given of methods of adjusting the frequency of oscillation of a maser to the frequency of a spectral line; effects by which the hyperfine structure and the nonuniformity of the radiation of the molecules along the length of the resonator influence the tuning by use of an external magnetic field are taken into account; it is shown that to adjust the frequency of a maser with an accuracy up to 10^{-10} it is necessary to compensate the frequency shift of the maser caused by nonuniform radiation of the molecules along the length of the resonator by using two equal beams moving in opposite directions; consideration is given to the possibility of tuning the frequency of the maser to that of a spectral line to accuracy $5 \cdot 10^{-10}$ by the method of two coupled resonators.

4. The effects of the waveguide system and of external power sources on the oscillation frequency of a maser are estimated; the dependence of the frequency on these factors is appreciable (10^{-8} to 10^{-9}); to eliminate this dependence it is necessary to use good decoupling valves.

5. Ways of further increasing the frequency stability of the oscillation of a maser are considered; it is shown that by using beams of "slow" molecules one hopes to obtain an absolute frequency stability of a maser better than 10^{-11}.

On the basis of the analysis of effects which influence the oscillation frequency of a maser, generators have been constructed which operate with the line J = 3, K = 2 of $N^{14}H_3$ and employ two molecular beams moving in opposite directions [51].

Experimental studies of the dependence of the frequency of such generators on the various parameters and of methods of tuning the frequency of the maser to that of a spectral line have confirmed the correctness of our calculations. By varying the pressure in the source of the molecular beam, and also by the method of an external magnetic field, it has been possible to tune the frequency of such a generator to that of a spectral line to an accuracy up to $1.5\text{-}1 \cdot 10^{-10}$ [51].*

The author is grateful to V. V. Nikitin, G. M. Strakhovskii, and I. V. Cheremiskin for making experimental material available before publication and for a discussion of the results, and to K. K. Svidzinskii for a helpful discussion.

The author is particularly grateful to Corresponding Member of the Academy of Sciences of the USSR N. G. Basov for his constant interest and for great assistance in this work.

*Papers have appeared [67, 68] which state that magnetic tuning of a two-beam maser operating with the line J = 3, K = 3 of $N^{15}H_3$ makes it possible to reproduce its frequency with an accuracy up to $\sim 3 \cdot 10^{-11}$; this is not absolute tuning, however, since the frequency obtained in this way depends on the pressure in the source of the molecular beam and on the field strength of the sorting system (cf. Part 2 of Section 3).

APPENDIX I

Derivation of the Formulas (2.7)

To derive formulas (2.7) we use Eq. (1.14):

$$\left.\begin{aligned}\left[\left(\frac{\partial}{\partial t}+\frac{\partial}{\partial \tau}\right)^2+\omega_{ab}\right]P(\tau,t)=-2\omega_{ab}\cdot\frac{|\mu_{ab}|^2}{\hbar}\mathscr{E}(t)\widetilde{R}(\tau,t),\\ \left(\frac{\partial}{\partial t}+\frac{\partial}{\partial \tau}\right)\widetilde{R}(\tau,t)=2\frac{1}{\hbar\omega_{ab}}\mathscr{E}(t)\left(\frac{\partial}{\partial t}+\frac{\partial}{\partial \tau}\right)P(\tau,t).\end{aligned}\right\}\quad (A.1)$$

Let

$$\mathscr{E}=\mathscr{E}_0\cos\omega t. \quad (A.2)$$

We then look for the solution of (A.1) in the form

$$\left.\begin{aligned}P(\tau,t)&=P_0'(\tau)\cos\omega t+P_0''(\tau)\sin\omega t,\\ \widetilde{R}(\tau,t)&=\widetilde{R}_0(\tau).\end{aligned}\right\}\quad (A.3)$$

Substituting (A.3) and (A.2) in (A.1), and neglecting the nonresonance terms proportional to $e^{\pm 2i\omega t}$, we get the following equations for the determination of P_0 and R_0:

$$\left.\begin{aligned}\frac{d^2P_0^{\pm}}{d\tau^2}\pm 2i\omega\frac{dP_0^{\pm}}{d\tau}+(\omega_{ab}^2-\omega^2)P_0^{\pm}=-2\omega_{ab}\frac{|\mu_{ab}|^2}{\hbar}\mathscr{E}_0\widetilde{R}_0,\\ \frac{d\widetilde{R}_0}{d\tau}=i\frac{\omega}{\hbar\omega_{ab}}\cdot\mathscr{E}_0(P_0^{+}-P_0^{-}),\end{aligned}\right\}\quad (A.4)$$

where

$$P_0^{+}=\frac{1}{2}(P_0'+iP_0''),\quad P_0^{-}=\frac{1}{2}(P_0'-iP_0'').$$

Equations (A.4) can be solved easily. We need to find the solution satisfying the initial condition $P_0^{\pm}(0)=0$, $\widetilde{R}_0(0)=1$. This solution is (up to terms in γ_{ab}/ω_{ab})

$$\left.\begin{aligned}P_0'(\tau)&=\left[\frac{\omega_{ab}-\omega}{\gamma_{ab}^2}\cdot\frac{|\mu_{ab}|^2}{\hbar}(1-\cos\gamma_{ab}\tau)\right]\mathscr{E}_0,\\ P_0''(\tau)&=\left[-\frac{|\mu_{ab}|^2}{\hbar\gamma_{ab}}\cdot\sin\gamma_{ab}\tau\right]\mathscr{E}_0.\end{aligned}\right\}\quad (A.5)$$

In these formulas $\gamma_{ab}=\frac{|\mu_{ab}|}{\hbar}\mathscr{E}_0$. Taking the average values of the expressions (A.5) over the time which the molecule spends in the resonator,

$$\left.\begin{aligned}P_0'&=\frac{1}{\tau}\int_0^{\tau}P_0'(\xi)\,d\xi,\\ P_0''&=\frac{1}{\tau}\int_0^{\tau}P_0''(\xi)\,d\xi,\end{aligned}\right\}\quad (A.6)$$

we get the required relations (2.7).

APPENDIX II

The equations obtained for the maser in Section 1, Eq. (1.18), depend in an essential way on the distribution of times of flight $(1/\bar{\tau})e^{-\tau/\bar{\tau}}$.

We can, however, write the equations for a maser in a closed form which is free from this restriction. In fact, using (1.3) and (1.12), we have

$$\frac{d^2\mathcal{E}_k}{dt^2}+\frac{\omega_k}{Q}\cdot\frac{d\mathcal{E}_k}{dt}+\omega_k^2\mathcal{E}_k=-4\pi N\cdot\frac{d^2}{dt^2}\int_0^\infty f(v)\,dv\cdot\int_0^{\lambda(t)}dz\iint_S dx\,dy\,\mathbf{P}\mathbf{E}_k(x,y,z), \tag{A.7}$$

$$\left[\left(\frac{\partial}{\partial t}+v\frac{\partial}{\partial z}\right)^2+\omega_{ab}^2\right]\mathbf{P}=-2\omega_{ab}\cdot\frac{|\mu_{ab}|^2}{\hbar}\cdot\mathbf{E}(x,y,z;t)\widetilde{R}, \tag{A.8}$$

$$\left(\frac{\partial}{\partial t}+v\frac{\partial}{\partial z}\right)\widetilde{R}=\frac{2}{\hbar\omega_{ab}}\cdot\mathcal{E}(x,y,z;t)\cdot\left(\frac{\partial}{\partial t}+v\frac{\partial}{\partial z}\right)\mathbf{P},$$

where v is the velocity of the molecule along the z axis (the axis of the resonator); $\lambda(t) = vt$ for $t < l/v$ and $\lambda(t) = l$ for $t \geq l/v$; and l is the length of the resonator. The weight function $f(v)$ can be any function, including that obtained from Eq. (2.18).

This system of equations does not assume that the process is quasi-stationary. This system of equations, however, one of which is an integrodifferential equation, is much more complicated to analyze than (1.18).

If the type of oscillations considered has a uniform distribution of the field in the resonator, Eq. (A.7) can be simplified and takes the form

$$\frac{d^2\mathcal{E}}{dt^2}+\frac{\omega_p}{Q}\frac{d\mathcal{E}}{dt}+\omega_p^2\mathcal{E}=-4\pi N\frac{d^2}{dt^2}\int_0^{\lambda(t)}F(\xi)\,P(\xi,t)\,d\xi, \tag{A.9}$$

where

$$\xi=\frac{z}{v},\quad F(\xi)=\int_0^l f\left(\frac{z}{\xi}\right)dz.$$

The solutions of (A.8) and (A.9) must satisfy definite boundary conditions at z = 0. In particular, for the usual maser the conditions at z = 0 are $P(0, t) = 0$, $\partial P(0,t)/\partial t = 0$ and $\widetilde{R}(0, t) = 1$.

LITERATURE CITED

1. Holloway, W. Neiberger, F. H. Reder, G. M. Winker, I. Essen, and J. W. L. Parry, Proc. IRE, 47, 1730 (1959).
2. N. G. Basov, I. D. Murin, A. P. Petrov, A. M. Prokhorov, and I. V. Shtranikh, Izv. vyssh. ucheb. zav. radiofizika, 2, 50 (1958).
3. I. D. Murin, Izv. vyssh. ucheb. zav., radiotekhnika, 1, 555 (1957).
4. R. C. Mockler, R. E. Beehler, and C. S. Snider, IRE Trans. Instr., 1, 120 (1960).
5. V. F. Lubentsov et al., Otchet VNIIFTRI (1959).
6. N. F. Ramsey, Phys. Rev., 78, 695 (1950).
7. N. F. Ramsey, Molecular Beams, Oxford Univ. Press (1956).
8. D. Kleppner, N. F. Ramsey, and P. Fjeldstadt, Phys. Rev. Letters, 1, 232 (1958).
9. H. M. Goldenberg, D. Kleppner, and N. F. Ramsey, Phys. Rev. Letters, 5, 361 (1960).
10. A. Scheibe, U. Adelsberger, G. Becker, G. Ohl, and R. Sus, Zs. angew. Phys., 11, 352 (1959).
11. A. H. Morgan and J. A. Barnes, Proc. IRE, 47, 1782 (1959).
12. J. R. Wittke, Proc. IRE, 45, 1782 (1959).
13. C. D. Alley, Abstracts of Reports at Conference on Quantum Electron-Resonance Phenomena, USA, September, 1959, p. 7.
14. N. G. Basov and A. M. Prokhorov, Uspekhi fiz. nauk, 7 (1955).
15. K. Shimoda, T. C. Wang, and C. H. Townes, Phys. Rev., 102, 1308 (1956).
16. K. Shimoda, J. Phys. Soc. Japan, 12, 1006 (1957); 13, 939 (1958).

17. J. C. Helmer, J. Appl. Phys., 28, 212 (1957).
18. J. Bonanomi and J. Herrmann, Helv. Phys. Acta, 29, 224 (1956).
19. J. Bonanomi, J. de Prins, J. Herrmann, and P. Kartaschoff, Helv. Phys. Acta, 30, 288 (1957).
20. F. S. Barnes, Proc. IRE, 47, 2085 (1959).
21. N. G. Basov and A. M. Prokhorov, Doklady Akad. nauk SSSR, 101, 47 (1955); Zhur éksptl. teoret. fiz., 30, 560 (1956) [Soviet Physics—JETP, 3, 426 (1956)].
22. V. Klimontovich and R. V. Khokhlov, Zhur. éksptl. teoret. fiz., 32, 1151 (1957) [Soviet Physics – JETP, 5, 937 (1957)].
23. V. S. Troitskii, Radiotekhnika i élektronika, 3, 98 (1958).
24. Slater, Ultrahigh Frequency Electronics [Russian translation], "Soviet Radio" Press (1948).
25. L. A. Vainshtein, Electromagnetic Waves, "Soviet Radio" Press (1957).
26. L. D. Landau and E. M. Lifshits, Quantum Mechanics, Gostekhizdat (1948), p. 170.
27. N. G. Basov and A. N. Oraevskii, Izv. vyssh. ucheb. zav., radiofizika, 1, No. 4, 63 (1958).
28. N. Bloembergen, Phys. Rev., 104, 324 (1956).
29. V. M. Fain, Zhur. éksptl. teoret. fiz., 33, 945 (1957) [Soviet Physics – JETP, 6, 726 (1958)].
30. S. M. Rytov, Zhur. éksptl. teoret. fiz., 29, 304 (1955) [Soviet Physics – JETP, 2, 217 (1956)].
31. I. G. Malkin, Some Problems of the Theory of Nonlinear Vibrations, Gostekhizdat (1956).
32. A. A. Andronov and A. A. Vitt, Zhur. tekh. fiz., 4, 122 (1934).
33. A. A. Andronov and A. A. Vitt, Zhur. tekh. fiz., 7, 3 (1930); see also A. A. Andronov, Collected Works, Izd. AN SSSR (1956), p. 70.
34. R. V. Khokhlov, Radiotekhnika i élektronika, 3, 566 (1958).
35. N. G. Basov and A. P. Petrov, Radiotekhnika i élektronika, 3, 198 (1958).
36. N. G. Basov and A. M. Prokhorov, Zhur. éksptl. teoret. fiz., 27, 431 (1954).
37. N. G. Basov, Doctoral Dissertation, FIAN (1956).
38. A. N. Oraevskii, Certificate Thesis, MFTI (1956).
39. G. R. Gunter-Mohr, C. H. Townes, and J. H. Van Vleck, Phys. Rev., 94, 1191 (1954).
40. J. P. Gordon, Phys. Rev., 99, 1253 (1955).
41. N. G. Basov and A. N. Oraevskii, Radiotekhnika i élektronika, 4, 1185 (1959).
42. C. Townes and A. Schawlow, Radiospectroscopy [Russian translation], IL (1959), pp. 282, 233.
43. A. R. Edmonds, Angular Momentum in Quantum Mechanics, Princeton University Press (1957).
44. N. G. Basov et al., Otchet FIAN (1959).
45. A. N. Oraevskii, Certificate Thesis, MFTI (1959).
46. V. S. Troitskii, Zhur. éksptl. teoret. fiz. (in press); A. I. Naumov, Zhur. éksptl. teoret fiz. (in press).
47. J. M. Jauch, Phys. Rev., 72, 720 (1947).
48. G. F. Hadley, Phys. Rev., 108, 2910 (1957).
49. N. G. Basov, Radiotekhnika i élektronika, 1, 51 (1956).
50. N. G. Basov, G. M. Strakhovskii, and V. Cheremiskin, Radiotekhnika i élektronika (in press).
51. N. G. Basov, V. V. Nikitin, and A. N. Oraevskii, Radiotekhnika i élektronika, 6, 796 (1961).
52. G. M. Strakhovskii and I. V. Cheremiskin, in this volume.
53. G. P. Gordon, H. J. Zeiger, and C. H. Townes, Phys. Rev., 99, 1264 (1955).
54. J. Bonanomi, J. de Prins, J. Herrmann, and P. Kartaschoff, Helv. Phys. Acta, 30, 492 (1957).
55. C. V. Heer, Abstracts of Reports at Conference on Quantum Electron-Resonance Phenomena, USA, September, 1959, p. 1.
56. A. R. Edmonds, Angular Momentum in Quantum Mechanics. Princeton University Press (1957).
57. J. Bonanomi, J. Herrmann, J. de Prins, and P. Kartaschoff, Rev. Sci. Instr., 28, 879 (1957).
58. N. N. Krylov, Theoretical Foundations of Radio Engineering, "Morskoi Transport" Press, Moscow-Leningrad (1953).
59. N. G. Basov and K. K. Svidzinskii, Izv. vyssh. uchebn. zav., radiofizika, 1, 89 (1958).
60. S. Gladstone and M. Edlund, Fundamental Theory of Nuclear Reactors [Russian translation], IL (1954).
61. K. F. Teodorchik, Zhur. tekh. fiz., 6, 845 (1946).
62. C. K. Jen, Phys. Rev., 74, 1396 (1948).
63. A. N. Oraevskii, Radiotekhnika i élektronika, 4, 718 (1959).

64. N. G. Basov and A. N. Oraevskii, Zhur. éksptl. teoret. fiz., 37, 1069 (1959) [Soviet Physics – JETP, 10, 761 (1959)].
65. N. G. Basov and A. N. Oraevskii, Izv. vyssh. uchebn. zav., radiotekhnika, 2, 3 (1959).
66. R. C. Mochler, Atomic Beam Frequency Standards, in Advances in Electronics and Electron Physics, 15, 1 (1962).
67. V. V. Nikitin, Radiotekhnika i élektronika, 7, No. 1 (1963).
68. J. A. Barnes, D. W. Allan, and A. E. Wainwright, IRE Trans. on Instr., No. 6, 26 (1962).

CHARACTERISTICS OF MASERS

(The J = 3, K = 3 Line of $N^{14}H_3$)

G. M. Strakhovskii and I. V. Cheremiskin

INTRODUCTION

At the present time, oscillators that make use of the natural frequencies of spectral lines of atoms or molecules are widely used as standards of frequency (time). Because of the high quality factor and good reproducibility of the frequencies of spectral lines these frequency standards can be used as primary standards; they do not require constant comparison against astronomical observations.

The first molecular frequency standards were used for stabilization of a radio-frequency oscillator by means of an absorption cell filled with ammonia, the spectral line of which served as a frequency discriminator [1, 2].

In order to determine the value of a stabilized frequency with high accuracy, it is necessary to use narrow lines characterized by high quality factors. The width of the spectral line of an isolated molecule (atom) in the centimeter range can be $\sim 10^{-5}$ cps, corresponding to a quality factor of $\sim 10^{15}$. However, under laboratory conditions the lines can be broadened by (1) collisions of molecules with molecules and with the walls of the container, (2) Doppler broadening, and (3) broadening due to saturation.

At low pressures there is very little interaction between molecules in a cell, and the Q of the line is determined primarily by the Doppler effect. This is of the order of $\sim 10^6$. A further reduction in pressure does not narrow the line any further. Dicke [3], in several experiments in microwave spectroscopy, has obtained narrower lines by diluting the gas in a gas which does not change the state of the molecule being studied. In this case collisions do not influence the time the molecule remains in the radiation field and the spectral line is not broadened by collisions; on the other hand, the Doppler width is reduced, since it is determined by the rate of diffusion of the test gas in the gas being used as a buffer.

A method of avoiding the limitations due to the Doppler effect is to use atomic and molecular beams moving at right angles to the direction of propagation of the electromagnetic energy. This method is used at the present time in most frequency standards of high stability. The first practical application of atomic beams in this field was the cesium frequency standard using the (F, m) transition $(4,0) \rightarrow (3,0)$ in the Cs atom [4] (the frequency of the transition $\gamma \cong 9192$ Mc). The best absolute stability obtained in a cesium device with two separated resonators [5,6] is of the order of $\pm 1.5 \cdot 10^{-10}$ [7]. At the present time, the Boulder Laboratories of the National Bureau of Standards, USA [8], are carrying out experiments with two cesium frequency standards which provide an accuracy of $\pm 2 \cdot 10^{-12}$ for measuring time periods ranging from one to several hours. The difference in the frequencies produced by the two standards has been maintained within $\pm 2 \cdot 10^{-12}$ over a period of ten months [8].

Zacharias [9] has shown that, if very slow molecules are used, it is possible to obtain very narrow lines and to provide frequency standards of still higher accuracy. He has proposed to project the molecular beam vertically upward and then to use only those molecules whose direction is reversed by the gravitational field of the earth. However, the slow molecules (atoms) comprise only a very small fraction of the molecules emitted from the source and are not sufficient for operation of the frequency standard.

Systems with energy-level populations not given by thermodynamic equilibrium made it possible in 1954-1955 to make a frequency standard using the radiation of the molecules directly. In the Soviet literature this device is called a molecular generator (MG); in the foreign literature it is called a maser [10-12]. The absolute stability* of a maser that has been obtained at the present time is of the order of 10^{-10}.

A further improvement in the frequency stability of a maser can be achieved by using slow molecules, because increasing the time of interaction between the molecule and radio-frequency field leads to a narrowing of the line. Calculations show that when this technique is used it may be possible to improve the stability of a maser by a factor of 10-100 [13, 14].

The question of high absolute stability is intimately related to the question of relative stability, i.e., the frequency stability over a given time period. A high relative stability is needed to obtain absolute stability, since tuning a maser to the frequency of the spectral line requires a definite time and the frequency of the maser during the tuning process can deviate only by an amount smaller than the absolute stability; in the second place, in a number of maser applications one wishes to measure a relative frequency change rather than absolute values of a frequency. Examples of this use of masers are found in experiments on verification of the general theory of relativity [15], our own work on the dependence of maser frequency on various parameters, etc. Hence, obtaining a high relative stability is an independent problem in its own right. At the present time it has been possible to develop masers in which a relative change of frequency (in several hours) of $\sim 10^{-11}$ has been achieved; it may be possible to achieve still higher values over times of the order of a second [16].

In 1960 a group of physicists under the direction of N. Ramsey built an atomic frequency standard using the transition between hyperfine levels in the hydrogen atom at frequencies of 1420 and 405 Mc [17]. The use of a storage volume in this generator made it possible to increase appreciably the time of interaction between the atoms and the radiation field [3, 18-21], and thus to obtain a narrow spectral line. The original width of the induced radiation line was found to be ~1 cps. The expected absolute stability of the atomic oscillator will be greater by far than any of the oscillators that has been proposed earlier.

At the present time one of the most precise frequency standards used in practice is the maser that makes use of one of the lines in the inversion spectrum of ammonia. Any attempt to increase the frequency stability of this standard, to improve its design, and to choose the optimum mode of operation would be impossible without a detailed investigation of the dependence of frequency on the various design parameters. A great deal of theoretical and experimental work has been done on this dependence [10, 22-28], but the results are far from complete. The absence of any sufficiently detailed experimental characteristics of the operation of a maser have made it impossible to carry out a quantitative comparison of theory and experiment and to understand all the basic effects that determine the frequency of a maser.

The present work is devoted to an experimental investigation of the characteristics of a maser and to certain problems associated with the design of a maser characterized by high relative stability.

Section 1

BRIEF SURVEY

The first masers that were built [11, 12, 29, 30] were based on the use of induced radiation from the inversion line J = 3, K = 3 of $N^{14}H_3$. Ammonia is used because the spectrum of this gas has been thoroughly investigated and this investigation has shown that the ammonia spectrum exhibits strong lines in the radio-frequency region.

Ammonia is a symmetric-top molecule.

The three hydrogen atoms form a triangle while the nitrogen atom oscillates with respect to the plane of this triangle, i.e., it executes an inversion. Because of the inversion, each rotational level of ammonia is split into two inversion levels and the transition between these levels falls in the microwave region.

---

*Absolute stability is defined as the accuracy with which it is possible to tune the oscillator frequency to the frequency of the spectral line. Mathematically it is characterized by the quantity

$$\delta = (\omega - \omega_l)/\omega_l \ .$$

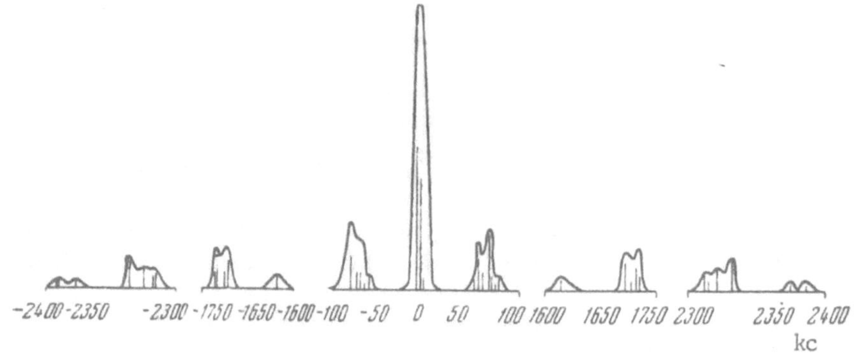

Fig. 1. Structure of the spectral line J = 3, K = 3 of $N^{14}H_3$.

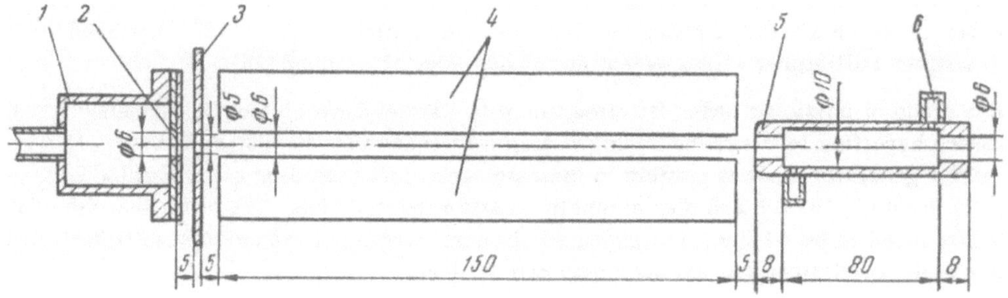

Fig. 2. Diagram of a maser. 1) Beam source; 2) grid; 3) liquid-air-cooled diaphragm; 4) electrodes of the quadrupole condenser; 5) cavity resonator; 6) waveguide.

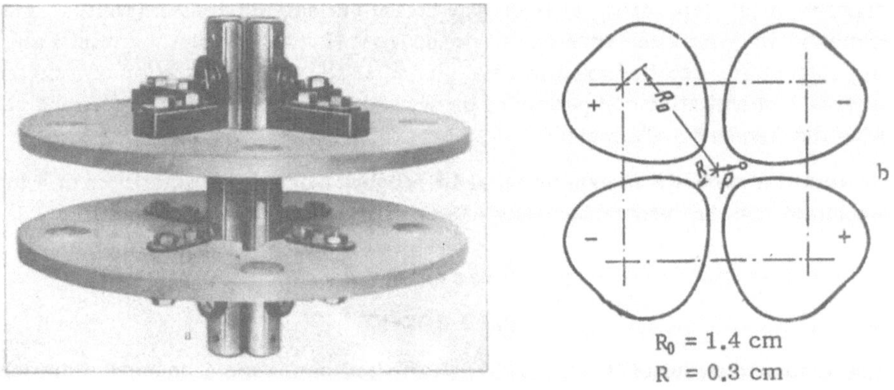

Fig. 3. Quadrupole condenser. a) General view; b) cross section.

The strongest line in the ammonia spectrum is characterized by the quantum numbers J = 3, K = 3, where J is the total rotational moment of the molecule and K is the projection on the axis of symmetry. If account is taken of the quadrupole interaction of the nitrogen nucleus N^{14} with the molecular field (of the order of 4 Mc [31]) and the magnetic interaction (of the order of 25-100 kc [31]) in the ammonia molecule, it is found that each inversion level is split into a series of sublevels corresponding to the quantum numbers I_H, I_N, J, F_1, and F.* Hence, the inversion lines consist of several components. Thus, the J = 3, K = 3 line for the transition with $\Delta F_1 = \Delta F = 0$

*$\mathbf{F}_1 = \mathbf{I}_N + \mathbf{J}$; $\mathbf{F} = \mathbf{F}_1 + \mathbf{I}_H$, where $\mathbf{I}_N$ is the spin of the nitrogen nucleus; $\mathbf{I}_H$ is the total spin of the hydrogen nuclei; $\mathbf{J}$ is the rotational moment of the molecule. The hyperfine structure of the J = 3, K = 3 line is shown in Fig. 1. The vector diagram showing the composition of the moments in $N^{14}H_3$ is given in Fig. 1 of [32].

used in the maser consists of twelve components corresponding to the twelve possible values of the quantum numbers F_1 and F (Fig. 1 of [32]). In the maser these components are not resolved and participate in the molecular radiation as a single spectral line.

The principle of operation of a maser is as follows: a beam of molecules from a molecular-beam source moves through a state-selector and enters a cavity resonator (Fig. 2). The system used to separate the states (Fig. 3) consists of an inhomogeneous electrostatic field which increases in the radial direction from the system axis. Molecules moving through this selection system are subject to a force

$$F = \pm 2\alpha E \frac{\partial E}{\partial r}, \quad (1)$$

where E is the strength of the electric field; $\partial E / \partial r$ is the derivative of the field strength in the radial direction. The plus sign refers to molecules in the lower inversion state, while the minus sign refers to molecules in the upper inversion state. Thus, molecules in the upper inversion state enter the region with minimum field and are focused at the axis of the selector system while molecules in the lower level are defocused and expelled from the selector system.

This separation of molecules according to energy levels leads to a condition of thermodynamic nonequilibrium [23, 29, 33, 34], since there are more molecules in the upper level of this transition than corresponds to thermodynamic equilibrium. The difference in the number of molecules in the upper and lower levels is called the number of active molecules. In the maser, the beam of active molecules enters a cavity resonator tuned to the frequency of the molecular transition and excites electromagnetic oscillations by virtue of the energy of the induced radiation of the molecules.

In [10, 22, 35-37] an investigation has been made of the steady-state oscillations in the maser, the condition for self-excitation has been obtained, and the amplitude and frequency of the oscillations have been computed.

The basic parameters that enter into the maser frequency ω are as follows: (1) the natural frequency of the cavity resonator; (2) the intensity of the molecular beam; and (3) the shape and width of the spectral line.

The most important factor is the natural frequency of the cavity resonator. If the natural frequency of the cavity deviates from the frequency of the spectral line by a half-width of the cavity passband, the generated frequency is shifted from the characteristic frequency of the molecular transition by the half-width of the spectral line.

The effect of the intensity of the molecular beam and of the shape and width of the spectral line can be given in terms of three effects [38]:

1. In the presence of unresolved hyperfine structure of the line a change in the intensity of the selection field causes a change in the shape and a shift of the peak of the spectral line. The 3, 2 line of $N^{14}H_3$ and the lines of $N^{15}H_3$ do not have hyperfine quadrupole structure, so that this effect does not appear in these molecules [39, 40].

2. When the magnitude of the selection field is changed there is a change in the distribution of molecules over time of flight through the cavity. This means that the generated frequency depends on the selection field even for lines that do not exhibit hyperfine structure if the natural frequency of the cavity resonator does not coincide with the peak of the spectral line.

3. Even when the resonator is tuned exactly to the frequency of the spectral line, the generated frequency does not coincide with the latter [22]. The reason is as follows: There is a nonuniformity in the radiation level of molecules along the resonator as a result of which a traveling wave appears; in addition to the primary wave there are additional oscillations in which the energy is propagated along the resonator. This leads to the appearance of a Doppler shift of the molecular transition even if use is made of the E_{010} mode, in which the Doppler shift should disappear for a parallel beam and for uniform radiation of the molecules along the resonator [42]. This effect, which we will call effect III, can be compensated to a large extent if the beam of molecules is admitted symmetrically from the two sides of the cavity resonator and if the power is extracted exactly in the center [22].

In [38] a formula was derived expressing the dependence of the maser frequency on the voltage U of the selection system and the saturation parameter γ (neglecting effect III):

$$\omega = \omega_l \left[1 + \frac{\omega_0 - \omega_l}{\omega_l} \cdot \frac{Q}{Q_l} \cdot G(\gamma, U) + \Delta(\gamma, U) \right], \tag{2}$$

where ω is the maser frequency; ω_0 is the natural frequency of the cavity resonator; ω_l is the frequency of the molecular transition; $\gamma = (d\mathcal{E}/\hbar)\tau$ is the saturation parameter; $\mathcal{E}$ is the amplitude of the field in the cavity; τ is the mean time of flight of molecules through the cavity; d is the dipole moment of the molecule; $2\pi h$ is Planck's constant.

The function Δ is related to the hyperfine structure of the spectral line. If the line is a single line, $\Delta = 0$. The function $G(\gamma, U)$ describes the dependence of the effective Q of the line on the voltage in the selection system U and the intensity of the molecular beam (through the saturation parameter γ). The frequency shift due to effect III had been computed in [43]. It appears additively in Eq. (2).

In order to obtain the maximum absolute stability it is necessary to tune the cavity precisely to the frequency of the spectral line. However, it is difficult in practice to determine the frequency of the loaded resonator with a sufficient degree of accuracy because of the distortion of the amplitude-phase characteristics of the cavity introduced by the waveguides connected to it. A more effective means of tuning the cavity is to use the dependence of the frequency of the maser oscillations on various parameters.

According to Eq. (2), which neglects effect III, the maser frequency is changed if one changes (1) the intensity of the molecular beam, (2) the voltage on the selector system, (3) the quality factor (width) of the spectral line Q_l.

The change in frequency is a minimum when the cavity is tuned to the frequency of the spectral line ($\omega_0 = \omega_l$).

The first estimates of the absolute stability of the maser using the 3,3 line were given by Helmer [25]. He measured the dependence of maser frequency on p and U as well as the magnetic field applied to the resonator. An analysis of the results obtained by this author led him to the conclusion that it is impossible to tune a maser using the 3,3 line of $N^{14}H_3$ to the frequency of the spectral line with an accuracy better than 10^{-8}.

Helmer considered three kinds of tuning:

1. Variation of the frequency as a function of pressure; the maser is tuned to a frequency satisfying the condition $\left(\frac{\partial\omega}{\partial p}\right)_{v,U} = 0$.

2. Variation of frequency as a function of the voltage applied to the selector system; in this case the condition $\left(\frac{\partial\omega}{\partial U}\right)_{v,p} = 0$ is satisfied.

3. Zeeman modulation. In this case we must satisfy the condition $\left(\frac{\partial\omega}{\partial H}\right)_{v,p,U} = 0$, where H is the magnetic field applied to the cavity, γ is the deviation of the cavity.

It is found that the frequency of a maser tuned by these three methods will differ by 100-200 cps.

Curves showing the dependence of frequency and amplitude of a maser oscillation for the 3,3 line of ammonia as functions of the voltage applied to the selector system and the pressure in the beam source are given in [23, 24]; this work was carried out in order to make comparisons with the theory of a single-velocity beam. The results of this work are shown in Fig. 4 and in Fig. 10 of [32]. However, conclusions as to the accuracy of tuning cannot be obtained from this work because of the inadequate detail in the curve showing the dependence of frequency on maser parameters. In [40] certain experimental data have been used to compute the theoretical functions G and Δ; Eq. (2) was then used to compute the dependence of the maser operating at the 3,3 line on pressure p and selector voltage U (neglecting effect III). The results are shown in Figs. 6 and 7 of [32]. A comparison of these curves and the curves obtained in [44, 45] shows that the hyperfine structure of the 3,3 line of ammonia (effect I) and the nonuniformity in the molecular radiation along the length of the resonator (effect III) actually lead to a shift in the maser oscillation frequency by an amount of the order of $5 \cdot 10^{-9}$ even when the cavity is tuned to the exact frequency of the spectral line. Hence, the frequency ω' obtained by tuning a maser

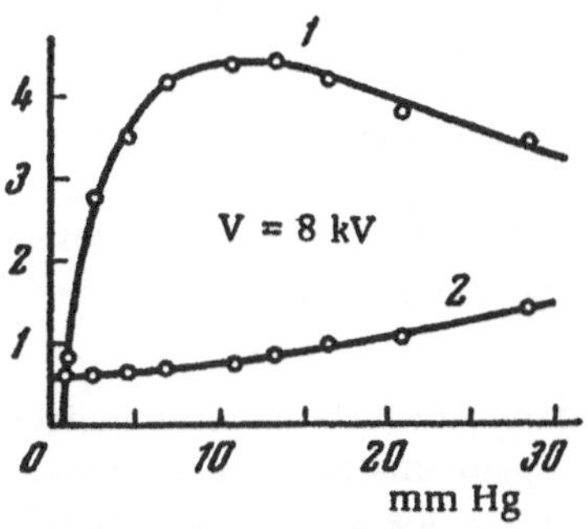

Fig. 4. Variation of relative amplitude (1) and pressure inside the vacuum system (2) as functions of the pressure in the molecular-beam source [24].

using the 3,3 line of $N^{14}H_3$ by one of the methods described above cannot coincide with the frequency of the spectral line ω_l if one of the parameters is changed. Moreover, ω' depends on the values of parameters which do not change in the tuning process. At different values of these parameters, in general, one obtains different oscillation frequencies. However, with a fixed design, if one establishes the frequency by one of the methods given above, it can be reproduced with a definite accuracy which is characterized by the reproducibility of the maser frequency:

$$\delta' = \frac{\omega - \omega'}{\omega'}.$$

If, by changing the pressure or selector voltage one tunes the cavity precisely to the frequency of the spectral line, then the frequency ω of a maser which uses two opposed symmetric beams and a single spectral line will coincide with ω_l [13], as follows from Eq. (2). In this case, ω does not depend on any of the parameters to within the accuracy with which Eq. (2) applies, i.e., $10^{-11}\,\omega_l$.

Experimental investigations have shown [40] that a maser using the 3,2 line with opposed beams (effect I vanishes and effect III is balanced) tuned by modulation of the pressure in the beam source with an accuracy of 1-1.5 cps will maintain its frequency value when the voltage applied to the selection system is changed from 20 to 30 keV. Hence, it may be stated that modulation of the pressure allows us to tune the oscillation frequency of a maser to the frequency of the spectral line. The frequency of a maser tuned in this way may be taken to be accurate to within 10^{-10}.

In [46] it has been proposed to tune a maser with two opposed beams (J = 3, K = 3 line of $N^{15}H_3$) by modulating the magnetic field and extrapolating the oscillation frequency to zero intensity of the molecular beam. The absolute stability obtained in this case is no worse than 10^{-10}.

The frequency of a maser tuned by modulating the external magnetic field without extrapolation to zero beam can be reproduced with an accuracy of 10^{-11} [47], but depends on the parameters of the system and can differ from the frequency of a maser tuned by changing the pressure by an amount of the order of 70 cps. In [48] this difference is attributed to the shift in the peak of the spectral line due to the application of the magnetic field.

The method of tuning by varying the selector voltage is usually not used because of its poor sensitivity.

As another method of tuning a maser to the frequency of a spectral line, in [38] it is proposed to use two coupled cavity resonators with a beam of active molecules passing through one. A theoretical calculation shows that this method should yield an absolute stability of the order of 10^{-11}.

It follows from Eq. (2) that the frequency stability of a maser increases as the quality factor of the line Q_l increases. One of the factors appearing in Q_l is the time that the molecule interacts with the radiation field. If the line with $\Delta\nu_0$ is determined only by the mean interaction time τ, then $\Delta\nu_0 = 1/\pi\tau$. It would appear that increasing the length of the cavity resonator could produce a narrower line and consequently enhance the stability of the maser. However, because of the nonparallel flow of the molecular beam, there exists an upper limit to τ determined by the diameter of the cavity resonator and the transverse velocities of the molecules in the selected beam. This quantity depends on the voltage applied to the selector system. Calculations carried out in [38] show that for typical maser parameters $\tau_{max} \cong 3 \cdot 10^{-4}$ sec, which gives an optimum resonator length $L_{opt} \cong 15$ cm.

It is possible to increase the time during which the molecules interact with the radiation field by using molecular beams in which the mean velocity of the molecules is appreciably smaller than the mean thermal velocity at typical temperatures. A number of methods of producing slow molecular beams have been given in [13, 14]: selection of the molecules by potential barriers, which could be inhomogeneous electric fields remaining constant in time; the use of combinations of inhomogeneous fixed and variable fields; retardation of the molecules of the working materials by collisions with a cold gas (helium). However, if only the slow molecules are used there is a strong reduction in the total number of molecules passing through the cavity resonator per unit time. For this reason masers using "slow molecules" have not been used at the present time.

It has been noted in [38] that the resonance properties of the waveguide system and changes in loading can have an appreciable effect on maser frequency. For example, if the coupling is changed by 10% the oscillation frequency is changed by $\sim 10^{-9}\ \omega_l$ cps. This strong dependence means a worsening of maser stability. The effect of the waveguide system can be reduced by reducing the coupling between the maser cavity and by using ferrite isolators.

In [49] it is proposed to use for this purpose two successive cavity resonators through which the same molecular beam would pass. In passing through the first resonator some of the molecules (approximately half) radiate and excite electromagnetic oscillations in the first resonator in exactly the same way as in a usual maser. In this case the ensemble of molecules forming the beam makes a transition into the superradiant state [49-52] in which molecules in the beam radiate with the same frequency with which they radiate in the first cavity. Thus, the oscillation frequency in the second cavity depends only on the characteristic frequency of the first cavity and is independent of the tuning of the second cavity. In this case, the waveguide system connected to the second cavity has no effect on the oscillation frequency.

It has been proposed to use spectral lines of other gases in addition to lines of ammonia; it was proposed to use these lines in masers in order to investigate the line shape and structure [53-56]. However, as frequency standards these are worse than the ammonia lines cited above because of hyperfine structure and because their intensity is usually smaller than that of the ammonia line. Thus, one would expect a maser using the $6_{-5} - 6_{-1}$ line of H_2O [57] to have a high stability; the hyperfine structure here has only magnetic satellites. However, calculations we have carried out show that because of the small intensity of this line, and because of the small dipole moment of the transition, to operate a maser of this kind it would be necessary to have a resonator with a quality factor of the order of 1,000,000, which would be impossible in practice.

In the maser one of the important parameters is the power delivered and the length of time of operation. These characteristics are directly related to the construction of the basic elements of the maser: the molecular-beam source, the selection system, and the cavity resonator. The power that can be delivered by a maser is estimated as 10^{-11}-10^{-10} W. Increasing the power of a maser means increasing the number of active molecules entering the cavity resonator. This number can be increased by increasing the total number of molecules from a source moving in the direction of the cavity or by increasing the molecular capture angle in the selector system.

Various kinds of molecular-beam sources of the "grid" and "channel" type are used in the maser [11, 24, 29, 54, 59]. It is known [60, 61] that the directivity of a molecular beam from a channel source is better and that, consequently, one can obtain a higher number of molecules in the beam for the same vacuum conditions in the system. However, the question of stability of the maser with such a source remains open because in a channel-type source the gas flow is not molecular but is a jet flow, causing an increase in the molecular velocity and, consequently, broadening of the radiation line. According to Eq. (2), this means a reduction in maser stability. At the present time the channel source is found to be advantageous for long-term operation.

For a given source the number of active molecules can be increased by using a higher efficiency in the selector system. Various kinds of selector systems have been proposed in [23, 29, 33, 34, 62, 63]:

1. Quadrupole condenser (and also multipole condensers) with rod profiles: (a) coincident with equipotential surfaces of four thin wires (cf. Fig. 3); (b) circular straight rods.
2. Selector systems made of rings.
3. Quadrupole and multipole condensers with focusing electrodes bent in a parabola along the axis of the selector system.

These selector systems have various focusing properties. It follows from [64] that the most effective system is a selector system with focusing electrodes bent in the form of a parabola. However, this system is more difficult to fabricate and requires more careful treatment of the rods, since the minimum distance between the rods is smaller than in other systems, and this tends to favor breakdown. In theoretical maser work one usually uses a selector system that employs a quadrupole condenser, since this provides the simplest form of potential energy,

$u \sim r^2$, making it possible to carry out the calculations completely. The characteristics of the cavity resonator in [22, 65] are introduced by the notion of a quantity M defined by the expression $M = \frac{LQ}{A} \cdot \left(\frac{8}{\pi^2}\right)^n$, where n is zero for uniform distribution of field along the axis of the resonator and unity if this distribution has a maximum at the center; A is the area of the cross section of the resonator; Q is the quality factor of the resonator; L is the length of the resonator. The kind of resonator that can be used is related to the minimum intensity of the molecular beam. The greater the value of M, the smaller the number of active molecules required to maintain maser oscillations. In terms of M the best cavity resonator is one operating in the E_{010} mode. A cavity using the H_{011} mode, such as that used by Gordon, Zeiger, and Townes [11], requires a beam intensity twice as high as that required for the E_{010} mode.

In [28] the characteristics of masers with resonators using the E_{011} and E_{012} modes have been given. These cavity resonators are less effective than resonators using the E_{010} mode. It is difficult to use resonators of this kind in actual masers because of the strong Doppler effect.

Although the output power of a maser is determined basically by the intensity of the molecular beam, to achieve the maximum efficiency it is necessary to increase the output coupling to the resonator. The optimum coupling [23] is determined by the condition

$$Q_0 \approx Q_1 \approx 2Q, \tag{3}$$

where Q_1 is the quality factor of the output coupling and Q_0 is the quality factor of the unloaded resonator. Actually, a coupling this strong causes an increase in the effect of the waveguide system on the tuning of the resonator and makes the stability of the maser worse. Hence, if the resonator is strongly coupled to the waveguide system it is necessary to take measures to provide good isolation.

In concluding this survey we may indicate that at the present time, of a number of masers that are already in operation, those with the best characteristics in terms of absolute and relative frequency stability and output power are those using the J = 3, K = 3 line of $N^{15}H_3$ with two opposed beams. A maser using the J = 3, K = 2 line of $N^{14}H_3$ is worse in terms of available power, and one using the J = 3, K = 3 line of $N^{14}H_3$ is worse in frequency stability. A disadvantage of a maser using the $N^{15}H_3$ molecule is the high cost of $N^{15}H_3$.

Section 2

DESCRIPTION OF THE APPARATUS

A general view of a maser using the 3,3 line is shown in Fig. 5. The molecular-beam source, the quadrupole condenser, and the cavity resonator are mounted on one frame and carefully aligned so as to be coaxial.

The molecular-beam source is a chamber with a volume of several cubic centimeters in which ammonia is admitted through a controlled leak. The beam is formed by a grid with square apertures 0.05 × 0.05 mm with a filling factor (transparency) of 0.25. The grid thickness is 0.05 mm. The diameter of the beam at exit is 6 mm.

Technical grade ammonia in a chamber at a pressure of 10 atm is used to produce the beam. The pressure is reduced by means of a reducer to 1.5-2.0 atm and then by means of the controlled leak, which is used to change the pressure inside the beam source to within the limits 0.01-1 mm Hg. Frequently, a controlled leak for ammonia is a section of vacuum tubing which is squeezed by a special clamp. Inside the vacuum tube there is a thin wire (diameter 0.5 mm). This controlled leak is completely satisfactory.

The gas pressure in the beam source is measured by means of a vacuum gauge VT-2 for an LT-4 vacuum gauge connected directly to the source chamber (Fig. 6). Between the beam source and the quadrupole condenser there is a diaphragm cooled by liquid air which selects a relatively narrow beam of molecules. In order to extend the time of continuous operation of the maser the diaphragm can be rotated in order to move apertures which are constricted by the freezing ammonia in approximately 3-5 hr. In addition to the molecular-beam source of the grid-type described above, use is made of sources in the form of a single channel 1-2 mm in diameter and 10 mm long. In this case the maser works well without a cooled diaphragm.

a

b

Fig. 5. One-beam maser using the J = 3, K = 3 line of $N^{14}H_3$. a) Maser with the front cover removed; b) general view of the installation comprising three masers.

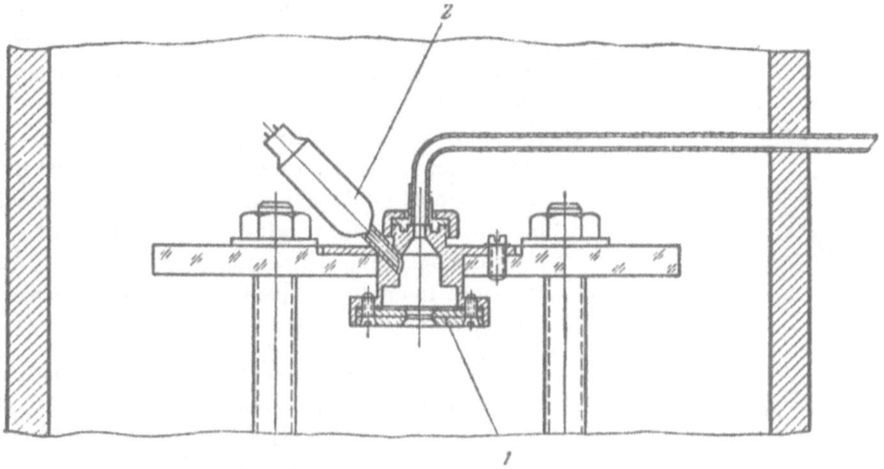

Fig. 6. Molecular-beam source (1) and vacuum gauge (2).

In normal operation of a maser it is necessary to maintain a vacuum of $(2\text{-}3) \cdot 10^{-5}$ mm Hg inside the vacuum chamber. The source produces a beam of 10^{17}-10^{18} molecules per second in the chamber; in order to maintain the required pressure the pumps must remove 10^{4}-10^{5} liters of gas per second. However, the required vacuum can be obtained more easily by freezing the ammonia with liquid air, since the vapor pressure of ammonia at the temperature of liquid air is 10^{-9} mm Hg. For this purpose, inside the vacuum chamber there is a copper jacket which is cooled by liquid air. The system is evacuated by means of a TsVL-100 pump. During operation of the maser a pressure of $(2\text{-}7) \cdot 10^{-6}$ mm Hg is maintained in the vacuum chamber, and it is only with large beams (greater than 1 mm Hg) that it increases to $1 \cdot 10^{-5}$ mm Hg.

The quadrupole condenser is 150 mm in length and the gap between plates is 2 mm. Figure 3a shows their general form; the inner portion of the electrodes is in the form of equipotential surfaces of a quadrupole condenser with infinitesimally thin rods. A cross section of the condenser is shown in Fig. 3b.

The selected beam of molecules enters a 80-mm-long cylindrical cavity resonator which oscillates in the E_{010} mode. At the ends of the resonator there are holes for entrance and exit of the molecular beam. In order to avoid radiation losses the holes are formed with waveguides 8 mm in length and 6 mm in diameter. By means of coupling apertures 2-2.5 mm in diameter, the resonator is coupled to the two waveguides. At one end the waveguides are terminated by plungers and at the other end extend through the frame of the maser. The vacuum seals in the waveguide are made by means of mica windows.

This design makes it possible to tune the resonator by passing through it the power from another klystron and to extract the power generated by the maser. The quality factor of the resonators, which are made from invar and silvered on the inside, is 6000-8000. The natural frequency of the cavity resonator can be shifted by several megacycles by means of rods 2 mm in diameter introduced through the side wall of the resonator to a depth of 1 mm; these rods have essentially no effect on the quality factor of the resonator. By moving the rod by 0.1 mm (ten divisions on a scale) it is possible to change the resonance frequency by approximately 0.5 Mc and to shift the generated frequency by 1000 cps on the average. The change in frequency and power as the resonator is adjusted are shown in Fig. 7.

The temperature of the resonator is held constant by means of a thermostat consisting of a sensing element, which determines the temperature of the cavity resonator, an amplifier, and a heater. The sensing element consists of several sections of bifilar windings of copper wire 0.07 mm in diameter mounted on the cavity. It is used as one of the arms of a resistance bridge. The other arms are wound of material having a low temperature coefficient of resistance (manganin or constantan). A 6-V d-c source is connected across one diagonal of the bridge. The voltage is balanced with the second arm of the bridge by means of a chopper which produces a 50-cps a-c voltage which is then amplified and mixed with a 50-cps stabilized reference voltage.

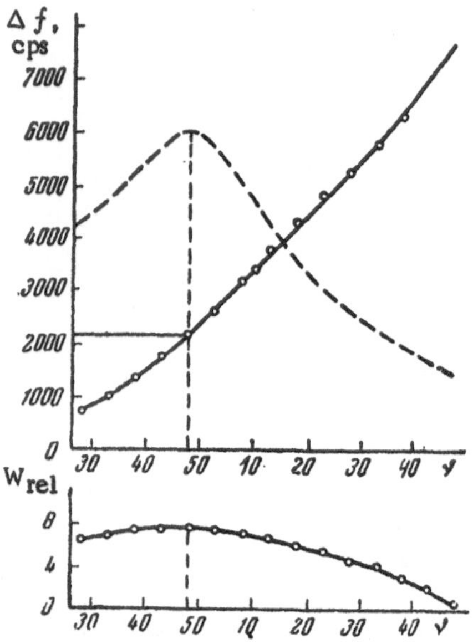

Fig. 7. The frequency and relative power of maser No. 3 as functions of the position of the tuning screw of the resonator.

With this arrangement the magnitude of the voltage at the amplifier output depends on the sign of the balance voltage of the bridge and, consequently, on whether the resonator is cold or hot. The phase of the reference voltage is chosen so that cooling of the resonator causes the amplifier output to increase. The balance voltage amplified by a factor of 10^4 is applied to the heater windings mounted on the cavity. The thermostat provides a temperature that is constant to within 0.01°C; the time required to establish the operating temperature (approximately 25°C) is 1-1.5 hr. A change in cavity temperature of 0.01°C changes the maser frequency by approximately 1 cps.

When the thermostat is first switched on, the temperature changes quite rapidly (approximately 1°C in 10 min). The time constant is approximately 1.5 hr. Six hours after the thermostat is switched off the temperature of the resonator oscillates; the maximum temperature change in 10 min is 0.1°C. The same measurements were carried out with cavities not connected directly with the walls of the vacuum chamber of the maser. The generated power was extracted by means of horns and the gap between the horns was 1-2 mm. In this case the temperature oscillations were also appreciable. This experiment shows that the heat loss from the resonator is to a considerable extent due to thermal radiation. Hence, in order to obtain a temperature constant to better than 0.01°C it is evidently necessary to use a double thermostat. The first is a jacket surrounding the resonator and the second is the resonator itself.

On the basis of these investigations of temperature stabilization of a resonator, we have developed a method for measuring the coefficient of linear expansion of a body over a small temperature range by use of a maser. The technique is as follows: A cavity resonator is made of the material to be investigated and the thermostat system described above is used.

The change in cavity frequency $\Delta\omega_0$ is related to the change in cavity diameter Δd by the expression $\Delta\omega_0/\omega_0 = \Delta d/d$, where d is the diameter of the cavity. If the change in characteristic frequency of the cavity for a temperature change of $\Delta t°$ is $\Delta\omega_0$, the coefficient of linear expansion of the material α is

$$\alpha = \frac{\Delta d}{d\Delta t^\circ} = \frac{\Delta\omega_0}{\omega_0\Delta t^\circ}. \tag{4}$$

According to Eq. (2) the change in frequency $\Delta\omega$ when the characteristic frequency of the cavity changes by $\Delta\omega_0$ is

$$\Delta\omega = \frac{QG}{Q_l}\cdot\Delta\omega_0. \tag{5}$$

Substituting $\Delta\omega_0$ from this expression in (4), we have

$$\alpha = \frac{Q_l}{QG\Delta t^\circ}\cdot\frac{\Delta\omega}{\omega_0}. \tag{6}$$

Using a tuning screw to adjust the characteristic frequency of the cavity ω_0 by 2-3 Mc and simultaneously measuring the change in maser frequency $\Delta\omega$, we can use Eq. (5) to determine the ratio Q_l/QG.

By switching off the thermostat it is also possible to measure the change in oscillation frequency of the maser $\Delta\omega$ due to a change of cavity temperature of $\Delta t°$. Substituting this quantity in Eq. (6), we obtain the coefficient of linear expansion of the material being investigated.

The change in maser frequency $\Delta\omega$ can be measured with an accuracy equal to the relative stability during

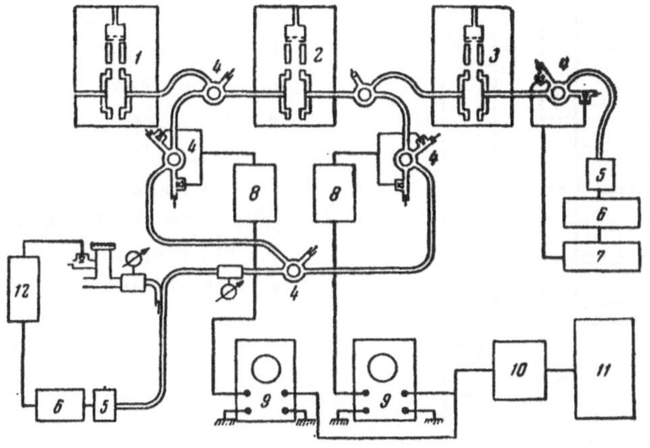

Fig. 8. Block diagram of the arrangement used to measure the frequency of a maser for the J = 3, K = 3 line. 1, 2, 3) Masers Nos. 1, 2, 3; 4) hybrid ring, balanced mixers; 5) marker heterodyne for klystron; 6) power supply for klystron; 7) intermediate-frequency amplifier and oscilloscope; 8) intermediate-frequency amplifier and second detector; 9) oscilloscope; 10) audio-frequency generator; 11) frequency measuring device; 12) 75-kc amplifier and discriminator.

the measurement time, i.e, 3-5 minutes. Within this time period it is a simple matter to obtain a relative stability of 10^{-11}. Under these conditions the error in the determination of α is approximately 10-15%.

This technique can be used to measure coefficients of linear expansion down to quantities of the order of $1 \cdot 10^{-8}$ deg^{-1}, a measurement that would be extremely difficult by ordinary techniques. In order to measure expansion coefficients smaller than 10^{-8} deg^{-1}, it is necessary to increase the relative stability of the maser. The use of this technique for measuring coefficients of the order of $1 \cdot 10^{-6}$ deg^{-1} is not convenient because of its complexity.

The measurement of the frequency deviation of a maser as its parameters are changed is carried out by making comparisons with a second maser whose frequency is held constant to within 2-5 cps. The change in power in relative units is measured simultaneously. A block diagram of the system used for these measurements is shown in Fig. 8.

With this scheme we can compare the frequencies of three masers in order as follows: No. 1 with No. 2 and No. 3 with No. 2. For example, the oscillations of masers 2 and 3, which are equal to within several hundred cycles, are mixed in a hybrid, which simultaneously provides good isolation, and applied to a second hybrid; this is a balanced mixer to which is also applied the power from a heterodyne klystron tuned to a frequency of 23830 Mc (i.e., 40 Mc below the frequency of the maser).

From the balanced mixer the power is applied to an if amplifier at 40 Mc with a band width of 2 Mc and a gain of 10,000. The heterodyne klystron is stabilized at frequencies up to 50 kc [66]. After the second detector at the output of the if amplifier, the signal is characterized by the difference Δf of the frequencies of masers 2 and 3 and is applied to the input of an EO-7 oscilloscope; simultaneously, a signal from a 3G-12 audio oscillator is applied to the input of the horizontal amplifier of the oscilloscope, and the frequency Δf is measured by means of the Lissajous pattern. The frequency Δf is measured with an accuracy of 1 cps by a frequency meter FM which makes use of counting units [67]; the time interval in this unit is determined by a quartz crystal.

The relative change in power is measured simultaneously with the frequency by taking part of the power from the resonator and amplifying it in a narrow-band amplifier (band width 70 kc) with a double frequency converter.

In order to evaluate certain effects in the characteristics, we have studied a maser with two opposed beams.

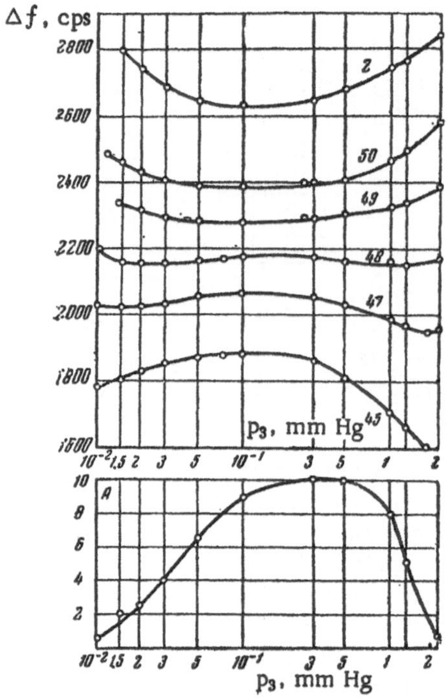

Fig. 9. Variation of frequency and relative power of maser No. 3 as functions of the pressure in the molecular-beam source with various deviations (the deviation is given in divisions on the screw scale); U_3=26 kV.

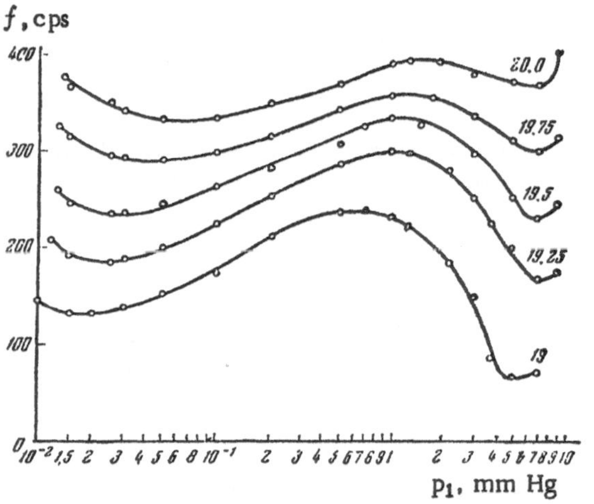

Fig. 10. The frequency of maser No. 1 as a function of pressure in the beam source. The region of small deviations is investigated in greater detail; U_1 = 26 kV.

The beam sources in this case were channels 5 mm in diameter and 10 mm long, but no cooling was used; selection was realized by means of quadrupole condensers as well as a system of rings 6 mm in diameter separated from each other by 3 mm. Voltages of the opposite polarities are applied to neighboring rings [62]. The two-beam maser was substituted for maser 3. The measurements were carried out in the same way as with the one-beam maser.

Section 3

MASER CHARACTERISTICS

1. Investigation of Maser Characteristics

Using the apparatus described above we have obtained three series of curves for masers with one molecular-beam source of the grid type:

1. The change in relative power W_{rel} and oscillation frequency Δf as functions of pressure in the beam source at various fixed resonator frequencies ν and voltages U on the quadrupole condenser: $W_{rel} = f(p)_{\nu,U}$; $\Delta f = f(p)_{\nu,U}$ (Figs. 9 and 10).
2. The change in relative power and oscillation frequency as functions of the voltage on the quadrupole condenser for various pressures and resonator frequencies: $W_{rel} = f(U)_{p,\nu}$; $\Delta f = f(U)_{p,\nu}$ (Fig. 11).
3. The change in relative power and oscillation frequency as the resonator frequency is changed: $W_{rel} = f(\nu)_{p,U}$; $\Delta f = f(\nu)_{p,U}$ for fixed pressures and voltages on the quadrupole condenser (Fig. 7).

In taking the maser characteristics we observed oscillation in the absence of voltage on the quadrupole condenser (with all four electrodes connected together and to ground). This effect is evidently due to the charge which accumulates on the frozen diaphragm. In the beginning of operation of the maser, as long as ammonia has not already frozen on the diaphragm, this effect is not observed. This kind of oscillation is similar to the anomalous oscillation observed earlier by Shimoda [23, 24]. Furthermore, it was noted that when a large amount of ammonia is frozen on the diaphragm the maser continues to operate even when covered by a heater. It is evident that some of the molecules evaporated from the diaphragm are selected by the quadrupole condenser and enter the resonator.

Using the same apparatus, we have taken similar curves for sources using a channel-type molecular beam with channel diameters of 1, 1.5, and 2 mm with a length of 10 mm. In this case the maser operates well without a cooled diaphragm. Typical curves of this kind are shown in Fig. 12. The best results in terms of extracted power are obtained with a channel 1.5 mm in diameter. In this case, the signal amplitude is 2-3 times greater than with a grid source. We then took the dependence of relative power and frequency of the maser similar to those described above for a two-beam maser with ring and quadrupole systems. The curves obtained in this work are shown in Figs. 13 and 14.

In order to compensate the effect due to nonuniform radiation of molecules along the length of the resonator (effect III), the vacuum gauges LT-2 for the upper and lower beam were specially calibrated at several points in the range of variation of pressure in the molecular-beam source. This provided sufficient accuracy to establish equal pressure in the upper and lower molecular-beam sources. Then the adjustment of the beams was verified by the simultaneous onset of oscillations due to the different beams. However, measurements show that the oscillation amplitude for the upper and lower beams differed by 10-20% at certain points with identical readings for both vacuum gauges. Evidently the basis for this discrepancy is the inaccuracy in adjustment.

To characterize the operation of the masers used in this work, in Fig. 15 we show the dependence of the voltage U_q required for quenching oscillation on pressure in the molecular-beam source. These functions have been computed theoretically in [43], neglecting the scattering of the molecular beam due to collisions with molecules of the residual gas in the system. It is evident from Fig. 15 that scattering appears even at source pressures of $(3\text{-}4) \cdot 10^{-2}$ mm Hg, in which case the vacuum in the system is maintained at $(3\text{-}5) \cdot 10^{-6}$ mm Hg. The scattering of the molecular beam can occur either in the resonator (the resonator operates as a storage volume) or in the quadrupole condenser close to the molecular-beam source.

Measurements carried out with a two-beam maser have shown that the first explanation cannot hold, because the oscillation amplitude increases when the second beam is switched on. If the beam were scattered in the resonator this growth in amplitude would not occur. Consequently, in the design of masers it is necessary to provide good evacuation of the space inside the selection system. In this respect the ring selection system and a system made up of thin rods has certain advantages as compared with the quadrupole (multipole) condenser system which has electrode cross sections in the form of equipotential surfaces.

2. Discussion of Experimental Characteristics

Above we have given the experimental dependence of maser frequency on pressure in the molecular-beam source, voltage applied to the quadrupole condenser, and characteristic frequency of the resonator (Figs. 7, 9-11). Figures 6 and 7 of [32] show curves obtained from a theoretical work in [40]. A comparison of the experimental curves with the calculated curves shows qualitative agreement.

A quantitative comparison between theory and experiment is difficult because, as shown in [40], the theoretical curves are computed as functions of the saturation parameter $\gamma^2 = (d^2 \mathcal{E}^2 / \hbar^2) \tau^2$, which is proportional to the maser power. However, the power is proportional to the pressure only at lower pressures in the molecular-beam source. At high pressures the power becomes a complicated function of pressure which can hardly be computed theoretically. Also, it is experimentally difficult to carry out absolute measurements of power with the required degree of accuracy. Hence, for a qualitative comparison of the theoretical and experimental work, we have measured the relative dependence of maser power (for a maser operating on the J = 3, K = 3 line) on pressure. This was used for plotting the calculated curves in [40].

Attention should be directed to the fact that for small deviations between the resonator frequency and the spectral line the experimental curves have a more complicated dependence of frequency on pressure than the theoretical curves without taking account of effect III – the nonuniformity in molecular radiation along the resonator (Fig. 6 of [32]). At small deviations there are four clearly defined different values of the pressure at which the maser frequency is the same (Figs. 9 and 10). Similar behavior is exhibited by the dependence of maser frequency on pressure when use is made of a beam source with a single channel (Fig. 12), except that the second frequency minimum and the amplitude maximum are shifted toward higher pressure.

This multiple-valued dependence of frequency on pressure is evidently due to the fact that at small deviations $\omega_0 - \omega_l$, in which case the term $\frac{Q}{Q_l} \cdot \frac{\omega_0 - \omega_l}{\omega_l} \cdot G(\gamma, U)$ is small, the maser frequency is determined by effect III [Eq. (2)]. This effect can be computed by the method suggested in [43]; the author divides the cavity

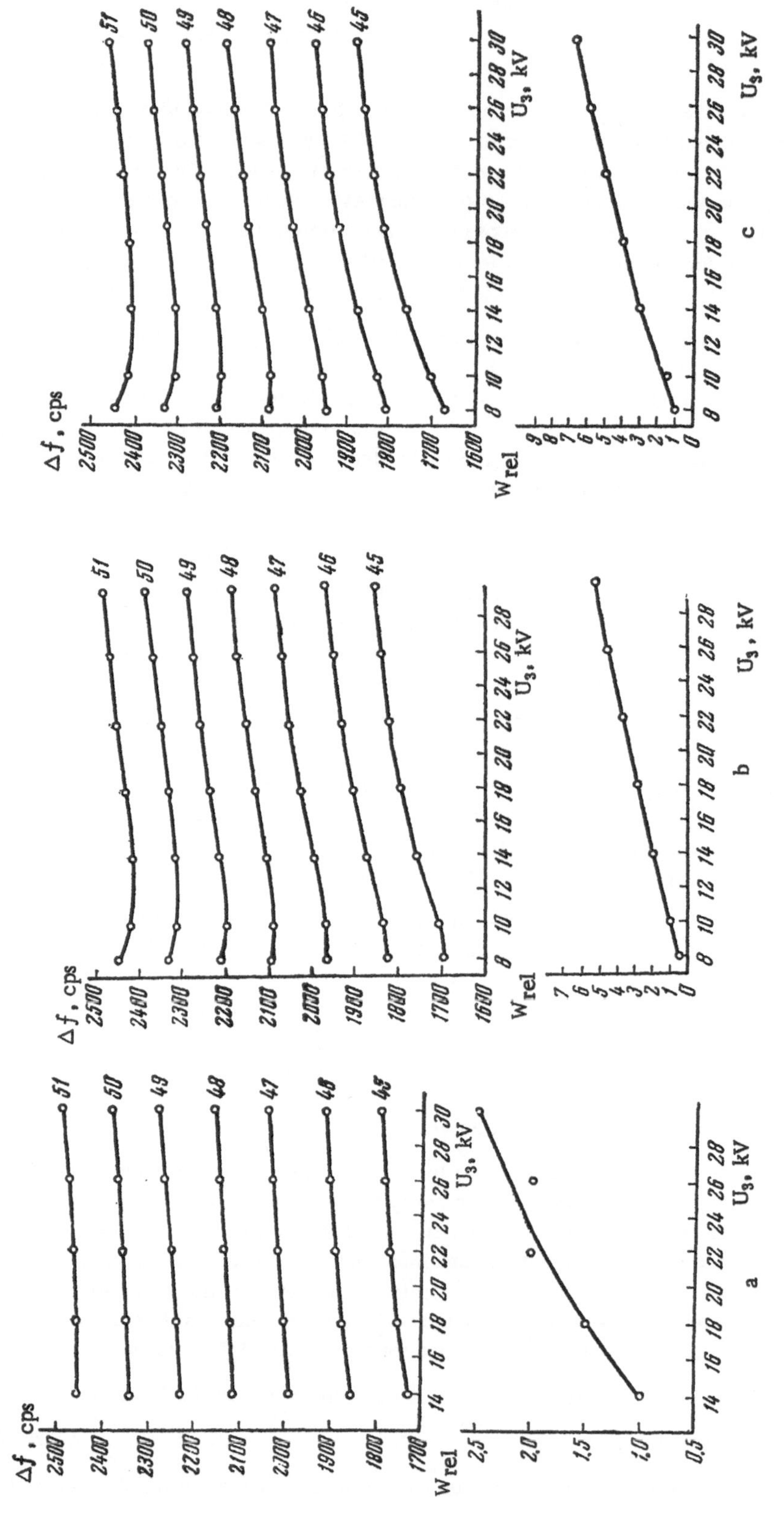

Δf, cps
W_{rel}
U_3, kV
45
46
47
48
49
50
51
a
b
c

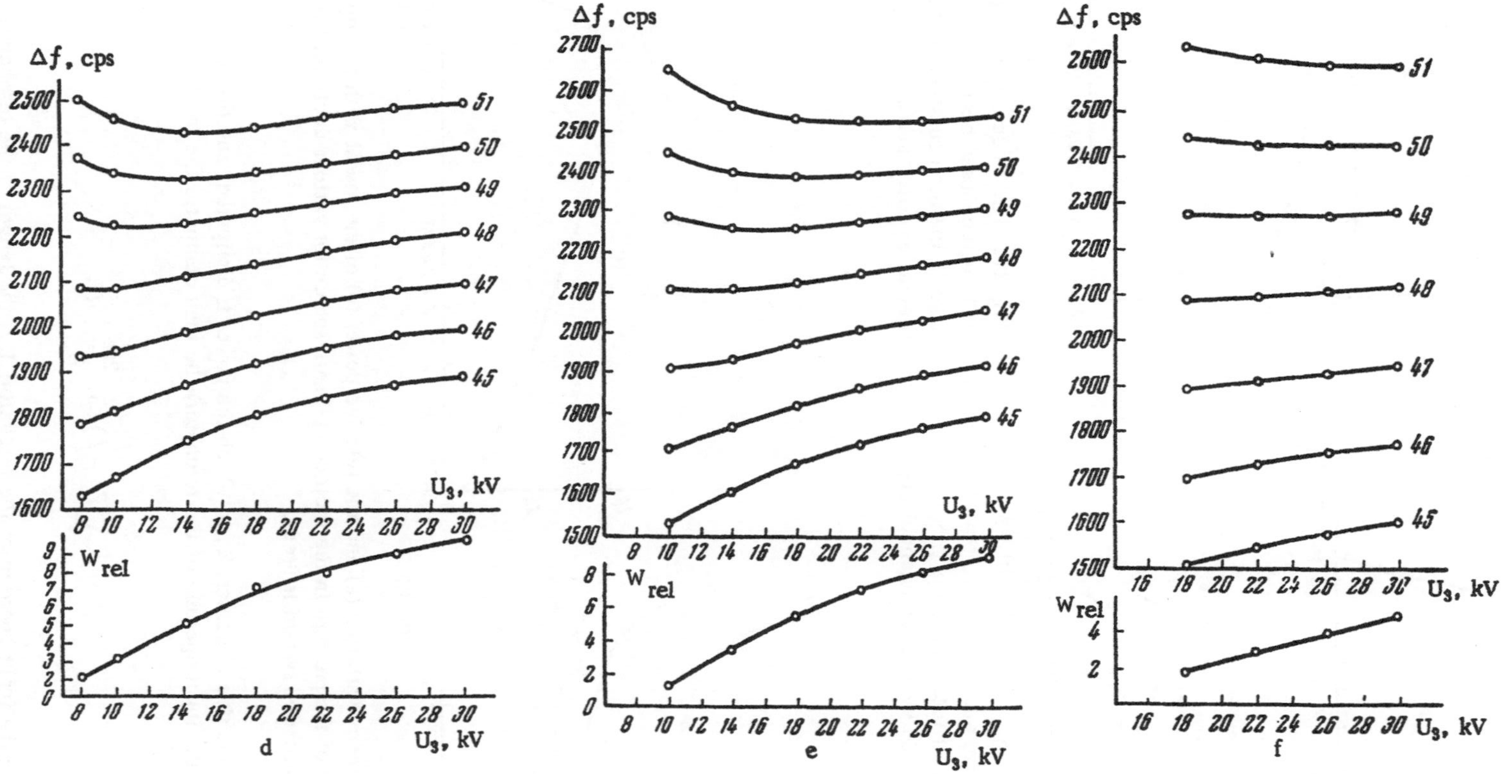

Fig. 11. The frequency and relative power of maser No. 3 as functions of the voltage on the quadrupole condenser for various pressures in the molecular-beam source with various deviations. a) $p = 1.5 \cdot 10^{-2}$ mm Hg; b) $p = 3 \cdot 10^{-2}$ mm Hg; c) $p = 5 \cdot 10^{-2}$ mm Hg; d) $p = 1 \cdot 10^{-1}$ mm Hg; e) $p = 5 \cdot 10^{-1}$ mm Hg; f) $p = 1$ mm Hg.

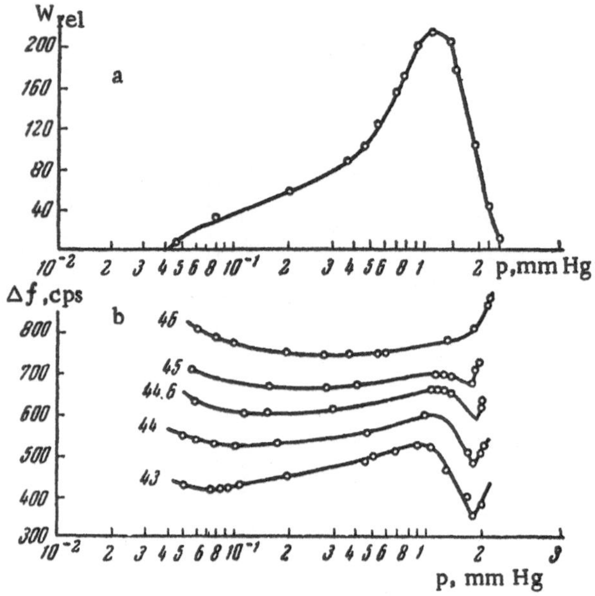

Fig. 12. The frequencies (b) and the relative power (a) of a maser as functions of pressure for the single-channel beam source; U_3 = 26 kV and the channel diameter is 1.5 mm.

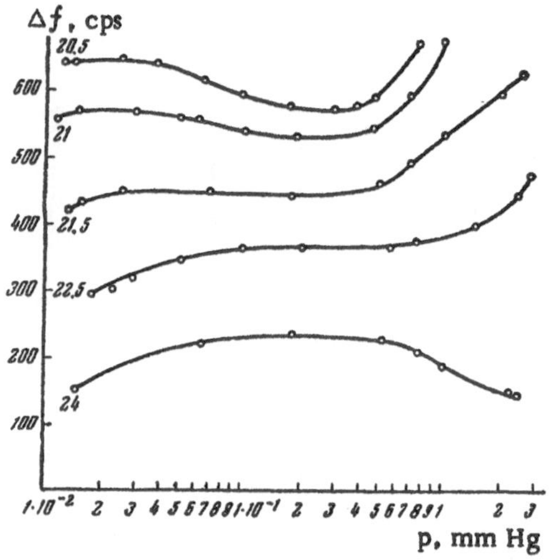

Fig. 13. The frequency variation of the maser with two opposed beams as a function of pressure in the molecular-beam source for various deviations (ring selection system).

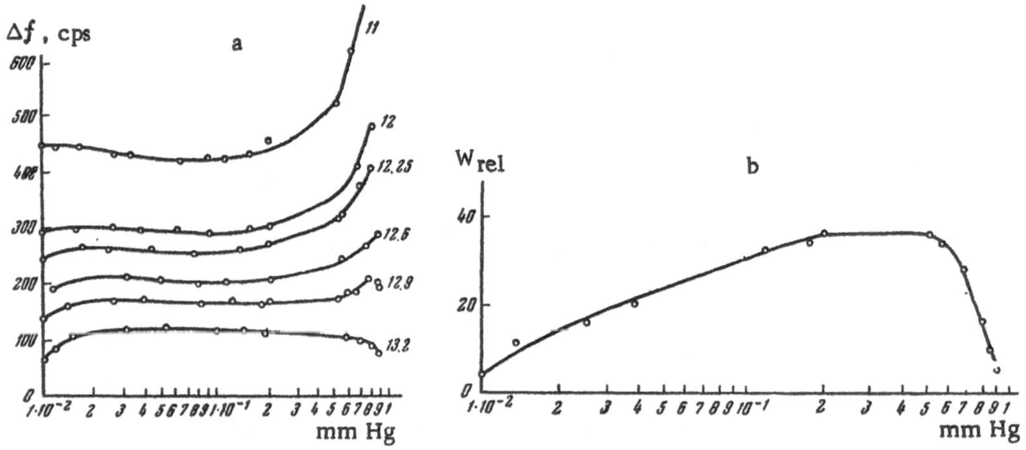

Fig. 14. The frequency variation (a) and the relative power (b) of the maser with two opposed beams as functions of the pressure in the molecular-beam source for various deviations (a quadrupole condenser selection system is used).

into two parts: the part to one side of center, Z, in which the power is dissipated, and the part on the other side of center (computing from the input aperture of the resonator); he then computes the total flux of power in the traveling-wave component,

$$p = K \int_0^L (Z - L) \sin \frac{2\theta z}{L} \cdot dz,$$

where z is the distance from the input aperture of the resonator; L is the length of the resonator; $\sin(2\theta z/L)dz$ is the probability of radiation by a molecule between z and z + dz in a E_{010} resonator; $\theta = (\mu L/2\hbar\bar{v})\mathcal{E}$ is a quantity analogous to the saturation parameter γ; $\bar{v}$ is the mean molecular velocity; $2\pi\hbar$ is Planck's constant, and K is a constant. In the derivation of this expression it is assumed that the characteristic frequency of the resonator is equal to the frequency of the molecular transition.

The traveling-wave component is expressed by means of Q_t, the quality factor of the resonator for traveling waves,

$$\frac{1}{Q_t} = \frac{P_t}{\omega_0 W} = \frac{1}{Q}\left\{\frac{Z-L}{L} + \frac{2\theta - \sin 2\theta}{2\theta(1-\cos 2\theta)}\right\}, \qquad (7)$$

where W is the energy stored in a resonator. If the frequency shift of the maser [22] is due to the traveling-wave effect,

$$\frac{\omega - \omega_l}{\omega_l} = \frac{4\pi^2 vL}{Q_t \lambda^2 \omega_l} \cdot \left[1 - \frac{1}{4} f_2(\theta)\right], \qquad (8)$$

where

$$f_2(\theta) = \frac{2\theta(1-\cos\theta)}{2\theta - \sin 2\theta},$$

then, substituting the value of Q_t from Eq. (7) in Eq. (8) we have

$$\frac{\omega - \omega_l}{\omega_l} = \frac{\pi}{2Q} \cdot \frac{v}{c} \cdot \frac{L}{\lambda} \cdot [4 - f_2(\theta)] \cdot \left[\frac{1}{f_2(\theta)} - d\right], \qquad (9)$$

where

$$d = \frac{L-Z}{L}.$$

According to Eq. (9) effect III can be computed for any value of θ. Qualitatively, taking account of effect III, the dependence of maser frequency on source pressure at small deviations can be explained as follows. At low beam intensities the maximum molecular radiation occurs at the end of the resonator, so that there is a traveling-wave component which propagates in the direction opposite to the velocity of the molecular beam. The frequency shift due to the Doppler effect is negative and increases with field amplitude. Correspondingly, the oscillation frequency of the maser is reduced.

As the intensity of the beam increases, the amplitude of the electromagnetic field in the resonator increases and the peak molecular radiation is shifted in the direction of the input aperture in the resonator. At some beam intensity the traveling wave changes direction, the frequency shift becomes positive, and the maser frequency increases with increasing field amplitude. On the curve showing the dependence of frequency on pressure in the molecular-beam source, this corresponds to the first minimum.

In the same way we explain the second minimum associated with the reduction in field amplitude due to the scattering of the molecular beam as the source pressure is increased. These minima are evident in Figs. 9 and 10. The magnitude of the frequency shift computed from Eq. (9) with θ = 1-2 is approximately 50-100 cps. This is in agreement with the change in frequency with pressure measured experimentally for the one-beam maser. This is also approximately 100 cps.

In addition to effect III, another factor that effects the dependence of maser frequency on source pressure is the Doppler effect associated with the fact that the mode in the resonator is not a pure E_{010} mode, but that other higher modes are present. These distortions can be due to the presence of the cutoff waveguides at the end faces of the resonator and the effect due to the tuning rod. The latter effect becomes greater as the rod penetrates deeper into the resonator, as is evident from a comparison of Figs. 9 and 10. In maser No. 1 the rod penetration is deeper than in maser No. 3.

An analysis of the characteristics of the maser with two opposed beams (effect III canceled) shows that the dependence of frequency on pressure of the molecular beam source is weaker (Figs. 13 and 14) than for the one-beam maser, but that this dependence again becomes appreciably strong at low and high beam intensities.

In part this can be explained by the difference in the adjustment of beams; the important role, however, is

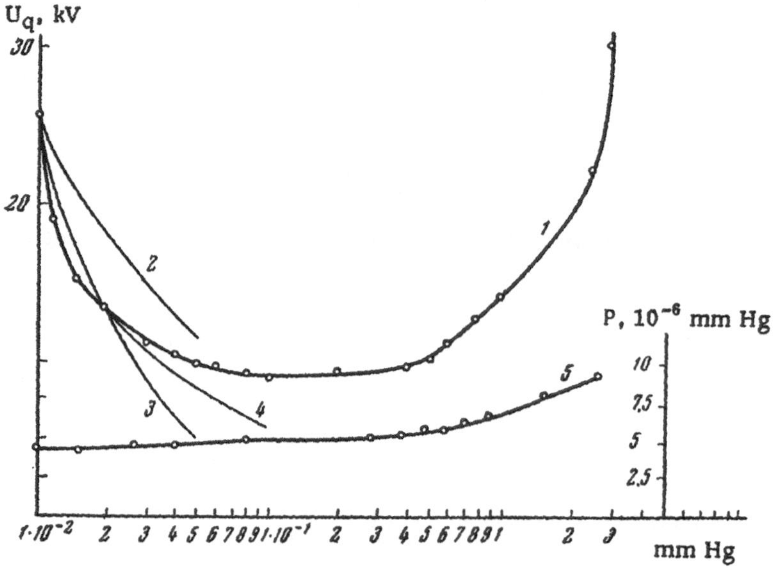

Fig. 15. The voltage for quenching oscillation U_q as a function of pressure in the molecular-beam source (curve 1). P is the pressure inside the vacuum system in mm Hg (curve 5).

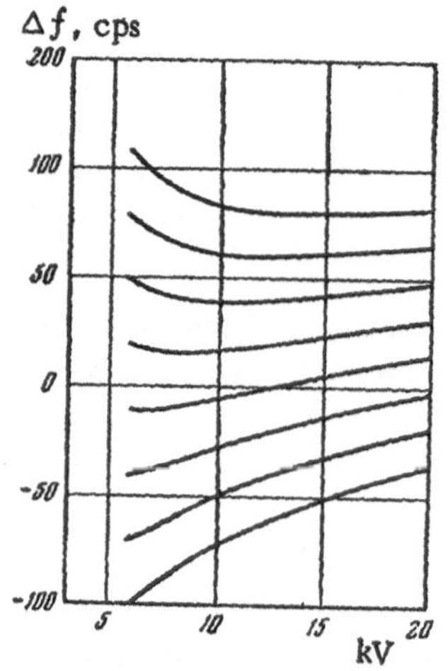

Fig. 16. The theoretical phase shift as a function of voltage on the octupole condenser [43]. The number of molecules in the beam is $1.1 \cdot 10^{18}$; d = 0.5.

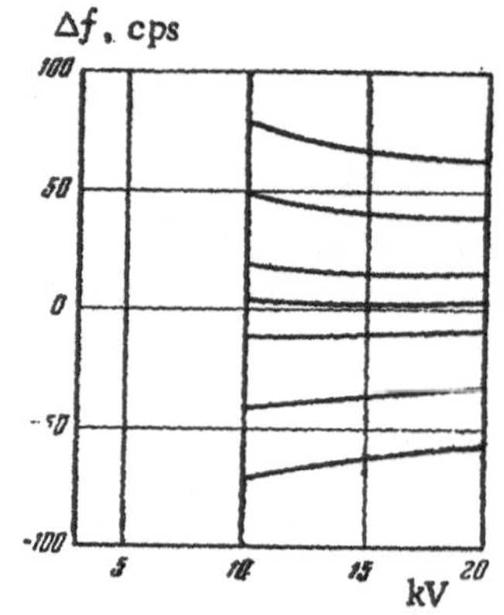

Fig. 17. The theoretical frequency shift as a function of voltage on the octupole condenser [43]. The number of molecules in the beam is $5.5 \cdot 10^{17}$; d = 0.5.

evidently that of the Doppler effect, because, at the extreme pressure points the molecules frequently radiate near the faces of the resonator where the resonator field is highly distorted. It is also possible that at high pressures, where the beam starts to scatter, some of the molecules experience collisions between themselves or with the selection system and all such molecules are selected and enter the resonator. In this case, only the part of the selection system near the resonator is actually effective and, to some extent, this is equivalent to a reduction in the voltage on the selection system. Taking account of the dependence of maser frequency on selection voltage it can be shown that, at high pressures in the molecular-beam source, an increase in pressure should cause an increase in maser frequency. This effect is observed in the two-beam maser (Fig. 14). A curve of the oscillation frequency corresponding to the value 12.9 for the scale of the tuning screw depends weakly on pressure in the beam source; at higher pressures, however, the oscillation frequency increases somewhat.

As is evident from Fig. 14, the use of two opposed beams means that the nonuniform molecular radiation in the resonator can be balanced in the operation of the maser. This reduces the effect of the pressure on maser frequency and makes it possible to increase stability.

The curve showing the maser frequency as a function of voltage applied to the quadrupole condenser is in qualitative agreement with the appropriate calculations carried out in [40]. The calculations were carried out assuming a Maxwellian velocity distribution for the beam molecules taking account of the hyperfine structure of the 3,3 line of $N^{14}H_3$. Effect III, associated with the nonuniform radiation of the molecules along the resonator, was not considered in this work.

The difference between the theoretical and experimental curves is especially marked in the region 8-12 kV, where the bend in the theoretical curves is much sharper than in the experimental curves. In this respect our experimental curves show good agreement with the calculations carried out in [58] for the 3,2 line of $N^{14}H_3$ or the 3,3 line of $N^{15}H_3$ for an octupole selection condenser. In this work the hyperfine structure of the line was not considered (for the 3,2 line of $N^{14}H_3$ and the 3,3 line of $N^{15}H_3$ there is no structure), but effect III was taken into account. Thus, it is possible to draw the conclusion that for a one-beam maser using the 3,3 line of $N^{14}H_3$ the oscillation frequency is determined primarily by effect III.

It is interesting to note that both for the experimental curves (Fig. 11), as well as the theoretical curves (Figs. 16 and 17 [43]), as the pressure in the molecular-beam source changes, the dependence of the maser frequency on quadrupole condenser voltage also changes. At low or high molecular-beam source pressures, i.e., with a small number of active molecules, the frequency is a weak function of voltage or is even independent of voltage. In this case the position of the curve satisfying the condition $\left(\frac{\partial f}{\partial U}\right)_{\nu,p} = 0$ is shifted as the pressure is changed. Near resonance, at optimum beam values, a change in voltage of the quadrupole condenser has a strong effect upon oscillator frequency. Thus, when the voltage is changed from 8 to 30 kV, the frequency is changed by approximately 100 cps.

In Figs. 9-11, in addition to the frequency functions, we show the dependence of relative maser power on pressure in the molecular-beam source and on quadrupole condenser voltage for various values of the cavity frequency. This can be explained on the basis of [22, 29]. In [29] there is an expression for the amplitude of the steady-state oscillations of the maser:

$$(\mathscr{E}_0)^2 = \frac{A\eta\tau^2\cdot\frac{1}{Q}\cdot\frac{\omega}{\omega_0}\cdot\left[\left(Q\cdot\frac{\omega}{\omega_0}\right)^2+1\right]-\tau^2(\omega_{mn}-\omega)^2-1}{\eta\tau^2}, \tag{10}$$

where

$$A = \frac{4\pi N_0\hbar}{Sl};$$

$$\eta = \frac{(d_{mn})^2}{\hbar^2};$$

N_0 is the number of active molecules entering the resonator per second; S is the cross-sectional area of the resonator; l is the length of the resonator; d_{mn} is the dipole moment matrix element; ω_0 is the natural frequency of the resonator; ω_{mn} is the frequency of the molecular transitions; ω is the oscillation frequency; Q is the quality factor of the resonator; τ is the mean molecular time of flight through the resonator field.

If it is assumed that $\omega_0 = \omega_{mn} \approx \omega$, then when $Q \gg 1$ Eq. (10) can be written in the form

$$(\mathscr{E}_0)^2 = \frac{4\pi N_0\hbar Q}{Sl} - \frac{\hbar^2}{(d_{mn})^2\tau^2}. \tag{11}$$

Using this expression we obtain the inequality that determines self-excitation of the oscillator [29]:

$$\beta = \frac{4\pi N_0}{S l \hbar} \cdot (d_{mn})^2 \cdot Q\tau^2 > 1, \qquad (12)$$

where β is the self-excitation parameter. Cutoff of oscillations occurs at $\beta = 1$.

Figure 15 shows the voltage applied to the quadrupole condenser U_q corresponding to the cutoff of oscillations as a function of pressure in the molecular-beam source. It is evident from this figure that for the portion for which $p = (1-2) \cdot 10^{-2}$ mm Hg and U_q varies within the limits 13-30 kV, p and U_q approximately satisfy $U_q p$ = const (Fig. 15, curve 3), i.e., the number of active molecules entering the resonator is proportional to the first power of the voltage on the quadrupole condenser rather than the square of the voltage, as is usually assumed in the theoretical calculations for a maser using the 3,3 line. Actually, on this portion of the curve the function $U_q = f(p)_\nu$ corresponds to curve 3 of Fig. 15 to within the experimental errors. In a quadratic molecule selector we would have $U_q^2 p$ = const (Fig. 15, curve 2). A relation of approximately this kind is observed in the region of $(2-10) \cdot 10^{-2}$ mm Hg, where the voltage varies from 8.5 to 13 kV (Fig. 15, curve 4).

Consequently, at these voltages on the quadrupole condenser the molecular selection process is approximately quadratic. Some deviation in the direction of a stronger dependence would appear to result because of the partial scattering of a beam due to molecular collisions. At source pressures greater than $1 \cdot 10^{-1}$ mm Hg, the beam scattering is so strong that an increase in the pressure in the beam source not only does not increase the number of active molecules entering the resonator, but can indeed reduce this number. Hence, the dependence of U_q on p is more complicated than in the preceding cases.

Thus, the analysis described here indicates that at voltages up to 13 kV there is approximately a quadratic selection of molecules in the field of the quadrupole condenser; at voltages above 13 kV the selection is approximately linear. In this connection we can analyze the dependence of the relative power on voltage applied to the quadrupole condenser U (Fig. 11). It follows from Eq. (11) that in linear selection there should be observed a linear dependence of the relative power W_{rel} on U. Actually, the experimental curves are basically straight, but at high voltages the increase in power is retarded as a result of a saturation of the energy stored in the resonator.

The saturation effect in the energy storage is due to the following effect. With molecular beams of low intensity the effective field in the resonator is small and molecules on the average will penetrate far into the resonator before undergoing a transition to the lower state. In the lower state they will spend only a relatively small fraction of their transit time in the resonator. With molecular beams of high intensity, however, the effective field in the resonator is high and molecules will, on the average, make a transition to the lower state immediately upon entering the resonator. Subsequently, the molecules in the lower state absorb energy. This absorption limits the power output of the resonator. An analysis of the experimental curves (Figs. 11d and 11e) supports this explanation. At high powers ($p = 1 \cdot 10^{-1}$, $5 \cdot 10^{-1}$ mm Hg) the retardation in the power increase with increase in voltage is emphasized.

Another effect that evidently leads to saturation is the following: With increasing capture angle for the molecules by the quadrupole condenser at higher voltages, not all molecules reach the resonator. Some of the molecules moving at large angles move through the cutoff waveguides of the resonator and do not participate in oscillation. This effect leads to a reduction in the power increase with increase in voltage. A deviation from the linear relation of W_{rel} on U was not observed from 8-13 kV.

From an examination of the dependence of relative power on source pressure (Figs. 9 and 12), it is evident that at low pressures and at a voltage on the quadrupole condenser of 26 kV, W_{rel}, in accordance with Eq. (10), is approximately proportional to the beam pressure. Then, at pressures $(3-4) \cdot 10^{-2}$ mm Hg the growth of power with increasing source pressure is retarded and W_{rel} reaches a maximum value at $(2-3) \cdot 10^{-1}$ mm Hg. A further increase in pressure leads to a reduction in the generated power and, at high beam intensities, to quenching of the oscillation.

This behavior of the dependence of relative power on source pressure is explained by the breakup of the beam due to collisions with molecules of the residual gas. As the beam intensity is increased the vacuum in the system becomes worse, although it still remains good enough so that molecules can move through distances greater

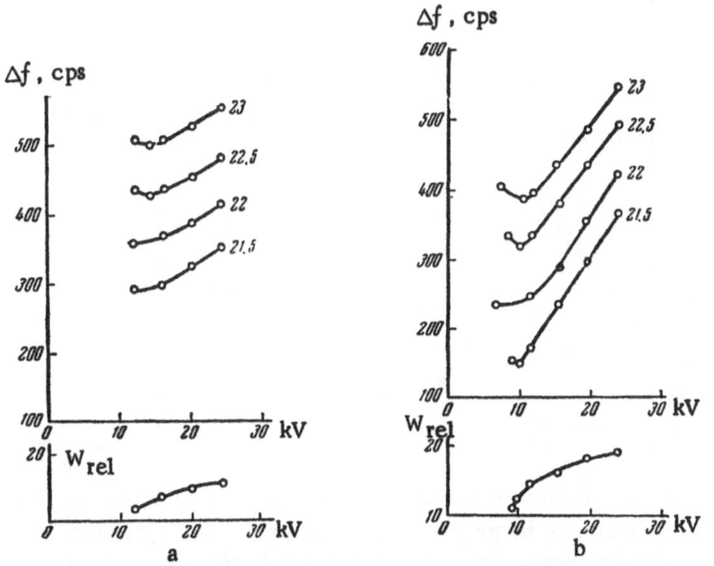

Fig. 18. The variation of frequency and relative power as functions of the voltage on a ring selection system for various deviations with the following pressures in the molecular beam source: a) $p = 5 \cdot 10^{-2}$ mm Hg; b) $p = 2 \cdot 10^{-1}$ mm Hg.

than the length of the quadrupole condenser without undergoing collisions. Evidently the vacuum inside the quadrupole condenser is worse than that in the rest of the system (at high beam intensities).

Figure 18 shows the dependence of oscillation frequency on voltage in a ring selection system. Analysis of these curves shows that for a ring system the dependence of frequency on voltage is much more sensitive than is the case for a quadrupole condenser. The frequency change is found to be twice as large for a change of 1 kV. With weak beams a change of voltage from 12-24 kV causes a frequency change of 50 cps. A further increase in beam leads to a stronger dependence of oscillation frequency on voltage. At source pressures of $2 \cdot 10^{-1}$ mm Hg a change of voltage from 12-24 kV causes the frequency to change by 160 cps.

As shown in [68], a ring selection system provides a higher molecular capture angle than a quadrupole system. For this reason a ring system with the same difference of potentials as a quadrupole condenser provides a greater number of active molecules in the beam and thus a higher field amplitude in the resonator. This then leads to a stronger dependence of oscillation frequency on the voltage applied to the electrodes of the selection system, since effect III and the functions G and Δ that appear in Eq. (2) depend on the strength of the field in the resonator and increase more rapidly with increasing field in the resonator.

3. Adjustment of Maser Frequency

The pressure in the molecular-beam source has a strong effect on oscillation frequency (Figs. 9, 10, 12). For this reason, a number of papers [22, 25, 29] propose that the change in oscillation frequency with pressure be used for tuning the resonator to the spectral line. As is evident from the figures given here, with this adjustment of the maser for the 3,3 line of $N^{14}H_3$ a different value of the pressure will correspond to a different oscillation frequency, because of the ambiguous condition $\left(\frac{\partial f}{\partial p}\right)_{v,U} = 0$ at small deviations.

The accuracy with which the maser can be tuned in this way is approximately 10^{-8}, and the frequency to which the maser is tuned depends on the voltage applied to the quadrupole condenser, as shown in Fig. 19. The change in oscillation frequency of the maser with change in selector voltage has also been proposed as a way for tuning the resonator in [25]. It follows from the experimental (Fig. 11) and theoretical (Fig. 7 [32]) curves that tuning the frequency by means of the condition $\left(\frac{\partial f}{\partial U}\right)_{v,p} = 0$ is also not very accurate. The oscillation frequency of a maser tuned by changing the voltage on the quadrupole selector will not, in general, coincide with the frequency obtained by tuning by varying the ammonia pressure in the molecular-beam source. This difference

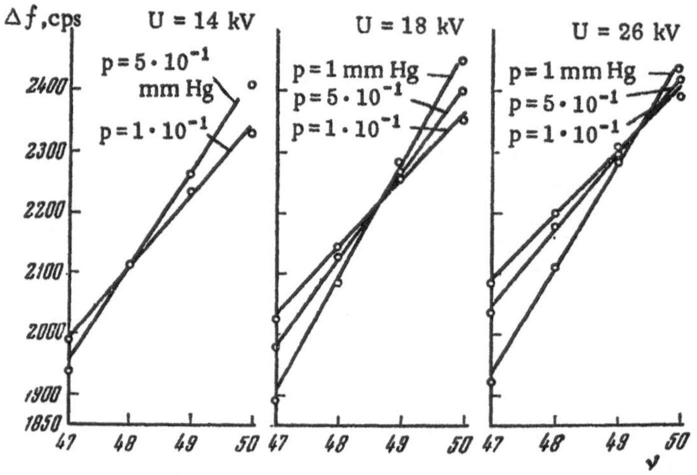

Fig. 19. The displacement of the intersection of the curve showing the dependence of frequency on deviation for various pressures in the molecular-beam source as a function of voltage on the quadrupole condenser.

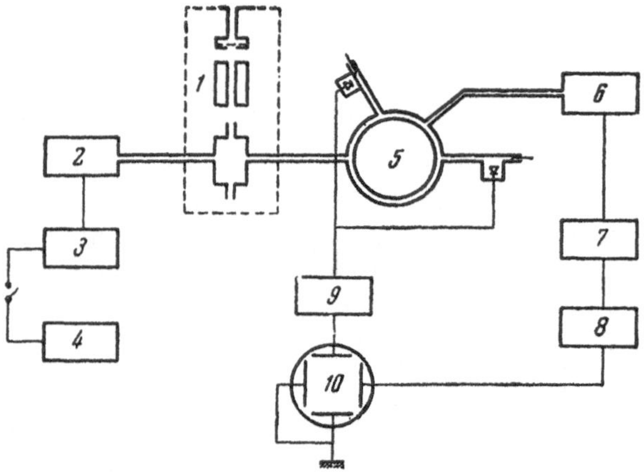

Fig. 20. Block diagram of the apparatus. 1) Maser; 2) pumping klystron; 3) power supply for klystron; 4) sawtooth voltage generator 1-10 kc; 5) balanced mixer; 6) heterodyne klystron; 7) klystron power supply; 8) sawtooth voltage generator 10-50 cycles; 9) intermediate-frequency amplifier (40 mc, bandwidth 70 kc); 10) oscilloscope.

is due to the fact that the voltage and pressure have different and independent effects on the oscillation frequency of the maser. In Eq. (2) of [13],

$$\omega = \omega_l \left[1 + \frac{Q}{Q_l} \cdot \frac{\omega_0 - \omega_l}{\omega_l} \cdot G(\gamma, U) + \Delta(\gamma, U) \right]$$

the pressure primarily effects $G(\gamma, U)$ and has little effect on $\Delta(\gamma, U)$ because of saturation, whereas the voltage has a strong effect on both the first and second terms.

In measurement of the dependence of frequency and amplitude of a maser a factor of primary interest is the maser characteristic in the region of the spectral line. In this connection a simple method for starting a maser and for preliminary tuning has been given in [69]. The basic steps in starting a maser are as follows: tuning the resonator to the absorption line; observation of induced radiation and oscillation; adjustment of the resonator to

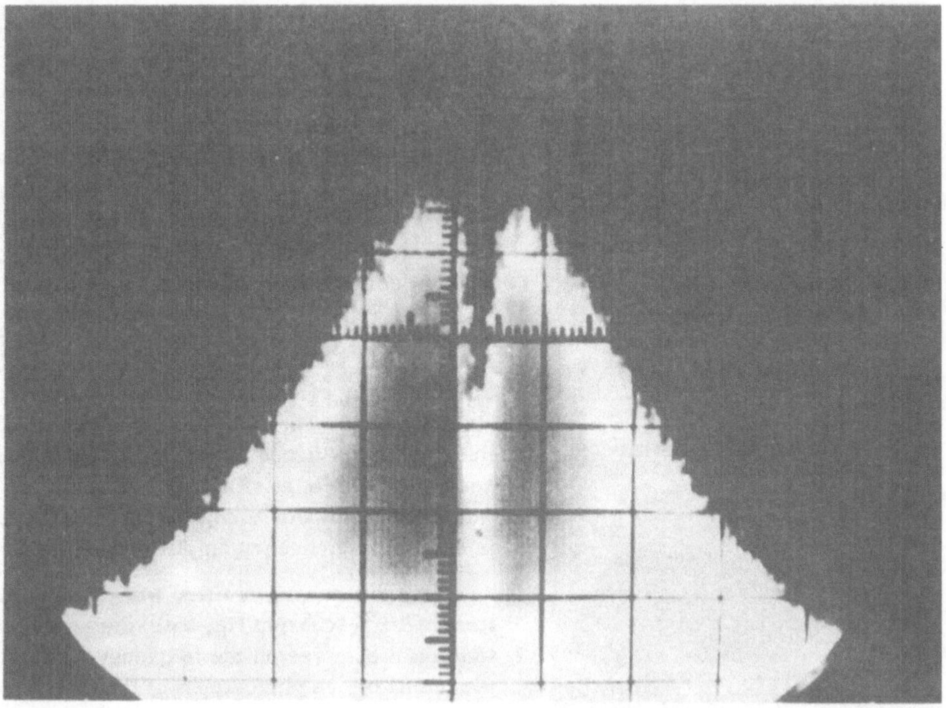

a

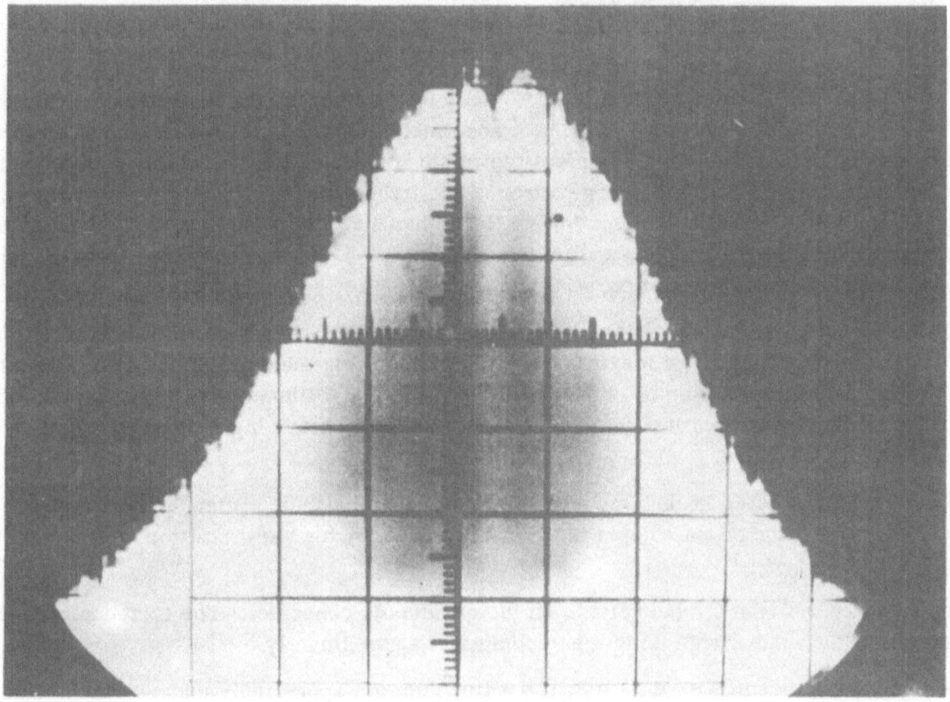

b

Fig. 21. The 3,3 absorption line of $N^{14}H_3$. a) At a pressure of $5 \cdot 10^{-4}$ mm Hg; b) at a pressure of $1 \cdot 10^{-4}$ mm Hg.

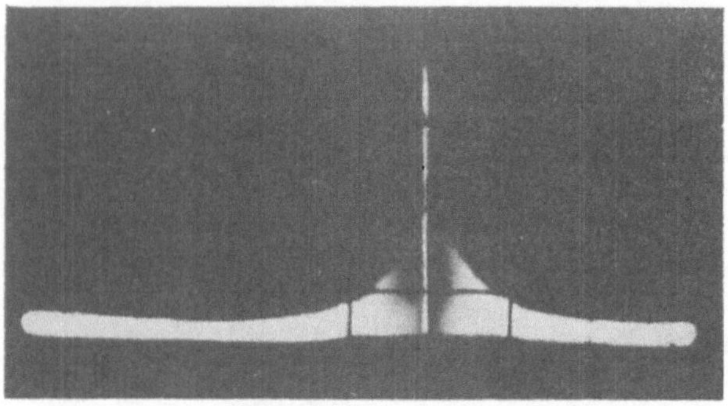

Fig. 22. Burst on the cavity resonance curve due to the maser oscillation.

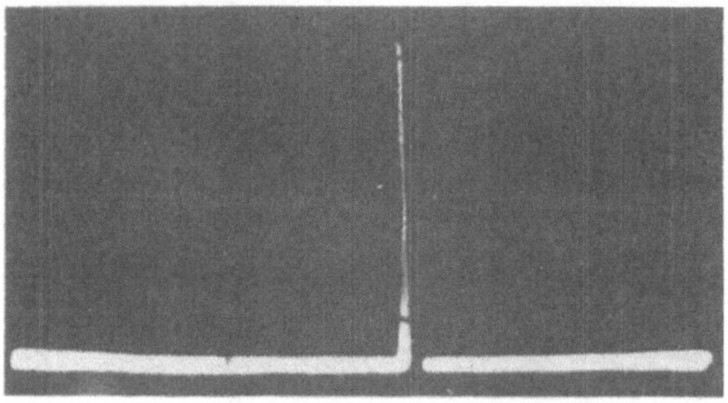

Fig. 23. The maser signal.

the spectral line. Usually the transition from one stage to another means a change in the electronic system. The method of adjustment proposed here, however, makes it possible, without any change in the electronics, to rapidly start the maser and adjust it to the frequency of the spectral line with a rather high degree of accuracy.

A block diagram of the system is shown in Fig. 20. In the closed position of the switch a sawtooth voltage at a frequency of several kilocycles is applied to the pump klystron. Thus, on the screen of the oscilloscope there appears a frequency characteristic of the resonator consisting of a series of narrow vertical lines, each of which represents the resonance curve of the narrow-band intermediate-frequency amplifier (band width 50-100 kc).

If the resonator is filled with ammonia, at pressures of 10^{-2}-10^{-4} mm Hg, a dip due to the ammonia absorption is observed on the frequency characteristic of the resonator. In photographs of the image on the oscilloscope screen the absorption line is observed up to pressures of several times 10^{-5} mm Hg (Fig. 21). As the vacuum is increased further and the selection system voltage switched on, the dip due to the absorption diminishes and a peak appears due to the beginning of induced radiation and then oscillation (Fig. 22).

By retuning the resonator it is possible to rather accurately move the oscillation line to the peak of the resonance curve of the resonator. If the high-frequency sawtooth is switched off, the maser signal itself is seen on the screen of the oscilloscope (Fig. 23). A comparison of the frequencies of two tuned masers (masers No. 1 and No. 3 of Fig. 8) with a third-reference maser (maser No. 2) has shown that the accuracy of tuning the maser frequency to the frequency of the spectral line with this technique is $\sim 5 \cdot 10^{-9}$, an accuracy that is completely adequate for this preliminary stage of operation.

Analysis of the curves shown at the beginning of this section shows that it is possible to tune the frequency of the maser by equalizing the minima (or maxima) on the curve showing the dependence of oscillation frequency on pressure. By this method a maser using one beam with the J = 3, K = 3 line can be tuned by equalizing the minima of frequency as functions of pressure (Figs. 9 and 10) with an accuracy higher than that obtained by using modulation of the pressure or voltage.

In Fig. 10 maser No. 1 is tuned in this way from the position of the tuning screw at 19.5, whereas the minima on the curves showing the pressure dependence of the frequency for screw positions 19.25 and 19.75 differ by 10 cps.

The tuning is carried out with a fixed voltage on the quadrupole condenser. The reproducibility of frequency for two masers of the same design using this technique is approximately $5 \cdot 10^{-10}$.

The use of two opposed beams with the J = 3, K = 3 line does not appreciably enhance the accuracy in the absolute tuning of the maser even when effect III is compensated, since the frequency of the 3,3 line of $N^{14}H_3$ depends on the saturation of the unresolved components of the hyperfine structure making up the line (the dependence of Δ on γ [40]). The existence of hyperfine structure in this case makes it difficult to adjust the frequency with an accuracy higher than $5 \cdot 10^{-10}$. In this case the requirements on beam adjustment are higher. Hence, to increase the absolute stability of a maser one should use ammonia lines which do not exhibit hyperfine structure, for example the J = 3, K = 2 line of $N^{14}H_3$ or lines of $N^{15}H_3$.

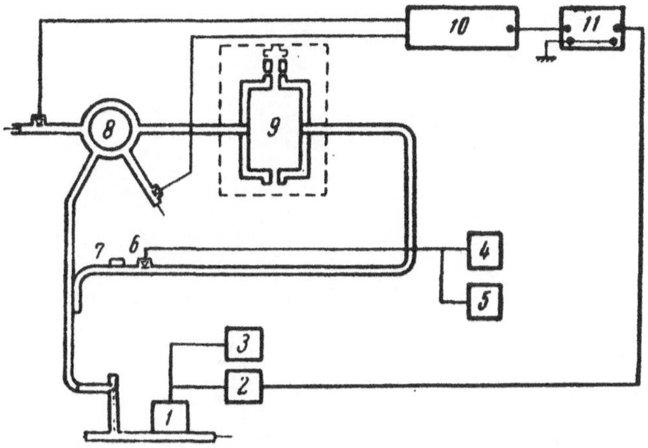

Fig. 24. Schematic diagram of the spectrometer. 1) Klystron; 2) klystron power supply; 3) sawtooth voltage generator (17 cycles); 4) sinusoidal voltage generator (40 mc); 5) ZG-12 oscillator; 6) mixer head; 7) attenuator; 8) hybrid ring, balanced mixer; 9) maser; 10) intermediate-frequency amplifier $f = 40$ Mc, $\Delta f = 2$ Mc; 11) EO-7 oscilloscope.

Section 4

RELATIVE STABILITY OF A MASER USING THE 3,3 LINE OF $N^{14}H_3$

It is evident from Eq. (2) that the relative stability of a maser depends on the quality factor Q_l of the line and on the degree of stabilization of the maser parameters: tuning of the resonator ω_0, the pressure in the molecular beam source p, and the voltage on the selector system U, since G and Δ are functions of p (through γ) and U.

In this connection we have investigated the dependence of the initial line width for induced radiation on intensity of the molecular beam for different source designs (grid, channel) in connection with the possibility of stabilizing the maser parameters.

1. Line Width for Induced Radiation of a Maser

Two different kinds of radio spectrometers have been used to measure line width. The first is a superheterodyne spectrometer of simple design. A diagram of this system is shown in Fig. 24. In this spectrometer the microwave signal is applied to a directional coupler, the main channel of which provides the transfer of heterodyne power to a hybrid ring. In this channel the fundamental frequency is mixed in a mixer crystal with a frequency of 40 Mc. The sideband that is obtained, $f_{het} + 40$ Mc $= f_{line}$, 23,870 Mc, i.e., the frequency of the ammonia line J=3, K = 3, is applied to the resonator of the maser as pump power, which causes induced radiation. Then this power is applied to a balanced mixer (hybrid ring), intermediate-frequency amplifier, and oscilloscope. Since a sawtooth voltage is applied to the repeller of the klystron, on the screen of the oscilloscope one observes the induced radiation line of ammonia.

In order to obtain frequency markers, to the mixer crystal in addition to the voltage at 40 Mc there is applied a voltage from a ZG-12 oscillator at 20 kc. This spectrometer is convenient for searching and for recording lines in the frequency region available by tuning the resonator and klystron. A shortcoming of this spectrometer is the necessity for having a klystron with a stability better than 10^{-7} over the period of a sweep on the oscilloscope.

By providing the klystron with a circulating water bath and using a fast sweep (10-50 cps) it is possible to photograph a radiation line on the screen of the oscilloscope in one passage of the beam. Figures 25a and 25b show the 3,3 line of $N^{14}H_3$ obtained with this spectrometer (the magnetic satellite lines are evident). However, the fast sweep requires a relatively wide-band detector and consequently lowers the sensitivity of the spectrometer.

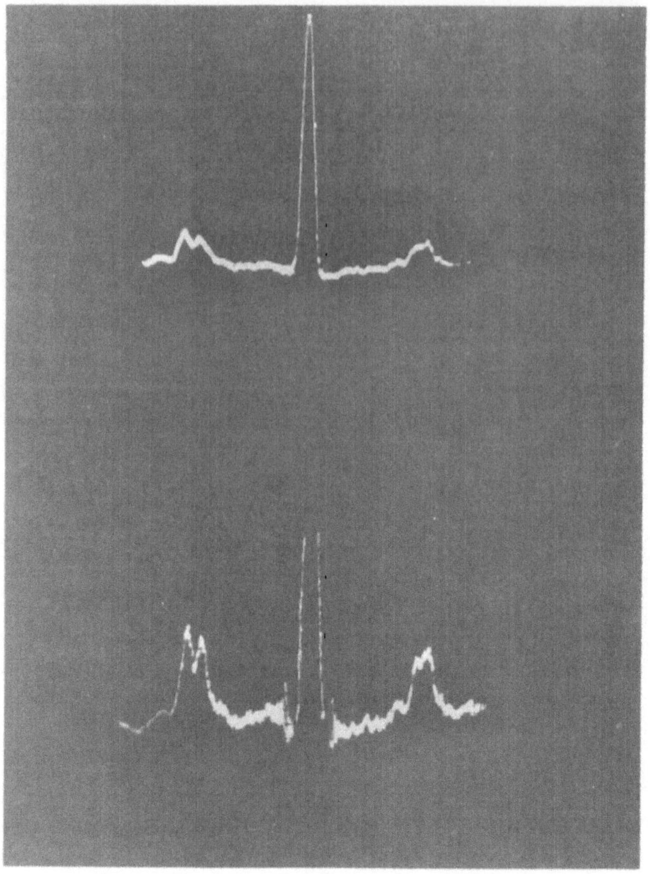

Fig. 25. a,b) Oscillogram showing the 3,3 induced radiation line of $N^{14}H_3$ (the magnetic satellite lines are evident) at two different values of the detector gain.

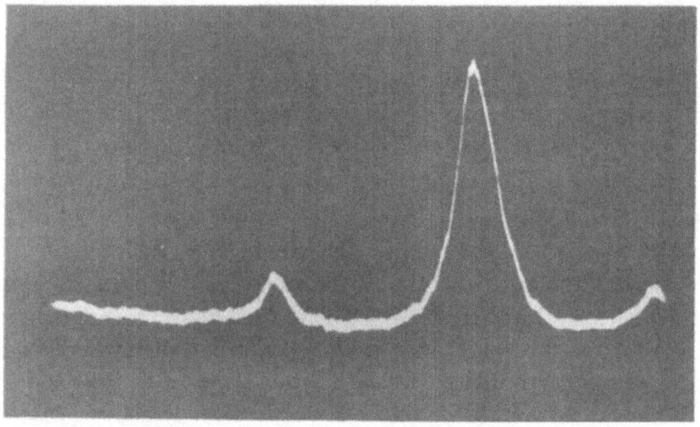

Fig. 26. Oscillogram of the 3,3 induced radiation line of $N^{14}H_3$. The distance between the frequency markers is 40 kc.

Better sensitivity is provided by a spectrometer which uses a quartz crystal multiplier as the pumping source. In order to sweep the frequency, a germanium diode, to which is applied a sawtooth voltage, is connected in parallel with the quartz crystal through a small capacity. The capacity of the diode changes in proportion to the voltage and correspondingly the pump frequency changes. Since the stability of a temperature-controlled quartz crystal over a period of several seconds is better than 10^{-8}, the time required for sweeping the line on the screen of the oscilloscope can be increased to several seconds. This allows us to reduce the band width of the detector and to increase the sensitivity of the spectrometer. The measurements of line width were carried out with this spectrometer.

In making measurements it should be noted that at low intensities of the beam the band width of the maser coincides with the width of the spectral line, but the observation of radiation and the measurement of line width are usually difficult because of the inadequate sensitivity.

At higher beam intensities the band width is reduced appreciably because of regeneration and direct measurements of band widths do not actually give the true widths of the spectral line.

In order to measure the width of the line it is convenient to use the following method. Having measured the dependence of band width on the self-excitation parameter in the range of beam intensities in which the possibility of observation is not limited by the detector sensitivity, we can extrapolate the obtained curve to zero beam intensity, thus obtaining the true value of the original width.

The band width of a maser α_0 exhibits the following dependence on τ, the mean time of flight of molecules through the resonator, and on β, the self-excitation parameter:

$$\alpha_0 = 2\frac{1-\beta}{\tau}. \tag{13}$$

This relation is obtained from the expression for the power gain:

$$k = k_0 \frac{\left(\frac{1}{\tau^2}+\alpha^2\right)^2}{\alpha^2\left(\frac{\beta}{\tau}\right)^2+\left[\alpha^2+\frac{1}{\tau^2}\cdot(1-\beta)\right]^2} \tag{14}$$

Here α is the deviation with respect to the frequency of the line peak, k_0 is the coefficient for power transfer through the resonator with no molecular beam. Equation (14) follows from the theory developed in [29] under the assumption that the quality factor of the line is appreciably greater than the quality factor of the resonator, that the deviation is not greater than several radiation line widths, and that there is no saturation.

In order to analyze the dependence of line width on the self-excitation parameter, in place of the gain k we introduce the new quantity $K = k - k_0$, i.e., a quantity which corresponds to the increment in gain due to the power radiated by the molecule in the resonator. In the absence of a molecular beam, $K = 0$. If we denote by K_{max} the value of K for a deviation $\alpha = 0$, the expression for α_0 can be obtained from the condition

$$\frac{K}{K_{max}} = \frac{1}{2}.$$

It is evident from Eq. (13) that the band width α_0 is a linear function of the self-excitation parameter β. When $\beta \to 1$, the band width $\alpha_0 \to 0$, i.e., when the self-excitation condition is satisfied, $\beta = 1$, the maser will radiate a monochromatic line, and when $\beta \to 0$ (consequently $N_0 \to 0$), we have

$$\alpha_0 = \frac{2}{\tau} \quad \text{or} \quad \Delta\nu_0 = \frac{1}{\pi\tau},$$

where $\Delta\nu_0$ is the radiation line width.

The mean transit time τ depends on the velocity distribution of the molecules, and this distribution does not remain fixed but can change somewhat as the selector voltage is changed, so that the linear dependence of α_0 on β will be violated. This effect has been noted in [13], where the change in the distribution of selected molecules by velocity was estimated from curves showing the dependence of maser frequency on selector voltage.

The quantity β can be easily obtained experimentally if we measure the maximum gain K_{max} (when $\alpha = 0$). It follows from Eq. (14) that

$$K_{max} = \frac{k_0}{(1-\beta)^2}.$$

The measurements were carried out with a maser using the 3,3 line of $N^{14}H_3$. With a grid source of thickness

0.05 mm and apertures 0.05 x 0.05 mm, the line width was found to be 5 kc (for a resonator length of 80 mm and T = 300°K), which is 2.5 times greater than the calculated value. In this case there is a linear compression of the band width with regeneration. It was possible to follow the compression of the line from 5 kc to 100 cps. The accuracy of the measurements was of the order of 15%.

An increase in line width over the estimated value is due partially to the unresolved hyperfine structure and the Doppler broadening, which evidently arises as a consequence of field distortion in the resonator due to the cutoff waveguides, the coupling holes, the tuning screw, and the divergence of the beam.

Measurements have shown that the width of the radiation line for a beam formed by a channel source (diameter 2 mm, length 10 mm) is increased appreciably (by a factor of 2) only at source pressures which provide the maximum output power. At low beam intensities the line widths are essentially the same for the channel and grid systems. Consequently, masers using these beam sources have essentially the same stability. This result has been verified by experimental investigations of stability.

Figure 26 shows an induced radiation line for a beam source pressure $p = 1 \cdot 10^{-1}$ mm Hg and a quadrupole condenser voltage U = 4 kV with the line width $\Delta\nu_0 \approx 5$ kc.

2. Stability of a Maser Using the 3,3 Line of $N^{14}H_3$

As we have indicated in the introduction, use of the maser frequently requires a high relative stability over a long period of time. From the results of the measurements of line width given in the previous part, it follows that in the design of a maser having this stability it is necessary to use as a molecular beam source either a single long channel (approximate dimensions: length 10 mm, diameter 1.5-2 mm) or a multichannel source consisting of a large number of thin long channels with a good transmission factor.

In this case one cannot use a cooled diaphragm, since molecules emitted from such a molecular-beam source have a rather high directivity and the vacuum inside the maser is not spoiled with an adequate number of molecules entering the resonator. The maser can operate continuously over a period of approximately one year [59, 70] so long as the layer of frozen ammonia on the cooled chamber wall does not reach a thickness of several millimeters.

An analysis of the experimental curves showing the dependence of frequency on maser parameters allows us to estimate the maximum possible long-term relative stability of a maser using the J = 3, K = 3 line. It follows from Figs. 9, 10, and 12 that a change in pressure from $1.5 \cdot 10^{-2}$ to $5 \cdot 10^{-2}$ mm Hg in a grid source, or from $1 \cdot 10^{-1}$ to $3 \cdot 10^{-1}$ mm Hg in a channel source causes a frequency change of approximately 15 cps. A change in voltage on the quadrupole condenser (Fig. 11) of 1 kV changes the frequency by approximately 5 cps. A superinvar resonator will change the frequency of the maser by approximately 20 cps for a temperature change of 1°C.

Thus, if the pressure in the molecular-beam source is stabilized with an accuracy of 1%, if the voltage on the quadrupole condenser is stabilized to 0.2%, and if the temperature of the invar resonator is maintained with an accuracy of 0.01°C, we can achieve a long-term relative stability of $\Delta f/f = 10^{-11}$. In this case, in order to avoid effects due to the waveguide system, it is necessary to provide good isolation. One must also take steps to provide mechanical stability for the maser elements, since a change in positions of the different parts can change the maser frequency.

In this work we have carried out measurements of the relative stability of the two masers described in Section 2. First in the maser we replaced the invar resonators by resonators made of superinvar, which has the lowest temperature coefficient of linear expansion ($\sim 3 \cdot 10^{-7}$ deg^{-1}).

The pressure in the molecular-beam source was held constant with the required degree of accuracy by fixing the pressure of the ammonia before the valve. For this purpose the ammonia from a chamber with a high capacity was admitted to the valve through a tube freely connected to the atmosphere. The ammonia pressure in the tube was monitored by a U-tube manometer filled with oil. The variation of pressure in operation was less than 1%. Stabilization of the high voltage was realized by stabilizing an a-c voltage which was used for the power supply of the high-voltage system. In this case use was made of an electronic stabilizer (SIP-1) which stabilizes the voltage with an accuracy of 0.1%.

A schematic diagram of the system used to measure frequency stability differs from that shown in Fig. 8

in that we used an oscilloscope with a post-acceleration intensifier and an automatic electronic potentiometer (EPP-09) was connected in parallel by means of which the beat frequencies of the masers were recorded on a strip chart. The initial beat frequency was established as 3-4 cps. The decoupling tubes and the mixer hybrid ring provided good decoupling of the masers, so that they were not pulled even when within tenths of a cycle of each other. Instead of the audio oscillator we used a low-frequency oscillator (NGPK-3).

As a result of the measurements carried out with these masers, we have obtained relative stabilities of 10^{-11} over a period of 4 hr.

CONCLUSION

1. In this work we have investigated experimentally the dependence of frequency and relative power of a maser using the J = 3, K = 3 line of $N^{14}H_3$ on the characteristic frequency of the resonator, the voltage applied to the quadrupole condenser (or to a ring-type selection system) and the ammonia pressure in the molecular-beam source. We have also investigated the dependence of frequency and relative power of a maser with two opposed beams on ammonia pressure in the molecular-beam source. The results can be explained theoretically on the basis of the work reported in [38, 40, 43].

2. Analysis of the characteristics has shown that the possible relative stability of a maser using the 3,3 line with one beam is 10^{-11} over an appreciable time period. Experimentally this stability has been obtained over a time period of 4 hr.

3. Rough adjustment of the characteristic frequency of the resonator of the maser to the frequency of the spectral line can be carried out by comparing the oscillation signal on the screen of an oscilloscope together with the peak of the resonance curve. The accuracy in tuning the maser in this way is $\sim 5 \cdot 10^{-9}$.

4. A maser can be tuned by equalizing the minima (or maxima) on the curve showing the dependence of maser frequency on ammonia pressure in the molecular-beam source. The reproducibility of a one-beam maser when tuned by this technique is $\sim 5 \cdot 10^{-10}$.

5. We have measured the widths of the spectral lines for induced radiation with different kinds of molecular-beam sources. It has been shown that for long-term operation of the maser one should use a channel source (or a multichannel source). However, the ammonia pressure in the molecular beam source in this case must be less than the pressure that provides maximum oscillation amplitude, because at high pressures one observes an appreciable broadening of the spectral line with a consequent deterioration of maser stability.

LITERATURE CITED

1. W. V. Smith, J. L. G. de Quevedo, R. L. Garter, and W. S. Bennett, J. Appl. Phys., 18, 1112 (1947).
2. N. A. Irisova, M. E. Zhabotinskii, and V. G. Veselago, Radiotekhnika, No. 4, 26 (1955).
3. J. P. Wittke and R. H. Dicke, Phys. Rev., 96, 530 (1954).
4. J. E. Sherwood, H. Lyons, R. H. McCracken, and P. Kusch, Bull. Am. Phys. Soc., 27(1), 43 (1952).
5. N. F. Ramsey, Phys. Rev., 78, 695 (1950).
6. N. F. Ramsey and H. B. Silslee, Phys. Rev., 84, 506 (1951).
7. Holloway, W. Neiberger, F. H. Reder, G. M. R. Winker, L. Essen, and J. W. L. Parry, Proc. IRE, 47, 1730 (1959).
8. Missiles and Rockets, 2/1, No. 1, 34 (1961).
9. K. F. Smith, Molecular Beams (Wiley, New York, 1955).
10. N. G. Basov and A. M. Prokhorov, Uspekhi Fiz. Nauk, 48, No. 5 (1955).
11. J. P. Gordon, H. J. Zeiger, and C. H. Townes, Phys. Rev., 99, No. 4, 1264 (1955).
12. N. G. Basov, The Maser, Radiotekh. i Elektron., 1, 51 (1956).
13. A. N. Oraevskii, Dissertation (1960).
14. N. G. Basov and A. N. Oraevskii, Zhur. Eksp. i Teoret. Fiz., 37, 1068 (1959).
15. N. G. Basov, O. N. Krokhin, A. N. Oraevskii, G. M. Strakhovskii, and B. M. Chikhaev, Uspekhi Fiz. Nauk, 75, No. 3 (1961).
16. N. G. Basov and A. P. Petrov, Radiotekh. i Elektron., 3, 298 (1958).
17. H. M. Goldenberg, D. Kleppner, and N. F. Ramsey, Phys. Rev. Lett., 5, 361 (1960).

18. D. Kleppner, N. F. Ramsey, and P. Fjeldstadt, Phys. Rev. Lett., 1, 232 (1958).
19. C. V. Heer, Report to the Conference on Electron Resonance Effects, USA, September, 1959, p. 1.
20. H. F. Ramsey, Rev. Sci. Instr., 28, 58 (1956).
21. R. H. Dicke, Phys. Rev., 89, 472 (1953).
22. K. Shimoda, T. C. Wang, and C. H. Townes, Phys. Rev., 102, 1308 (1956).
23. K. Shimoda, J. Phys. Soc. Japan, 12, 1006 (1957).
24. K. Shimoda, J. Phys. Soc. Japan, 13, 939 (1958).
25. J. C. Helmer, J. Appl. Phys., 28, 212 (1957).
26. J. Bananomi and J. Herrmann, Helv. Phys. Acta, 29, 224 (1956).
27. J. Bananomi, J. de Prins, J. Herrmann, and P. Kartaschoff, Helv. Phys. Acta, 30, 288 (1957).
28. F. S. Barnes, Proc. IRE, 47, 2085 (1959).
29. N. G. Basov, The Maser, Doctoral Dissertation, FIAN, 1956.
30. J. P. Gordon, H. J. Zeiger, and C. H. Townes, Phys. Rev., 95, 284 (1954).
31. J. P. Gordon, Phys. Rev., 99, 1253 (1955).
32. A. N. Oraevskii (see p. 1 of this collection).
33. H. G. Bennewitz and W. Paul, Ch. Schlier, Z. Physik, 141, 6 (1955).
34. F. O. Vonbun, J. Appl. Phys., 29, 632 (1958).
35. N. G. Basov and A. M. Prokhorov, Doklady Akad. Nauk SSSR, 101, 47 (1955).
36. N. G. Basov and A. M. Prokhorov, Zhur. Eksp. i Teoret. Fiz., 30, 560 (1955).
37. V. Klimantovich and R. V. Khokhlov, Zhur. Eksp. i Teoret. Fiz., 32, 1151 (1957).
38. N. G. Basov and A. N. Oraevskii, Radiotekh. i Elektron., 4, 1185 (1959).
39. J. Bonanomi, J. de Prins, J. Herrmann, and P. Kartaschoff, Rev. Scient. Instr., 28, 879 (1957).
40. N. G. Basov, V. V. Nikitin, and A. N. Oraevskii, Radiotekh. i Elektron., 6, 796 (1961).
41. N. G. Basov, Pribory i Tekh. Eksp., No. 1, 71, 77 (1957).
42. N. G. Basov and A. M. Prokhorov, Zhur. Eksp. i Teoret. Fiz., 27, 431 (1954).
43. K. Shimoda, J. Phys. Soc. Japan, 16, 1728-1739 (1961).
44. Absolute and Relative Stability of the Frequency of the Maser [in Russian] (FIAN Report, 1959).
45. N. G. Basov, G. M. Strakhovskii, and I. V. Cheremiskin, Radiotekh. i Elektron., 6, 1020 (1961).
46. J. de Prins, Menoud, and Kartaschow, Helv. Phys. acta, 34, No. 5, 438 (1961).
47. V. V. Nikitin, Radiotekh. i Elektron. (in press).
48. V. V. Nikitin and A. N. Oraevskii, Radiotekh. i Elektron. (in press).
49. W. H. Higa, Rev. Sci. Instr., 28, 726 (1957).
50. R. H. Dicke, Phys. Rev., 93, 99 (1954).
51. W. H. Wells, J. Appl. Phys., 29, 714 (1958).
52. F. H. Reder and C. J. Bickart, Rev. Sci. Instr., 31, 1164 (1960).
53. N. G. Basov and K. K. Svidzinskii, Izvestiya Vysshikh Ucheb. Zavedenii, Radiofiz., 1, No. 2, 89 (1959).
54. N. G. Basov and V. S. Zuev, Pribory i Tekh. Eksp., No. 1, 120 (1961).
55. J. Phys. Soc. Japan, 16, No. 2, 309 (1961).
56. Rev. Sci. Instr., 32, No. 10, 1083 (1961).
57. N. G. Basov and A. N. Oraevskii, Izvestiya Vysshikh Ucheb. Zavedenii, Radiofiz., No. 1, 3 (1959).
58. N. G. Basov, I. D. Murin, A. P. Petrov, A. M. Prokhorov, and I. V. Shtranikh, Izvestiya Vysshikh Ucheb. Zavedenii, Radiofiz., 2, 50 (1958).
59. V. F. Lubentsov et al., VNIIFTRI Report, 1959.
60. N. Ramsey, Molecular Beams, Oxford University Press, 1956.
61. I. A. Giordmaine and T. C. Wang, J. Appl. Phys., 31, 463 (1960).
62. A. F. Krupnov, Izvestiya Vyssikh Ucheb. Zavedenii, Radiofiz., 2, 658 (1959).
63. H. G. Bennewitz and W. Paul, Z. Physik, 139, 489 (1954).
64. J. C. Helmer, F. B. Jacobus, and P. A. Sturrock, J. Appl. Phys., 31, 458 (1960).
65. J. Singer, Masers, Wiley, New York, 1959.
66. V. G. Veselago and N. A. Irisova, Radiotekh. i Elektron., 2, 4 (1957).
67. I. D. Murin, Izvestiya Vysshikh Ucheb. Zavedenii, Radiofiz., 1, No. 5, 555 (1957).
68. Design of Ring Selectors, FIAN Report, 1961.

69. G. P. Lyubimov, G. M. Strakhovskii, and I. V. Cheremiskin, Bulletin of Moscow State University, No. 1, 79 (1961).
70. IRE Nat. Conv. Rec., No. 3, 78 (1960).
71. R. H. Dicke and J. R. Wittke, Phys. Rev., 103, 620 (1956).
72. W. Gordy et al., Microwave Spectroscopy, Wiley, New York, 1953.
73. P. Thaddeus and J. Lousber, Nuovo cimento, 13, No. 5, 1060 (1959).
74. C. H. Townes and A. L. Schawlow, Microwave Spectroscopy of Gases, McGraw-Hill, New York, 1953.

THEORY OF THE HYPERFINE STRUCTURE OF THE ROTATIONAL SPECTRA OF MOLECULES

K. K. Svidzinskii

INTRODUCTION

The investigations of the hyperfine structure (hfs) of molecular spectra is an extremely effective way to study the multipole moments of nuclei and intramolecular fields. This is of great importance in nuclear physics, in the study of the nature of the chemical bond and electronic structure of molecules, and in a number of problems of molecular physics.

The high resolution necessary for the investigation of hfs in molecular spectra has been achieved through the high degree of development of radiospectroscopic techniques [1], beginning with the well-known work of Ramsey and Rabi [2] with molecular beams. For example, such techniques have been used to measure the overwhelming majority of nuclear dipole and quadrupole moments as well as for the precise measurement of the Lamb shift in hydrogen [34] and the anomalous magnetic moment of the electron [4], which have had such great significance for contemporary physics.

Up to the present time, hfs has been investigated in a very large number of molecules. The most complete survey of these investigations is contained in the aforementioned fundamental monographs [1,2].

In recent years there has occurred a much wider development and application of methods of creating systems of energy levels with populations not in thermodynamic equilibrium; this has led to the development of a new field of study – quantum radiophysics [9]. In principle, the methods of quantum physics open up new possibilities for increasing both the resolution and sensitivity of spectroscopes simultaneously. In the first work of this kind [6], the hfs of the inversion spectrum of NH_3 was successfully studied with line widths of 2-3 kc and a precision of frequency measurement of the order of tens of cycles per second. Similar methods can be used to study a wide range of substances.

Some work recently completed at Columbia University in the United States of America can be cited as an example of an application of these methods [5, 7]. At the present time, there is in the Oscillation Laboratory at FIAN (Physical Institute, Academy of Sciences, USSR) an apparatus [10] which will measure the inversion spectrum of ND_3 with a resolution of about 300 cps and a precision not lower than 10 cps. Such high precision requires, as a rule, the calculation of an entire series of hyperfine interactions of a different kind. Because of this, the theoretical interpretation of the spectrum is an extremely difficult task even in the case of relatively simple molecules.

For the calculation of the hfs in molecular spectroscopy it is usual to apply direct methods based on the well-known formulas for the matrix elements of vector operators as given, for example, in the famous book by Condon and Shortley [11].

Since the current precision of measurement requires the simultaneous calculation of different interactions of the tensor type, such direct methods lead to exceedingly cumbersome expressions, particularly in those cases in which the hfs is caused by interaction of several nuclei having large spins. In addition, such calculations, as for example in [12], have to a significant extent a specific character, which complicates the theoretical interpretation of hfs in each new case.

The deficiency of such direct methods lies in the fact that they do not utilize in a clear way the rotational invariance of the hyperfine interaction operators and the rotational state vectors of the molecules. Meanwhile, particularly in the last two decades, the mathematical apparatus of the theory of groups of three-dimensional rotations has been well developed [14-16], based mainly on the work of Wigner [13] and Racah [14]. This apparatus, associated with the application of irreducible tensor operators and Wigner 3nj-symbols, has become an organic part of the modern theory of nuclear and atomic spectra, the theory of nuclear reactions and angular correlations of radiation, and a number of other problems in which the rotational invariance is an essential feature of the interactions.

From this point of view, the calculation of hfs is a typical problem in which the application of rotational-invariance methods can be extremely effective. Such a problem for molecular spectra was contemplated in a fundamental paper by Van Vleck [17] as early as 1951.

In the present paper an attempt will be made to complete this task, using as a basis the ideas set forth by Van Vleck in the aforementioned work. This apparatus has been applied to the analysis of the hfs of atomic spectra in [18].

However, the calculation of hyperfine interactions in molecules, as distinct from atoms, is made extremely complicated by the absence in molecules of the central symmetry characteristic of atoms. The essential difficulty is caused by our lack of accurate knowledge of the electronic structure of molecules.

In fact, all that can be established firmly for further consideration are the following general properties of molecules:

1. Division of the states of the molecule into electronic, vibrational, and rotational (adiabatic approximation).

2. Separation of the spin and "orbital" states of the electrons (Russel-Saunders approximation).

3. Mutual compensation of the electronic spins in the formation of stable chemical bonds.

4. The completely general properties of invariance of the electronic Hamiltonian under such operations as time reversal, parity, and the point group of the molecule.

In spite of the fact that the properties enumerated are completely general in character, it turns out that they are all that are necessary for the solution of the central problem of this paper, namely, the extraction of the dependence of the hyperfine interaction energy on the rotational quantum numbers and nuclear spins to an accuracy of not less than 10 cps, which is currently possible experimentally. Of course, not being in possession of concrete information about the electronic states of the molecules, we shall leave aside the complex and independent problem of calculating the molecular constants of the hyperfine interactions, which at the present time is still very far from a satisfactory solution [1].

The basic idea of the paper is to attain the maximum simplification of the hfs calculation by making the fullest possible use of symmetry without sacrificing the generality and completeness of the treatment.

In this connection group-theoretical methods adequate for the conceptual statement of the problem are widely employed. The treatment is limited to that specific class of molecules in which the ground electronic state is nondegenerate. To this class belong symmetric-top molecules in the ground electronic $^1\Sigma$ state and asymmetric-top molecules whose ground electronic states are also singlets.

It should be noted, however, that this limitation is not a very strong one, since this class encompasses the overwhelming majority of polyatomic molecules. Only a few paramagnetic molecules, such as O_2 or NO, are excluded.

Section 1 is concerned with the apparatus of the theory, based on the application of irreducible tensor operators and the Wigner 3nj-symbols. Although this apparatus has received an excellent exposition in the book by Fano and Racah [14] and in the book by Edmonds [15], we have decided to give a summary of it here. This decision was inspired by the desire not to have an excessive number of references to formulas in these books as well as countless statements about choice of phase conventions, normalization, etc.

Particular attention in this section is drawn to the connection between the wave functions of the symmetric top and the matrices of the finite rotations of Wigner. The treatment presented is convenient since it allows, in the first place, the fullest possible description of the properties of the wave functions of the symmetric top and, in the second place, it automatically takes into account the change in sign in the commutation relations of the "inner" momenta in the rotating system of coordinates [17].

The calculation of the matrix elements for the case of a large number of interacting nuclei is considered in Section 2. Basic to this section is the application of the method of genealogical coefficients for the substantial simplification of the matrix operator of the hyperfine interactions of several identical nuclei. For the case of three identical nuclei the symmetrized state vectors and reduction coefficients are calculated.

Section 3 presents an analysis of the symmetry properties of the molecular wave functions, and the matrix elements of various intramolecular field operators are calculated from the rotational wave functions. The selection rules which emerge from the symmetry properties of the Hamiltonian are considered for the matrix elements based on the electronic wave functions (i.e., for the hfs constant).

Section 4 is devoted to the calculation of the hfs of the rotational spectra of molecules. In order to provide an accuracy of calculation of hyperfine effects not lower than 10 cps, octupole and hexadecapole moments are included in the expansion of the hyperfine interaction energy, besides the usual dipole and quadrupole moments.

In the calculation of the magnetic interactions an important role is played by electronic states excited by rotational-electronic perturbation. The effect of this perturbation is accounted for by setting up a so-called effective Hamiltonian which then serves as the initial operator for the energy of the hyperfine interactions.

In Section 5, the theory is applied to the concrete case of calculating the hfs of the inversion transition $J = K = 6$ in ND_3.

This calculation, presented here as an illustration, has practical interest of its own. It is one of the experiments [10, 19, 20] carried out at FIAN with the aim of investigating the possibility of using a beam of ND_3 molecules as a frequency standard.

Section 1

THE MATHEMATICAL APPARATUS

1. The Matrices of Finite Rotations and Wave Functions for the Symmetric Top

Transformations of Finite Rotations. As is well known, the commutation relations for the components of angular momentum

$$[J_x, J_y] = iJ_z; \quad [J_y, J_z] = iJ_x; \quad [J_z, J_x] = iJ_y \tag{1.1}$$

allow the determination of a set of $2j + 1$ eigenvectors $|jm)$ [21] of the operators $\mathbf{J}^2 = j(j+1)$ and $J_z = m$, where j is a positive integer or half-integer, $m = -j, -j + 1, \ldots, j$, such that the eigenvectors $|jm)$ satisfy the equations

$$\left.\begin{aligned} J_z |jm) &= m|jm), \\ (J_x \pm iJ_y)|jm) &= [j(j+1) - m(m\pm 1)]^{1/2} |jm\pm 1). \end{aligned}\right\} \tag{1.2}$$

The vectors $|jm)$ describe a quantum mechanical system with a specific value of the square of the momentum $\mathbf{J}^2 = j(j + 1)$. For an infinitesimally small rotation of the system in three-dimensional space by an angle $\delta\varphi$ about an axis given by the direction of a unit vector $\mathbf{n}$, the vectors $|jm)$ transform according to the law [22]

$$|jm) \to D(\delta\varphi\mathbf{n})|jm) = [1 + i\delta\varphi\mathbf{n}\mathbf{J}]|jm). \tag{1.3}$$

It follows from (1.3) that the operator

$$D(\varphi\mathbf{n}) = \sum_{s=0}^{\infty} \frac{(i\varphi\mathbf{n}J)^s}{s!} \equiv \exp\{i\varphi\mathbf{n}J\} \tag{1.4}$$

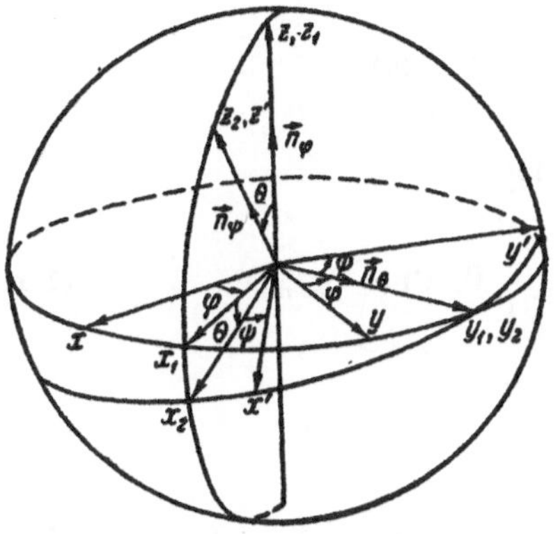

Fig. 1. Eulerian rotations.

corresponds to rotation by a finite angle φ.

Eulerian Rotations. Any finite rotation can be represented as a set of three Eulerian rotations carried out in the following way (Fig. 1):

1. A rotation by φ about the z axis carries the system (xyz) into ($x_1y_1z_1$) (Fig. 1).

2. A rotation by θ about the y_1 axis takes the system ($x_1y_1z_1$) into ($x_2y_2z_2$).

3. A rotation by ψ about the z_2 axis carries system ($x_2y_2z_2$) into (x'y'z').

If we agree that the operator corresponding to the complete rotation $D(\psi, \theta, \varphi)$ act from the left [i.e., on a ket-vector $|jm)$], it has the form

$$D(\psi, \theta, \varphi) = e^{i\psi J_z} e^{i\theta J_y} e^{i\varphi J_z}. \tag{1.5}$$

Matrices of Finite Rotations. Let us describe the initial system by the umprimed vectors $|jm)$ and the rotated one with primed vectors $|jm')$. Then the finite rotation operator $D(\psi, \theta, \varphi)$ acts in this way:

$$|jm) \to |jm') = D(\psi, \theta, \varphi)\,|jm). \tag{1.6}$$

Hence, the finite rotation matrix

$$D^{(j)}_{m'm}(\psi, \theta, \varphi) \equiv (jm'|e^{i\psi J_z} e^{i\theta J_y} e^{i\varphi J_z}|jm) \tag{1.7}$$

gives us the relation

$$(jm'| = \sum_{m=-j}^{j} D^{(j)}_{m'm}(\psi, \theta, \varphi)\,(jm| \tag{1.8}$$

or

$$|jm) = \sum_{m'=-j}^{j} |jm')\,D^{(j)}_{m'm}(\psi, \theta, \varphi). \tag{1.9}$$

Thus, according to these symbols, the matrix $D^{(j)}_{m'm}(\psi, \theta, \varphi)$ gives the bra-vectors of the rotated system in terms of the bra-vectors of the original system (i.e., it operates from the left). On the other hand, if the matrix $D^{(j)}_{m'm}(\psi, \theta, \varphi)$ operates from the right, it gives the ket-vectors of the original system in terms of the ket-vectors of the rotated system.

The matrix (1.7) is unitary, since the operators J_z and J_y are Hermitian. The Hermitian conjugate operator $D(\psi, \theta, \varphi)^+$ acts from the right, and the Hermitian conjugate matrix corresponding to it is

$$\begin{aligned} D^{(j)}_{m'm}(\psi, \theta, \varphi)^+ &= (jm|e^{-i\varphi J_z} e^{-i\theta J_y} e^{-i\psi J_z}|jm) = \\ &= D^{(j)}_{m'm}(-\varphi, -\theta, -\psi) = [D^{-1}]^{(j)}_{m'm}(\psi, \theta, \varphi), \end{aligned} \tag{1.10}$$

which gives, obviously, the inverse rotation.

Group-Theoretical Properties. From the point of view of group theory, the matrices $D^{(j)}_{mm'}$ constitute an irreducible unitary representation of the group of three-dimensional rotations of rank j [22]. The unitary property is expressed by the relation (1.10). The fundamental group properties are fulfilled by the fact that the result of two consecutive rotations $D(\psi_1, \theta_1, \varphi_1)$ and $D(\psi_2, \theta_2, \varphi_2)$ is a new rotation specified by the angles ψ, θ, φ.

$$D^{(j)}_{m'm}(\psi,\ \theta,\ \varphi) = \sum_{m''=-j}^{j} D^{(j)}_{m'm''}(\psi_2,\ \theta_2,\ \varphi_2)\, D^{(j)}_{m''m}(\psi_1,\ \theta_1,\ \varphi_1), \tag{1.11}$$

and that there exists an inverse rotation, leading to the identity transformation

$$\sum_{m''=-j}^{j} D^{(j)}_{m'm''}(-\varphi,\ -\theta,\ -\psi)\, D^{(j)}_{m''m}(\psi,\ \theta,\ \varphi) = D^{(j)}_{m'm}(0,\ 0,\ 0) = \delta_{m'm}. \tag{1.12}$$

The irreducibility of these representations is expressed by the property [3]

$$\int_0^{2\pi} d\varphi \int_0^{\pi} d\theta \sin\theta \int_0^{2\pi} d\psi\, D^{(j_1)^*}_{m_1 m}(\psi,\ \theta,\ \varphi)\, D^{(j_2)}_{m_2 m_2'}(\psi,\ \theta,\ \varphi) = \frac{8\pi^2}{2j_1+1}\delta_{j_1, j_2}\delta_{m_1, m_2}\delta_{m_1', m_2'}. \tag{1.13}$$

<u>The Operators of Infinitesimal Rotations in the Coordinate Representation.</u> The projection of the momentum operator $\mathbf{J}$ on an arbitrary axis in space, $\mathbf{n}$, is the operator of an infinitesimal rotation relative to this axis (multiplied by Planck's constant $\hbar$, which we shall always set equal to unity) and in the coordinate representation is expressed as a differential operator with respect to the angle of rotation α about the axis $\mathbf{n}$:

$$\mathbf{nJ} = -i\frac{\partial}{\partial\alpha}. \tag{1.14}$$

The infinitesimal rotation operators relative to different axes $\mathbf{n}_1\mathbf{J}$ and $\mathbf{n}_2\mathbf{J}$ are related to each other as the unit vectors $\mathbf{n}_1$ and $\mathbf{n}_2$ are. The vectors of the axes of the three Eulerian rotations $\mathbf{n}_\varphi$, $\mathbf{n}_\theta$, and $\mathbf{n}_\psi$, as can be seen from Fig. 1, are related to the Cartesian vectors by the following equations:

$$\left.\begin{aligned} \mathbf{n}_\varphi &= \mathbf{n}_z, \\ \mathbf{n}_\theta &= -\mathbf{n}_x \sin\varphi + \mathbf{n}_y \cos\varphi, \\ \mathbf{n}_\psi &= \mathbf{n}_x \sin\theta\cos\varphi + \mathbf{n}_y \sin\theta\cdot\sin\varphi + \mathbf{n}_z\cos\theta. \end{aligned}\right\} \tag{1.15}$$

Considering (1.14), it is easy to obtain from (1.15) expressions for the operators J_x, J_y, and J_z in terms of the Eulerian angles:

$$\left.\begin{aligned} J_z &= -i\frac{\partial}{\partial\varphi}, \\ J_x \pm iJ_y &= -ie^{\pm i\varphi}\left[\frac{1}{\sin\theta}\left(\frac{\partial}{\partial\psi} - \cos\theta\frac{\partial}{\partial\varphi}\right) \pm i\frac{\partial}{\partial\theta}\right]. \end{aligned}\right\} \tag{1.16}$$

In analogous fashion we obtain the expressions for the Cartesian components of the operator $\mathbf{J}$ in the rotated system of coordinates:

$$\left.\begin{aligned} J_{z'} &= -i\frac{\partial}{\partial\psi}, \\ J_{x'} \pm iJ_y &= +ie^{\pm i\psi}\left[\frac{1}{\sin\theta}\left(\frac{\partial}{\partial\varphi} - \cos\theta\frac{\partial}{\partial\psi}\right) \pm i\frac{\partial}{\partial\theta}\right]. \end{aligned}\right\} \tag{1.17}$$

<u>Differential Properties of the Finite Rotation Matrices.</u> If the rotation $D(\psi, \theta, \varphi)$ transforms the unit vector $\mathbf{n}$ into the vector $\mathbf{n}'$, then the operators $\mathbf{n}J$ and $\mathbf{n}'J$, acting respectively in the initial and rotated systems, are related by

$$Dn\,JD^{-1} = \mathbf{n}'\mathbf{J} \tag{1.18}$$

or

$$D\mathbf{nJ} = \mathbf{n}'JD. \tag{1.19}$$

Calculating the matrix elements of both sides of (1.19) with the aid of (1.2) and (1.7), we obtain the differential equations which are obeyed by the matrix elements $D^{(j)}_{m'm}(\psi, \theta, \varphi)$:

$$\left.\begin{aligned}
&\frac{\partial}{\partial\varphi} D^{(j)}_{m'm} = imD^{(j)}_{m'm};\\
&\frac{\partial}{\partial\psi} D^{(j)}_{m'm} = im'D^{(j)}_{m'm};\\
&e^{\pm i\varphi}\left[\frac{1}{\sin\theta}\left(\frac{\partial}{\partial\psi} - \cos\theta\frac{\partial}{\partial\varphi}\right) \pm i\frac{\partial}{\partial\theta}\right] D^{(j)}_{m'm} =\\
&= i\sqrt{j(j+1) - m'(m' \pm 1)}\, D^{(j)}_{m', m\pm 1};\\
&e^{\pm i\psi}\left[\frac{1}{\sin\theta}\left(\cos\theta\frac{\partial}{\partial\psi} - \frac{\partial}{\partial\varphi}\right) \pm i\frac{\partial}{\partial\theta}\right] D^{(j)}_{m'm} =\\
&= i\sqrt{j(j+1) - m'(m' \pm 1)}\, D^{(j)}_{m'\pm 1, m},
\end{aligned}\right\} \tag{1.20}$$

and which take care of all the differential properties of D-matrices.

Analytical Expression. Analytical expressions of the dependence of the matrix elements $D^{(j)}_{m'm}$ on the parameters of the rotation ψ, θ, φ, are called Wigner functions. The functions $D^{(j)}_{m'm}(\psi, \theta, \varphi)$ can be found by solving the system of the first three equations of (1.20), taking the normalization condition (1.12) into account; this gives

$$D^{(j)}_{m'm}(\psi, \theta, \varphi) = e^{i\psi m'} d^{(j)}_{m'm}(\theta)\, e^{i\varphi m}, \tag{1.21}$$

where

$$d^{(j)}_{m'm}(\theta) = \sum (-1)^s \frac{[(j+m')!\,(j-m')!\,(j+m)!\,(j-m)!]^{1/2}}{(j-m'-s)!\,(j+m-s)!\,s!\,(s+m'-m)!} \times$$
$$\times (\cos {}^1\!/_2\,\theta)^{2j-m'+m-2s} (\sin {}^1\!/_2\theta)^{2s+m'-m}, \tag{1.22}$$

in which the summation is over all integers s for which the arguments of all the factorials are not negative.

Note that for the particular choice of Euler rotations we have made (second rotation about the y axis), the matrix $d^{(j)}_{m'm}(\theta)$ is real. Moreover,

$$D^{(j)}_{m'm}(0, \pi, 0) = d^{(j)}_{m'm}(\pi) = (-1)^{j-m'} \cdot \delta_{m', -m}. \tag{1.23}$$

Transformation of a Vector Sum of Moments. The properties of finite rotation matrices will be completed with a discussion of the decomposition of the direct product of matrices. It is known [11] that the basis of the direct product $|j_1m_1)|j_2m_2) = |j_1m_1j_2m_2)$ decomposes according to the eigenvectors $|jm)$ of the basis of the total momentum $\mathbf{j} = \mathbf{j}_1 + \mathbf{j}_2$ by means of the transformation of the vector sum of momenta:

$$|j_1m_1\ j_2m_2) \to |j_1j_2jm) = C\,|j_1m_1j_2m_2). \tag{1.24}$$

Since the eigenvectors $|j_1j_2jm)$ of the total momentum again form a basis for the irreducible representation $D^{(j)}$, then according to (1.24)

$$CD^{(j_1)} \times D^{(j_2)}C^{-1} = D^{(j)}. \tag{1.25}$$

For the matrix elements of the transformation C we adopt the symbol $C^{jm}_{j_1m_1,\, j_2m_2} = (j_1j_2jm|j_1m_1j_2m_2)$, so that

$$|j_1m_1j_2m_2) = \sum_{j=|j_1-j_2|}^{j_1+j_2} |j_1j_2jm)\,(j_1j_2m|j_1m_1j_2m_2). \tag{1.26}$$

The relation inverse to (1.25),

$$D^{(j_1)}\times D^{(j_2)} = C^{-1}D^{(j)}C, \tag{1.27}$$

written in matrix form,

$$D^{(j_1)}_{m_1'm_1}(\psi,\ \theta,\ \varphi)\,D^{(j_2)}_{m_2'm_2}(\psi,\ \theta,\ \varphi) =$$

$$= \sum_{j=|j_1-j_2|}^{j_1+j_2} (j_1m_1'j_2m_2' \mid j_1j_2jm')\,D^{(j)}_{m'm}(\psi,\ \theta,\ \varphi)\,(j_1j_2jm \mid j_1m_1j_2m_2), \tag{1.28}$$

is called the Clebsch-Gordan series. The matrix elements $(j_1j_2jm\mid j_1m_1j_2m_2)$ are therefore frequently called Clebsch-Gordan coefficients. The transformation C is unitary and can be chosen to be real. These properties are expressed by the relations

$$\sum_{jm} (j_1m_1j_2m_2 \mid j_1j_2jm)\,(j_1j_2jm \mid j_1m_1'j_2m_2') = \delta_{m_1m_1'}\delta_{m_2m_2'}, \tag{1.29}$$

$$\sum_{m_1m_2} (j_1j_2jm \mid j_1m_2j_2m_2)\,(j_1m_1j_2m_2 \mid j'm') = \delta_{j,\,j'}\cdot\delta_{m,\,m'}, \tag{1.30}$$

$$(j_1j_2jm \mid j_1m_1j_2m_2) = (j_1m_1j_2m_2 \mid j_1j_2jm). \tag{1.31}$$

Besides this, the coefficients $(j_1j_2jm\mid j_1m_1j_2m_2)$ are zero unless

$$m = m_1 + m_2. \tag{1.32}$$

$$\delta(j_1,\ j_2,\ j) \neq 0. \tag{1.33}$$

The latter inequality is a symbolic way of writing the condition obeyed by the sides of a triangle of lengths j_1, j_2, and j, i.e., each of them is greater than (or equal to) the difference between the two others, but less than (or equal to) their sum.

Relation to the Spherical Harmonics. The angular part of the wave function of a particle in a state with a well-defined value for the magnitude of the orbital momentum l and its projection m on the z axis is described by the matrices of the transition from the coordinate representation given by the spherical angles θ and φ to the representation lm:

$$(\theta\varphi \mid lm) = Y_{lm}(\theta,\ \varphi). \tag{1.34}$$

The same state can be described by the system of vectors $|lm')$ turned relative to the initial set by the rotation $D(\psi = 0, \theta, \varphi)$. The particle is then found on the z' axis of the rotated system of coordinates and will have the coordinates $\theta' = \varphi' = 0$. According to (1.9) we have

$$(\theta\varphi \mid lm) = \sum_{m'} (00 \mid lm')\,D^{(l)}_{m'm}(0,\ \theta,\ \varphi). \tag{1.35}$$

But if a particle is situated on the z' axis, then the projection of its orbital momentum can equal only zero, in virtue of which

$$(00 \mid lm') = C\delta_{m',\,0}, \tag{1.36}$$

and then (1.35) gives

$$(\theta\varphi \mid lm) = CD^{(l)}_{0m}(0,\ \theta,\ \varphi). \tag{1.37}$$

The proportionality coefficient C is determined by the normalization condition

$$\int (lm \mid \theta\varphi) \sin\theta \, d\theta \, d\varphi \, (\theta\varphi \mid l'm') = \delta_{l,l'} \delta_{m\,m'} \tag{1.38}$$

with the aid of (1.13), which gives, finally,

$$(\theta\varphi \mid lm) = Y_{lm}(\theta, \varphi) = \left(\frac{2l+1}{4\pi}\right)^{1/2} D^{(l)}_{0m}(0, \theta, \varphi). \tag{1.39}$$

Representation of the Symmetric Top. The state of a rotating system of particles can be described in a representation in which are simultaneously given the magnitude of the angular momentum J and the projections M and K of the vector **J** respectively on the z and z' axes of a fixed (laboratory) and a proper (rotating together with the system of particles) system of coordinates. We shall call this representation the JKM representation, or the symmetric top representation. The vectors of the state |JKM) in the coordinate representation given by the Eulerian angles φ, θ, ψ can be found in the same way that was used in the preceding part. Performing the rotation $D(\psi, \theta, \varphi)$ on the system of vectors |JKM), we obtain

$$(\varphi\theta\psi \mid JKM) = \sum_{M'=-J}^{J} (000 \mid JKM') D^{(J)}_{M'M}(\psi, \theta, \varphi). \tag{1.40}$$

The matrix (000|JKM') describes the state in which the proper and laboratory coordinate systems coincide, by virtue of which

$$(000 \mid JKM') = C\delta_{K,M'}. \tag{1.41}$$

The normalization condition

$$\int_0^{2\pi}\int_0^{2\pi}\int_0^{2\pi} d\varphi \, d\psi \sin\theta \, d\theta \, (J'K'M' \mid \varphi\theta\psi)(\varphi\theta\psi \mid JKM) = \delta_{J'J}\delta_{K'K}\delta_{M'M}, \tag{1.42}$$

with (1.13) gives us, finally,

$$(\varphi\theta\psi \mid JKM) = \left(\frac{2J+1}{8\pi^2}\right)^{1/2} D^{(J)}_{KM}(\psi, \theta, \varphi). \tag{1.43}$$

The functions $(\varphi\,\theta\,\psi$|JKM) are frequently (though not completely accurately) called the wave functions of the symmetric top. Their properties are completely indicated by (1.43). They also obey Eq. (1.20), in which it is necessary to make the replacement $j \to J$, $m' \to K$, $m \to M$. The solution of this system can be represented also in the form

$$(\varphi\theta\psi \mid JKM) = e^{i(M\varphi+K\psi)} \cdot \frac{(-1)^{J+M}}{\pi \cdot 2^J} \cdot \left[\frac{(2J+1)(J-M)!}{(J+M)!\,(J+K)!\,(J-K)!\,8}\right]^{1/2} \times$$
$$\times \frac{(\sin\theta)^M}{(\mathrm{tg}\,{}^1/_2\,\theta)^K} \left\{\frac{\partial}{\partial(\cos\theta)}\right\}^{J+M} \cdot [(\sin\theta)^{2J} (\mathrm{tg}\,{}^1/_2\theta)^{2K}]. \tag{1.44}$$

Formula (1.43) makes possible the calculation of the matrix elements of the D-matrix in the JKM representation. Applying (1.28) and (1.13), we obtain

$$(JKM \mid D^{(j)}_{m'm} \mid J'K'M') = \left(\frac{2J'+1}{2J+1}\right)^{1/2} (J'jJK \mid J'K'jm') \cdot (J'jJM \mid J'M'jm). \tag{1.45}$$

Change of Sign in the Commutation Relations of the Components of Momentum. It is well known that the components of angular momentum in the rotated system of coordinates obey commutation relations that differ from the usual ones (1.1) by the change in sign in front of the imaginary units [17]:

$$[J_{x'}, J_{y'}] = -iJ_{z'}; \quad [J_{y'}, J_{z'}] = -iJ_{x'}; \quad [J_{z'}, J_{x'}] = -iJ_{y'}. \tag{1.46}$$

We note here that this circumstance is a formal consequence of the adopted means of description, i.e., the choice of the JKM representation, in which the projections of J on the z and z' axes of different systems are given simultaneously. In fact, by writing (1.43) in the form

$$(\varphi\theta\psi \mid JKM) = \left(\frac{2J+1}{8\pi^2}\right)^{1/2} (JK \mid \exp\{i\psi J_z + i\theta J_y + i\varphi J_z\} \mid JM), \quad (1.47)$$

it is easily seen that with the simultaneous specification of the projections M and K the corresponding vectors of the states are transformed like $(JK|$ and $|JM)$, i.e., contragrediently. The transition to the scheme of description in which $|jm)$ is replaced by $(\overline{jm}|$ and thereby $|jm)$ and $(\overline{jm}|$ transform cogrediently is attained by means of the conjugation operator K [14], which consists of Hermitian conjugation $(jm|^{+} = |jm)$ and transformation of the transition to contragredience $U|jm) = |jm)$, so that

$$(jm \mid K = U \mid jm). \quad (1.48)$$

This operation is formally the same as the operator T of weak (Wigner) time reversal [23]. In particular, it changes the sign of the momentum

$$KJK^{-1} = -\mathbf{J} \quad (1.49)$$

and the sign in front of the imaginaries in the commutation relations for the components of momentum.

The relation (1.49) justifies the artificial procedure (Van Vleck) [17] consisting of changing the signs of all internal momenta of the rotating system in order to maintain the correct relation between the moments in going over to the rotating system of coordinates. However, if all magnitudes and relations characterizing the internal interactions in the system are projected always in the laboratory coordinate system, then the necessity for applying this rule naturally falls away.

2. Irreducible Tensors

The set of 2j + 1 quantities $a_m^{(j)}$ (m = −j, −j + 1, . . . , j), which transform under rotations like the eigenvectors $|jm)$, i.e., according to the law (1.9),

$$a_m^{(j)} = \sum_{m'} a_{m'}^{(j)} D_{m'm}^{(j)} (\psi, \theta, \varphi), \quad (1.50)$$

is called an irreducible tensor of rank j. The designation is due to the fact that the representation of the rotation group realized by the matrices $D_{m'm}^{(j)}$ is irreducible. In (1.50) the quantities $a_m^{(j)}$ pertain to the initial coordinate system and $a_{m'}^{(j)}$ to the rotated one. The inverse transformation, which gives the $a_{m'}^{(j)}$ in terms of the $a_m^{(j)}$, has the form

$$a_{m'}^{(j)} = \sum_{m} a_m^{(j)} D_{mm'}^{-1(j)} (\psi, \theta, \varphi). \quad (1.51)$$

Contragredience. It can be seen from the transformation (1.50) that the irreducible tensor $a_m^{(j)}$ transforms like a row. The irreducible tensor $\bar{a}_m^{(j)}$, which transposes like a column by means of a matrix that is the transpose of the inverse $\tilde{D}^{-1(j)}$,

$$\bar{a}_m^{(j)} = \sum_{m'} D_{mm'}^{-1(j)} \bar{a}_{m'}^{(j)}, \quad (1.52)$$

transforms contragrediently with respect to $a_m^{(j)}$. The transformation of the transition to contragredience U means

$$\bar{a}_\mu^{(j)} = \sum_{m} a_m^{(j)} U_{m\mu}^{(j)}. \quad (1.53)$$

Since the matrix $D_{m'm}^{(j)}$ is unitary,

$$UD^{(j)}U^{-1} = \tilde{D}^{-1(j)} = D^{(j)*}. \quad (1.54)$$

As a matrix U having the property (1.54) it is possible and convenient [14] to choose the real matrix (1.23):

$$U^{(j)}_{m'm} = D^{(j)}_{m'm}(0, \pi, 0) = (-1)^{j-m'}\delta_{m',-m}. \tag{1.55}$$

Then, according to (1.53),

$$a^{(j)}_m = (-1)^{j-m}\bar{a}^{(j)}_{-m};\ \bar{a}^{(j)}_m = (-1)^{j+m}a^{(j)}_{-m}. \tag{1.56}$$

The matrix $U^{(j)}_{m'm}$ is unitary and has the properties

$$\tilde{U}^{(j)}_{m'm} = U^{(j)}_{mm'} = (-1)^{2j}U^{(j)}_{m'm}, \tag{1.57}$$

i.e., it is symmetric for integral j and antisymmetric for half-integral j.

Scalar Formation. The expression

$$(a^{(j)}\cdot b^{(j)}) = \sum_m a^{(j)}_m \bar{b}^{(j)}_m \tag{1.58}$$

means the product of a row and a column; this gives a number, i.e., it is invariant under rotation (scalar). In order to make a scalar out of the two irreducible tensors $a^{(j)}$ and $b^{(j)}$, which transform cogrediently, it is necessary to apply the operation U, i.e.,

$$(a^{(j)}\cdot b^{(j)}) = \sum_{m\mu} a^{(j)}_m U^{(j)}_{m\mu} b^{(j)}_\mu = \sum_m a^{(j)}_m (-1)^{j-m} b^{(j)}_{-m}. \tag{1.59}$$

Conjugation Operator. The complex conjugate irreducible tensor

$$a^{(j)*}_m = K_0 a^{(j)}_m, \tag{1.60}$$

as follows from (1.54), transforms contragrediently with respect to the original $a^{(j)}$, i.e., like $\bar{a}^{(j)}$. The operator of conjugation of an irreducible tensor $K = K_0U$ is invariant under rotation, since the conjugate tensor

$$Ka^{(j)}_m = K_0 U a^{(j)}_m = \bar{a}^{(j)*}_m$$

transforms cogrediently with $a^{(j)}_m$. This explains its important role, since the operators K_0 and U separately are not invariants of the rotation.

Self-Conjugate Tensors. Irreducible tensors with the property

$$Ka^{(j)}_m = a^{(j)}_m, \tag{1.61}$$

are called self-conjugate. This means that

$$a^{(j)*}_m = \bar{a}^{(j)}_m = (-1)^{j-m}a^{(j)}_{-m}, \tag{1.62}$$

since, from $K = K_0U = 1$ follows the equivalence of the operators U and K_0. For the common tensor with real Cartesian components, $K_0 = U = 1$. If a unitary substitution is used to make from them an irreducible tensor, it will be self-conjugate, since the property K = 1 is an invariant of a unitary transformation. Therefore, physical quantities can correspond only to self-conjugate irreducible tensors.

It follows from the equivalence of K_0 and U that $U^*U = 1$, which is satisfied only for integral j, because by (1.55), $U^*U = (-1)^{2j}$. Hence, self-conjugate irreducible tensors can be only of integral rank j. The converse is also true: for an irreducible tensor $a^{(j)}_m$ of integral rank, one can always choose the phase factor (which depends only on j) such that it satisfies the condition (1.61). Irreducible tensors of half-integral rank cannot be self-conjugate and, therefore, there is no unitary substitution that will transform them to purely real components.

Composition of Irreducible Tensors. A linear combination of the components of two irreducible tensors $a^{(j_1)}_{m_1}$, $b^{(j_2)}_{m_2}$ formed by means of Clebsch-Gordan coefficients,

$$[a^{(j_1)}_{m_1} \times b^{(j_2)}_{m_2}]^{(j)}_{(m)} = \sum_{m_1 m_2} (j_1 j_2 j m \mid j_1 m_1 j_2 m_2)\, a^{j_1}_{m_1} b^{(j_2)}_{m_2}, \tag{1.63}$$

transform, by virtue of (1.25), by means of a matrix $D^{(j)}_{m'm}$, i.e., it is an irreducible tensor of rank j. Note that the composition of self-conjugate tensors is again a self-conjugate tensor, since the self-conjugate property is invariant under a unitary transformation such as (1.63) is. The rank j can run over the values $j = |j_1 - j_2|$, $|j_1 - j_2| + 1, \ldots, j_1 + j_2$. The composition of irreducible tensors corresponds in quantum mechanics to the well-known process of coupling of momenta $\mathbf{j}_1 + \mathbf{j}_2 = \mathbf{j}$. In particular, if $j_1 = j_2$, then one can form a scalar ($j = 0$) by means of (1.63). It is clear that the corresponding matrix $(j_1 j_2 0 0 \mid j_1 m_1 j_2 m_2)$ should be proportional to the matrix $U^{(j_1)}_{m_1 m_2}$. In fact [4],

$$(j_1 j_2 00 \mid j_1 m_1 j_2 m_2) = (-1)^{j_1 - m_1} \delta_{m_1, -m_2} \delta_{j_1, j_2} (2j_1 + 1)^{1/2}, \tag{1.64}$$

so that

$$(a^{(j)} \cdot b^{(j)}) = (2j + 1)^{1/2} [a^{(j)} \times b^{(j)}]^{(0)}. \tag{1.65}$$

The Wigner 3j-Symbol. By applying (1.63) twice, we can obtain from three irreducible tensors the invariant

$$(-1)^{j_3 + j_2 - j_1} [(a^{(j_1)} \times b^{(j_2)})^{(j_3)} \times c^{(j_3)}]^{(0)} = \sum_{m_1 m_2 m_3} \begin{pmatrix} j_1 j_2 j_3 \\ m_1 m_2 m_3 \end{pmatrix} a^{(j_1)}_{m_1} b^{(j_2)}_{m_2} C^{(j_3)}_{m_3}. \tag{1.66}$$

The coefficients of this transformation are called Wigner 3j-symbols and, as follows from (1.63) and (1.64), are equal to

$$\begin{pmatrix} j_1 & j_2 & j_3 \\ m_1 & m_2 & m_3 \end{pmatrix} = (-1)^{j_1 - j_2 - m_3} (2j_3 + 1)^{-1/2} (j_1 j_2 j_3 - m_3 \mid j_1 m_1 j_2 m_2). \tag{1.67}$$

The Wigner 3j-symbols are very convenient in calculations, since they have high symmetry with respect to transposition of its arguments. Specifically, it is unchanged for an even permutation of its columns and is multiplied by $(-1)^{j_1+j_2+j_3}$ in an uneven permutation. Furthermore,

$$\begin{pmatrix} j_1 & j_2 & j_3 \\ m_1 & m_2 & m_3 \end{pmatrix} = (-1)^{j_1+j_2+j_3} \begin{pmatrix} i_1 & i_2 & j_3 \\ -m_1 & -m_2 & -m_3 \end{pmatrix}. \tag{1.68}$$

The symbol is zero unless $m_1 + m_2 + m_3 = 0$ and unless j_1, j_2, and j_3 obey the "triangular" rule (1.33). The formation of an invariant in the case when one of the tensors transforms contragrediently with respect to the two others, for example $[[Ua^{(j_1)} \times b^{(j_2)}]^{(j_3)} \times C^{(j_3)}]^{(0)}$, is carried out by means of the quantity

$$\sum_{m_1'} U^{(j_1)}_{m_1 m_1'} \begin{pmatrix} j_1 & j_2 & j_3 \\ m_1 & m_2 & m_3 \end{pmatrix} = (-1)^{j_1 - m_1} \begin{pmatrix} j_1 & j_2 & j_3 \\ -m_1 & m_2 & m_3 \end{pmatrix}. \tag{1.69}$$

Transformation of Coupling Schemes. 3nj-Symbols. In the construction of an irreducible tensor from several irreducible tensors, the order of composition or coupling scheme of the irreducible tensors plays a very important role. For example, to the tensor $[(a^{(j_1)} \times b^{(j_2)})^{(j_{12})} \times c^{(j_3)}]^{(j)}$ belongs the coupling scheme $\mathbf{j}_1 + \mathbf{j}_2 = \mathbf{j}_{12}$, $\mathbf{j}_{12} + \mathbf{j}_3 = \mathbf{j}$, or $((j_1 j_2) j_{12}, j_3) j$, and to the tensor $[a^{(j_1)} \times (b^{(j_2)} \times c^{(j_3)})^{(j_{23})}]^{(j)}$ corresponds $\mathbf{j}_2 + \mathbf{j}_3 = \mathbf{j}_{23}$, $\mathbf{j}_1 + \mathbf{j}_{23} = \mathbf{j}$ or $[j_1, (j_2 j_3)\, j_{23}] j$. The transition between these schemes is accomplished by the coupling transformation matrix $(j_{12}, j_3 \mid j_1, j_{23})^{(j)}$, e.g.,

$$[(a^{(j_1)} \times b^{(j_2)})^{(j_{12})} \times c^{(j_3)}]^{(j)}_m = \sum_{j_{23}} (j_{12}, j_3 \mid j_1, j_{23})^{(j)} [a^{(j_1)} \times [b^{(j_2)} \times c^{(j_3)}]^{(j_{23})}]^{(j)}_m, \tag{1.70}$$

where the summation index j_{23} runs over all values permitted for the given coupling scheme, i.e., from $\max\{|j_2 - j_3|, |j - j_1|\}$ to $\min\{j_1 + j_2, j_1 + j\}$. The elements $(j_{12}, j_3 \mid j_2, j_{23})^{(j)}$ of this matrix are expressed by the so-called 6j-symbol

$$(j_{12}, j_3, | j_1, j_{23})^{(j)} = (-1)^{j_1+j_2+j_3+j} [(2j_{12}+1)(2j_{23}+1)]^{1/2} \begin{Bmatrix} j_1 j_2 j_3 \\ j_3 j j_{23} \end{Bmatrix}. \tag{1.71}$$

The 6j-symbol reverts to zero if the triangle rule is not fulfilled for the four triads (j_1, j_2, j_{12}), (j_2, j_3, j_{23}), and (j_1, j_{23}, j), (j_{12}, j_3, j), since only in this case can there be a vector coupling of the indicated momenta. The symmetry properties of 6j-symbols can be written down in the form

$$\begin{Bmatrix} j_1 & j_2 & j_3 \\ l_1 & l_2 & l_3 \end{Bmatrix} = \begin{Bmatrix} j_\alpha & j_\beta & j_\gamma \\ l_\alpha & l_\beta & l_\gamma \end{Bmatrix} = \begin{Bmatrix} l_\alpha & l_\beta & j_\gamma \\ j_\alpha & j_\beta & l_\gamma \end{Bmatrix}, \tag{1.72}$$

where (α, β, γ) is an arbitrary permutation of the numbers (1, 2, 3). When one of the tensors is a scalar, the transformation is trivial. For example, when $j_2 = 0$, obviously $j_{12} = j_1$, $j_{23} = j_3$, and $(j_{12}, j_3 | j_1, j_{23})^{(j)} = 1$.

From (1.71) and (1.72) it is easy to obtain

$$\begin{Bmatrix} l_1 & l_2 & l_3 \\ 0 & l_3 & l_2 \end{Bmatrix} = (-1)^{l_1+l_2+l_3} [(2l_3+1)(2l_3+1)]^{-1/2}. \tag{1.73}$$

Changes in the order of coupling of four irreducible tensors require 9j-symbols, which are defined in the following way:

$$(j_{12}, j_{34} | j_{13} j_{24})^{(j)} = [(2j_{12}+1)(2j_{34}+1)(2j_{13}+1)(2j_{24}+1)]^{1/2} \begin{Bmatrix} j_1 & j_2 & j_{12} \\ j_3 & j_4 & j_{34} \\ j_{13} & j_{24} & j \end{Bmatrix}, \tag{1.74}$$

9j-symbols do not change in an even permutation of rows or columns and in reflections across two of its diagonals. For an uneven permutation of rows or columns, the 9j-symbol is multiplied by $(-1)^\sigma$, where σ is the sum of all its arguments.

From the meaning of the transformation matrix (1.74) it is clear that the 9j-symbol is zero if the three arguments in any row or column do not fulfill the triangular rule. If one of the recoupled tensors is a scalar, then the matrix (1.74) reduces to the matrix (1.71). In analogy to (1.73), it is not difficult to find

$$\begin{Bmatrix} j_1 & j_2 & j_3 \\ j_5 & j_4 & j_3 \\ j_6 & j_6 & 0 \end{Bmatrix} = (-1)^{j_2+j_3+j_5+j_6} [(2j_3+1)(2j_6+1)]^{-1/2} \begin{Bmatrix} j_1 & j_2 & j_3 \\ j_4 & j_5 & j_6 \end{Bmatrix}. \tag{1.75}$$

Generalized 3nj-symbols arise in the recoupling of the product of n + 1 irreducible tensors [15].

However, for our purposes we shall require only 3j-, 6j-, and 9j-symbols.

Calculation of 3nj-Symbols. The generalized 3nj-symbols are complicated algebraic functions of their arguments and appear as extremely cumbersome sums. We shall not present here explicit expressions for the 3nj-symbols, which can be found in [15] and [16]. The latter is especially concerned with the technique of application of the 3nj-symbols and the quantities associated with them.

In practice, one encounters mostly cases in which at least one of the arguments of a 3nj-symbol does not take on too large a value. For these cases there are tables of formulas for 3j- and 6j-symbols [15] and the Clebsch-Gordan coefficients associated with them [16] and the Racah coefficients. There exist also decimal tables of these coefficients for a wider range of variation of the arguments.

These tables provide practically all the cases encountered in hfs calculations. Certain relations between 6j- and 9j-symbols which are used below will be found in Appendix A.

3. Spherical Harmonics

Scalar Spherical Harmonics. In the present work we will make use of scalar spherical harmonics defined in the following way:

$$Y_m^{(l)}(\theta, \varphi) = i^l D_{0m}^{(l)}(0, \theta, \varphi). \tag{1.76}$$

They differ from those generally used,(1.39) [22],by a factor $i^l\left(\frac{4\pi}{2l+1}\right)^{\frac{1}{2}}$, which is selected in accordance with the condition for self-conjugation (1.62):

$$Y_m^{(l)*}(\theta, \varphi) = (-1)^{l-m} Y_{-m}^{(l)}(\theta, \varphi). \qquad (1.77)$$

The formula for the scalar product

$$(Y^{(l)}(\theta, \varphi)\cdot Y^{(l)}(\theta', \varphi')) = \sum_m Y_m^{(l)*}(\theta, \varphi)\cdot Y_m^{(l)}(\theta', \varphi') = P^{(l)}(\cos\omega), \qquad (1.78)$$

where ω is the angle between directions (θ, φ) and (θ', φ'), follows from (1.11) if one considers that the Legendre polynomial equals

$$P^{(l)}(\cos\omega) = D_{00}^{(l)}(0, \omega, 0) = i^l Y_0^{(l)}(\omega, 0). \qquad (1.79)$$

It follows from (1.78) that $Y_m^{(l)}(\theta, \varphi)$ are unit irreducible tensors

$$(Y_m^{(l)}(\theta, \varphi)\cdot Y_m^{(l)}(\theta, \varphi)) = \sum_m |Y_m^{(l)}(\theta, \varphi)|^2 = 1. \qquad (1.80)$$

Canonical Basis. In the analysis of a vector field we will use a canonical self-conjugate basis $\mathbf{e}_\mu^{(1)}$ defined in accordance with (1.62). If $\mathbf{e}_x, \mathbf{e}_y, \mathbf{e}_z$ are the usual Cartesian unit vectors,

$$\mathbf{e}_1^{(1)} = \frac{1}{\sqrt{2}}(\mathbf{e}_y - i\mathbf{e}_x), \quad \mathbf{e}_0^{(1)} = i\mathbf{e}_z, \quad \mathbf{e}_{-1}^{(1)} = \frac{1}{\sqrt{2}}(\mathbf{e}_y + i\mathbf{e}_x). \qquad (1.81)$$

The components of the irreducible tensor of the first rank formed from the Cartesian components of a vector are defined by the expansion

$$\mathbf{A} = \sum_\mu (\mathbf{A}\cdot\mathbf{e}_\mu^{(1)})\,\mathbf{e}_\mu^{(1)*} = \sum (-1)^{1-\mu}\mathbf{e}_{-\mu}^{(1)} A_\mu^{(1)}, \qquad (1.82)$$

where $A_\mu^{(1)} = \mathbf{A}\mathbf{e}_\mu^{(1)}$, i.e.,

$$A_1^{(1)} = \frac{1}{\sqrt{2}}(A_y - iA_x), \quad A_0^{(1)} = iA_z, \quad A_{-1}^{(1)} = \frac{1}{\sqrt{2}}(A_y + iA_x). \qquad (1.83)$$

It is not difficult to see that if the components of the vector are real, the tensor $A_\mu^{(1)}$ is self-conjugate, i.e.,

$$A_\mu^{(1)*} = (-1)^{1-\mu} A_{-\mu}^{(1)}.$$

For the decomposition of tensors of the second rank, we use a basis constructed according to the rule (1.63):

$$e_\mu^{(\lambda)} = \sum_{\mu_1,\mu_2} (11\lambda\mu\,|\,1\mu_1 1\mu_2)\,\mathbf{e}_{\mu_1}^{(1)}\mathbf{e}_{\mu_2}^{(1)}, \qquad (1.84)$$

where $\lambda = 0, 1, 2$. In particular, a second-rank tensor $B_\mu^{(2)}$ can be constructed from the components of the second-rank tensor B_{ik}:

$$\left.\begin{aligned} B_0^{(2)} &= \frac{1}{\sqrt{6}}[B_{ii} - 3B_{zz}], \\ B_{\pm 1}^{(2)} &= \frac{1}{2}[\pm(B_{xz} + B_{zx}) + i(B_{yz} + B_{zy})], \\ B_{\pm 2}^{(2)} &= \frac{1}{2}[B_{yy} - B_{xx} \pm i(B_{xy} + B_{yx})]. \end{aligned}\right\} \qquad (1.85)$$

Vector Spherical Harmonics. We will use the vector spherical harmonics

$$\mathbf{Y}_{\nu}^{(l\lambda)}(\theta, \varphi) = [Y^{(l)}(\theta, \varphi) \times \mathbf{e}^{(1)}]_{\nu}^{(\lambda)} = \sum_{m\mu} (l1\lambda\nu \mid lm1\mu) Y_{m}^{(l)}(\theta, \varphi) \mathbf{e}_{\mu}^{(1)}, \tag{1.86}$$

which are defined so that they are self-conjugate irreducible unit tensors of rank λ:

$$\mathbf{Y}_{\nu}^{(l,\lambda)*}(\theta, \varphi) = (-1)^{\lambda-\nu} \mathbf{Y}_{-\nu}^{(l,\lambda)}(\theta, \varphi), \tag{1.87}$$

$$(\mathbf{Y}^{(l,\lambda)}(\theta, \varphi) \cdot \mathbf{Y}^{(l,\lambda)}(\theta, \varphi)) = \sum_{\nu=-\lambda} (-1)^{\lambda-\nu} \mathbf{Y}_{-\nu}^{(l,\lambda)}(\theta, \varphi) \cdot \mathbf{Y}_{\nu}^{(l,\lambda)}(\theta, \varphi) = 1. \tag{1.88}$$

For a given value of λ it is possible to construct three vector spherical harmonics corresponding to the possible values $l = \lambda, \lambda \pm 1$.*

The vector spherical harmonics form a complete system and are normalized according to the condition

$$\int_0^{\pi} \int_0^{2\pi} \sin\theta \, d\theta \, d\varphi \mathbf{Y}_{\nu}^{(l,\lambda)} \cdot (\theta, \varphi) \mathbf{Y}_{\nu'}^{(l',\lambda')} = \frac{4\pi}{2l+1} \delta_{l,l'} \delta_{\lambda,\lambda'} \delta_{\nu,\nu'}. \tag{1.89}$$

We present here some properties of the spherical harmonics [15] which we shall make use of later:

$$\frac{\mathbf{r}}{r} Y_{\nu}^{(\lambda)}(\theta, \varphi) = \frac{[(\lambda+1)(2\lambda+3)]^{1/2}}{2\lambda+1} \mathbf{Y}_{\nu}^{(\lambda+1,\lambda)}(\theta, \varphi) + \frac{[\lambda(2\lambda-1)]^{1/2}}{2\lambda+1} \mathbf{Y}_{\nu}^{(\lambda-1,\lambda)}(\theta, \varphi). \tag{1.90}$$

The following properties are associated with the application of the gradient operator ∇ to the product of an arbitrary scalar function $f(r)$ and a spherical harmonic:

$$(\mathbf{r} \times \nabla) f(r) Y_{\nu}^{(\lambda)}(\theta, \varphi) = [\lambda(\lambda+1)]^{1/2} f(r) \mathbf{Y}_{\nu}^{(\lambda,\lambda)}(\theta, \varphi), \tag{1.91}$$

$$\nabla f(r) Y_{\nu}^{(\lambda)}(\theta, \varphi) = \frac{[(\lambda+1)(2\lambda+3)]^{1/2}}{2\lambda+1} \left(\frac{d}{dr} - \frac{\lambda}{r}\right) f(r) \mathbf{Y}_{\nu}^{(\lambda+1,\lambda)}(\theta, \varphi) +$$

$$+ \frac{[\lambda(2\lambda-1)]^{1/2}}{2\lambda+1} \left(\frac{d}{dr} + \frac{\lambda+1}{r}\right) f(r) \mathbf{Y}_{\nu}^{(\lambda-1,\lambda)}(\theta, \varphi), \tag{1.92}$$

$$\nabla f(r) \mathbf{Y}_{\nu}^{(\lambda,\lambda)}(\theta, \varphi) = 0, \tag{1.93}$$

$$\nabla \times f(r) \mathbf{Y}_{\nu}^{(\lambda,\lambda)}(\theta, \varphi) = \frac{[\lambda(2\lambda+3)]^{1/2}}{2\lambda+1} \left(\frac{d}{dr} - \frac{\lambda}{r}\right) f(r) \mathbf{Y}_{\nu}^{(\lambda+1,\lambda)}(\theta, \varphi) -$$

$$- \frac{[(\lambda+1)(2\lambda-1)]^{1/2}}{2\lambda+1} \left(\frac{d}{dr} + \frac{\lambda+1}{r}\right) f(r) \mathbf{Y}_{\nu}^{(\lambda-1,\lambda)}(\theta, \varphi). \tag{1.94}$$

Expansion of Field Potentials in Spherical Harmonics. The well-known expansion of the Green's function of Poisson's equation

$$\frac{1}{|\mathbf{r}-\mathbf{r}'|} = \sum_{l=0}^{\infty} r_<^l r_>^{-(l+1)} P^{(l)}(\cos\omega) = \sum_l r_<^l r_>^{-(l+1)} (Y^{(l)}(\theta, \varphi) Y^{(l)}(\theta', \varphi')) \tag{1.95}$$

allows one to carry out the expansion of this equation:

$$\varphi(\mathbf{r}) = \int \frac{\rho(\mathbf{r}')}{|\mathbf{r}-\mathbf{r}'|} dv' \tag{1.96}$$

*An exception is the case $\lambda = 0$, when only one function $\mathbf{Y}^{(1,0)}(\theta, \varphi) = \frac{1}{\sqrt{3}} (Y^{(1)}(\theta, \varphi) \cdot \mathbf{e}^{(1)}) = \frac{1}{\sqrt{3}} \frac{\mathbf{r}}{r}$ is possible.

and

$$\mathbf{A}(\mathbf{r}) = \int \frac{\mathbf{j}(\mathbf{r}')}{|\mathbf{r}-\mathbf{r}'|}\, dv', \tag{1.97}$$

where $\rho(\mathbf{r})$ and $\mathbf{j}(\mathbf{r})$ are charge and current densities. The expansion formulas have the form

$$\varphi(r, \theta, \varphi) = \sum_{\lambda\mu} Y_\mu^{(\lambda)*}(\theta, \varphi)\,[r^\lambda q_\mu^{(\lambda)}(r) + r^{-(\lambda+1)} Q_\mu^{(\lambda)}(r)] \tag{1.98}$$

and

$$\mathbf{A}(r, \theta, \varphi) = \sum_{\lambda\nu}\sum_{l} \mathbf{Y}_\nu^{(l,\lambda)*}(\theta, \varphi)\,[r^l m_\nu^{(l,\lambda)}(r) + r^{-(l+1)} M_\nu^{(l,\lambda)}(r)]. \tag{1.99}$$

The expansion coefficients in Eqs. (1.98) and (1.99) are called multipole moments and are defined by the relations

$$q_\mu^{(\lambda)}(r) = \int_r^\infty r'^{-(\lambda+1)} \rho(r', \theta', \varphi')\, Y_\mu^{(\lambda)}(\theta', \varphi')\, dv', \tag{1.100}$$

$$Q_\mu^{(\lambda)}(r) = \int_0^r r'^\lambda \rho(r', \theta', \varphi')\, Y_\mu^{(\lambda)}(\theta', \varphi')\, dv', \tag{1.101}$$

$$m_\nu^{(l,\lambda)}(r) = \int_r^\infty r'^{-(l+1)} \mathbf{j}(r', \theta', \varphi') \cdot \mathbf{Y}_\nu^{(l,\lambda)}(\theta', \varphi')\, dv', \tag{1.102}$$

$$M_\nu^{(l,\lambda)} = \int_0^r r'^l \mathbf{j}(r', \theta', \varphi') \cdot \mathbf{Y}_\nu^{(l,\lambda)}(\theta', \varphi')\, dv'. \tag{1.103}$$

The quantities $q_\mu^\lambda(r)$ and $m_\nu^{(l,\lambda)}(r)$ characterize, respectively, the distribution of charge and current outside a sphere of radius r, and the quantities $Q_\mu^{(\lambda)}(r)$ and $M_\nu^{(l,\lambda)}$, inside this sphere.

Note that if in the expansion (1.99) all quantities $m_\nu^{(l,\lambda)}$ and $M_\nu^{(l,\lambda)}$ with $l = \lambda \pm 1$ are equal to zero, then the vector potential satisfies the Coulomb gauge condition

$$\operatorname{div} \mathbf{A} = 0 \tag{1.104}$$

by virtue of (1.93).

General Properties of Multipole Moments. The multipole moments given by Eqs. (1.100)-(1.103) are self-conjugate irreducible tensors of rank λ, i.e., they have the properties (1.62) and transform according to (1.51) under rotation.

Multipole moments also have a specific parity. The parity of the quantities $q_\nu^{(\lambda)}$ and $Q_\nu^{(\lambda)}$ is determined by the parity of the spherical harmonic $Y_\nu^{(\lambda)}$, which equals $(-1)^\lambda$ [17]. The parity of the quantities $m_\nu^{(l,\lambda)}$ and $M_\nu^{(l,\lambda)}$ equals the product of the parity of the spherical harmonic $Y_m^{(l)}$ and the parity of the current vector, which is 1, i.e., it equals $(-1)^{l+1}$. Hence, the multipoles $m_\nu^{(\lambda,\lambda)}$ and $M_\nu^{(\lambda,\lambda)}$ have parity $(-1)^{\lambda+1}$, and the multipoles $m_\nu^{(\lambda\pm1,\lambda)}$ and $M_\nu^{(\lambda\pm1,\lambda)}$ have the parity $(-1)^\lambda$. The first of these is therefore called magnetic, and the second electric.

4. Irreducible Tensor Operators

The set of $2\lambda + 1$ operators $A_\mu^{(\lambda)}$ ($\mu = -\lambda, -\lambda + 1, \ldots, \lambda$) is called an irreducible tensor operator of rank λ, if under the rotations $D(\psi, \theta, \varphi)$ the operators $A_\mu^{(\lambda)}$ transform according to the law

$$D(\psi, \theta, \varphi)\, A_\mu^{(\lambda)} D^{-1}(\psi, \theta, \varphi) = \sum_{\nu=-\lambda}^{\lambda} A_\nu^{(\lambda)} D_{\nu\mu}^{(\lambda)}(\psi, \theta, \varphi). \tag{1.105}$$

We shall label the components of the operator acting in the original system of coordinates with the index μ and the components of the operator in the rotated system with the index ν. Thus, according to (1.50), we have

$$A^{(\lambda)}_{\mu} = \sum_{\nu} A^{(\lambda)}_{\nu} D^{(\lambda)}_{\nu\mu}(\psi, \theta, \varphi). \tag{1.106}$$

Conjugation and Time Reversal Operators. All the relations presented in Part 2 above remain in force for irreducible tensor operators, with the condition that the operator of complex conjugation is replaced by the operator of Hermitian conjugation. The operator of conjugation of an irreducible tensor operator is defined as a combination of Hermitian conjugation and transition to the contragredient

$$KA^{(\lambda)}_{\mu} = (-1)^{\lambda-\mu} A^{(\lambda)+}_{-\mu}. \tag{1.107}$$

For the reasons stated in Part 2, only irreducible tensor operators of integral rank λ can correspond to physical quantities. The phase factors of such operators can always be chosen conveniently so that they satisfy the self-conjugate condition $KA^{(\lambda)}_{\mu} = A^{(\lambda)}_{\mu}$ or

$$A^{(\lambda)+}_{\mu} = (-1)^{\lambda-\mu} A^{(\lambda)}_{-\mu}, \tag{1.108}$$

since the composition of only such operators is again a self-conjugate operator.

The conjugation operator K is the same, to within a sign, as the weak (Wigner) time reversal operator T [23], so that $K^2 = T^2$ and $K = \pm T$. For self-conjugate tensors the operation of time reversal leads to

$$TA^{(\lambda)}_{\mu} = \pm A^{(\lambda)}_{\mu}, \tag{1.109}$$

and the sign is chosen in accordance with the correspondence principle, i.e., according to whether the sign of the physical quantity changes with time reversal in classical physics. For example, for self-conjugate operators of electrical multipole moments (1.100),

$$Tq^{(\lambda)}_{\mu}T^{-1} = q^{(\lambda)}_{\mu}, \tag{1.110}$$

and for magnetic multipoles (1.102),

$$Tm^{(l,\lambda)}_{\mu}T^{-1} = -m^{(l,\lambda)}_{\mu}. \tag{1.111}$$

Henceforth, we shall use only self-conjugate irreducible tensor operators.

The Wigner-Eckart Theorem. The construction of the matrix element $(jm|A^{(\lambda)}_{\mu}|j'm')$ corresponds to the construction of the invariant of the components from the three irreducible tensors $(jm|$, $A^{(\lambda)}_{\mu}$, and $|j'm')$. The first of these transforms contragrediently with respect to the other two. Hence, the construction of a scalar should be carried out by means of the 3j-symbol (1.69), and we can write

$$(jm|A^{(\lambda)}_{\mu}|jm') = (-1)^{j-m}\begin{pmatrix} j & \lambda & j' \\ -m & \mu & m' \end{pmatrix}(j\|A^{(\lambda)}\|j') = \\ = (-1)^{j-m}\begin{pmatrix} j' & j & \lambda \\ m' & -m & \mu \end{pmatrix}(j\|A^{(\lambda)}\|j'). \tag{1.112}$$

The quantity $(j\|A^{(\lambda)}\|j')$, which is called the reduced matrix element, does not depend on the orientation of the physical system in space, i.e., on the quantum numbers m, μ, and m'. The fundamental relation (1.112), namely the Wigner-Eckart theorem, shows that the process of calculating a matrix element of a tensor operator geometrically amounts to a coupling of momenta $\mathbf{j}' + \boldsymbol{\lambda} = \mathbf{j}$. To emphasize this fact, let us write (1.112) in the form

$$(jm|A^{(\lambda)}_{\mu}|j'm') = \langle j\|A^{(\lambda)}\|j'\rangle (j'\lambda jm|j'm'\lambda\mu), \tag{1.113}$$

where

$$< j \| A^{(\lambda)} \| j' > = (2j+1)^{-1/2} (j \| A^{(\lambda)} \| j'). \tag{1.114}$$

The relation

$$(j \| A^{(\lambda)+} \| j') = (-1)^{\lambda+j'-j} (j' \| A^{(\lambda)} \| j) \tag{1.115}$$

is true for irreducible matrix elements of conjugate operators, as follows from (1.112) and (1.68).

Matrix Elements of the Product of Operators. Consider two irreducible tensor operators $A^{(\lambda_1)}_{\mu_1}$ and $B^{(\lambda_2)}_{\mu_2}$ which act on different eigenvectors $|\alpha_1 j_1 m_1)$ and $|\alpha_2 j_2 m_2)$, which describe two interacting subsystems (a molecule and a nucleus say). Then in the representation $\alpha_1 j_1 \alpha_2 j_2 jm$, where $\mathbf{j} = \mathbf{j}_1 + \mathbf{j}_2$, we have for the operator $[A^{(\lambda_1)} \times B^{(\lambda_2)}]^{(\lambda)}_{\mu}$, according to the Wigner-Eckart theorem,

$$\begin{aligned}(\alpha_1 j_1,\ \alpha_2 j_2,\ jm \,|\, [A^{(\lambda_1)} \times B^{(\lambda_2)}]^{(\lambda)}_{\mu} \,|\, \alpha_1' j_1',\ \alpha_2' j_2',\ j'm') = \\ = (-1)^{j-m} \begin{pmatrix} j & \lambda & j' \\ -m & \mu & m' \end{pmatrix} (\alpha_1 j_1,\ \alpha_2 j_2 j \,\|\, [A^{(\lambda_1)} \times B^{(\lambda_2)}]^{(\lambda)} \,\|\, \alpha_1' j_1',\ \alpha_2' j_2',\ j').\end{aligned} \tag{1.116}$$

The calculation of the given matrix element corresponds to the coupling

$$\mathbf{j}_1' + \mathbf{j}_2' = \mathbf{j}', \quad \boldsymbol{\lambda}_1 + \boldsymbol{\lambda}_2 = \boldsymbol{\lambda}, \quad \mathbf{j}' + \boldsymbol{\lambda} = \mathbf{j}.$$

To separate variables, it is necessary to transform to the coupling scheme

$$\mathbf{j}_1' + \boldsymbol{\lambda}_1 = \mathbf{j}_1, \quad \mathbf{j}_2' + \boldsymbol{\lambda}_2 = \mathbf{j}_2, \quad \mathbf{j}_1 + \mathbf{j}_2 = \mathbf{j}.$$

Consequently, the calculation of the matrix element involved in (1.113) amounts to a transition between these two coupling schemes, and we can write

$$\begin{aligned}< (\alpha_1 j_1,\ \alpha_2 j_2)\, j \,\|\, [A^{(\lambda_1)} \times B^{(\lambda_2)}]^{(\lambda)} \,\|\, (\alpha_1' j_1',\ \alpha_2' j_2')\, j' > = ((j_1' j_2')\, j', (\lambda_1 \lambda_2))\, \lambda \,|\, (j_1' \lambda_2)\, j_1, \\ (j_2' \lambda_2)\, j_2)^{(j)} < \alpha_1 j_1 \,\|\, A^{(\lambda_1)} \,\|\, \alpha_1' j_1' > < \alpha_2 j_2 \,\|\, B^{(\lambda_2)} \,\|\, \alpha_2' j_2' >.\end{aligned} \tag{1.117}$$

With the aid of (1.74) and (1.114) we obtain, finally,

$$\begin{aligned}(\alpha_1 j_1,\ \alpha_2 j_2,\ j \,\|\, [A^{(\lambda_1)} \times B^{(\lambda_2)}]^{(\lambda)} \,\|\, \alpha_1' j_1',\ \alpha_2' j_2',\ j') = \\ = [(2j+1)(2j'+1)(2\lambda+1)]^{1/2} \begin{Bmatrix} j_1 & j_2 & j \\ j_1' & j_2' & j' \\ \lambda_1 & \lambda_2 & \lambda \end{Bmatrix} (\alpha_1 j_1 \,\|\, A^{(\lambda_1)} \,\|\, \alpha_1' j_1')\, (\alpha_2 j_2 \,\|\, B^{(\lambda_2)} \,\|\, \alpha_2' j_2').\end{aligned} \tag{1.118}$$

We shall use this general formula to calculate spin-spin interaction.

The hyperfine interaction energy is a scalar and is expressed as the scalar product of irreducible tensors (1.59). Setting, in (1.118), $\lambda = 0$ ($\lambda_1 = \lambda_2 = \lambda$) and making use of (1.65) and (1.75), we obtain the important formula (Racah, 1942),

$$\begin{aligned}(\alpha_1 j_1,\ \alpha_2 j_2,\ j \,|\, (A^{(\lambda)} B^{(\lambda)}) \,|\, \alpha_1' j_1',\ \alpha_2' j_2',\ j) = (-1)^{\lambda+j+j_1+j_2} \times \\ \times \begin{Bmatrix} \lambda & j_1 & j_1' \\ j & j_2' & j_2 \end{Bmatrix} (\alpha_1 j_1 \,\|\, A^{(\lambda)} \,\|\, \alpha_1' j_1')\, (\alpha_2 j_2 \,\|\, B^{(\lambda)} \,\|\, \alpha_2' j_2').\end{aligned} \tag{1.119}$$

Setting $B^{(\lambda_2)}_{\mu_2} \equiv 1$ ($\lambda_2 = 0$) in (1.118), we obtain the useful formula

$$(\alpha_1 j_1, \alpha_2 j_2, j \| A^{(\lambda)} \| \alpha_1' j_1', \alpha_2 j_2, j') = (-1)^{j_1+j_2+j'+\lambda} \times$$
$$\times [(2j+1)(2j'+1)]^{1/2} \begin{Bmatrix} \lambda & j_1 & j_1' \\ j_2 & j' & j \end{Bmatrix} (\alpha_1 j_1 \| A^{(\lambda)} \| \alpha_1' j_1'). \qquad (1.120)$$

Similarly,

$$(\alpha_1 j_1, \alpha_2 j_2, j \| B^{(\lambda)} \| \alpha_1 j_1, \alpha_2' j_2', j') = (-1)^{j_1+j_2'+j+\lambda} \times$$
$$\times [(2j+1)(2j'+1)]^{1/2} \times \begin{Bmatrix} \lambda & j_2 & j_2' \\ j_1 & j' & j \end{Bmatrix} (\alpha_2 j_2 \| B^{(\lambda)} \| \alpha_2' j_2'). \qquad (1.121)$$

We will encounter cases in which both operators $A^{(\lambda_1)}_{\mu_1}$ and $B^{(\lambda_2)}_{\mu_2}$ act not on different, but on the same state vector $|\alpha jm)$. For this case we have

$$< \alpha j \| [A^{(\lambda_1)} B^{(\lambda_2)}]^{(\lambda)} \| \alpha' j' > = \sum_{\alpha'', j''} (j_1' (\lambda_1 \lambda_2) \lambda | (j' \lambda_2) j'', \lambda_1)^{(j)} \times$$
$$\times < \alpha j \| A^{(\lambda_1)} \| \alpha'' j'' > < \alpha'' j'' \| B^{(\lambda_2)} \| \alpha' j' >, \qquad (1.122)$$

from which, using (1.71) and (1.114), we find

$$(\alpha j \| [A^{(\lambda_1)} \times B^{(\lambda_2)}]^{(\lambda)} \| \alpha' j') = (-1)^{\lambda+j+j_1+j_2'} \times$$
$$\times \begin{Bmatrix} \lambda & j_1 & j_1' \\ j & j_2' & j_2 \end{Bmatrix} (\alpha_1 j_1 \| A^{(\lambda)} \| \alpha_1' j_2')(\alpha_2 j_1 \| B^{(\lambda)} \| \alpha_2' j_2'). \qquad (1.123)$$

Section 2

CALCULATION OF THE MATRIX ELEMENTS OF THE ENERGY OF INTERACTION OF SEVERAL NUCLEI

The equations of Part 3 of Section 1 allow the direct calculation of the matrix elements of the interaction of two systems that have well-defined momenta j_1 and j_2 (for example, molecules and nuclei). However, the usual hfs of molecular spectra is caused by the interaction of several nuclei in the molecule. In this connection, special attention will be given in this section to the application of the apparatus set forth in the preceding section to the calculation of the matrix elements of the energy of several interacting nuclei.

1. Several Interacting Nuclei in the Molecule

Interactions Differing in Magnitude. Imagine a hfs caused by the interaction of n different nuclei with spins $j_1, j_2, \ldots, j_n$ with the field of a molecule in the state $|\alpha J)$, where J is the angular momentum of the molecule and α is the set of remaining quantum numbers. The energy of the hyperfine interaction can be represented in the form of a sum of scalar products of irreducible tensor operators of the form $(A^{(\lambda)}(J) \cdot B^{(\lambda)}(j_k))$, where $B^{(\lambda)}(j_k)$ is the multipole moment of the k-th nucleus and $A^{(\lambda)}(J)$ is the corresponding entity for the molecule.

If the coupling energies differ significantly for the different nuclei, then the most convenient representation is naturally $\alpha J_1 j_1 F_1, j_2 F_2, \ldots, j_n F$, which corresponds to the coupling scheme

$$\mathbf{J} + \mathbf{j}_1 = \mathbf{F}_1;\ \mathbf{F}_1 + \mathbf{j}_2 = \mathbf{F}_2;\ \mathbf{F}_{n-1} + \mathbf{j}_n = \mathbf{F}, \qquad (2.1)$$

where the nuclear spins are enumerated in the order of decreasing coupling energy with the molecular field. Then the principal part of the hfs energy is given by the matrix element diagonal in this representation, which corresponds to first-order perturbation theory.

The matrix element of the interaction energy of the k-th nucleus does not depend on the quantum numbers standing after the quantum number F_k in the sequence $\alpha J j_1 F_1, \ldots, F_{k-1} j_k, \ldots, F$. According to (1.119),

$$(\alpha J \ldots F_{k-1} j_k F \mid (A^{(\lambda)}(y) \cdot B^{(\lambda)}(j_k)) \mid \alpha' Y' \ldots F'_{k-1} j_k F_k) =$$
$$= (-1)^{F'_{k-1}+j_k+F_k+\lambda} \begin{Bmatrix} \lambda & F_{k-1} & F'_{k-1} \\ F_k & j_k & j_k \end{Bmatrix} (\alpha J \ldots F_{k-1} \| A^{(\lambda)}(J) \| \alpha' J' \ldots F'_{k-1}) \times$$
$$\times (j_k \| B^{(\lambda)}(j_k) \| j_k). \tag{2.2}$$

The matrix element of the operator $A^{(\lambda)}(J)$ is calculated by the successive application of Eq. (1.120). For example,

$$(\alpha J \ldots F_{k-2} j_{k-1} F_{k-1} \| A^{(\lambda)}(J) \| \alpha' J' \ldots F'_{k-2} j_{k-1} F'_{k-1}) = (-1)^{F_{k-2}+j_{k-1}+F_{k-1}+\lambda} \times$$
$$\times [(2F_{k-1}+1)(2F'_{k-1}+1)]^{1/2} \begin{Bmatrix} \lambda & F_{k-2} & F'_{k-2} \\ j_{k-1} & F'_{k-1} & F_{k-1} \end{Bmatrix} \times$$
$$\times (\alpha J \ldots j_{k-2} F_{k-2} \| A^{(\lambda)}(J) \| \alpha' J' \ldots j_{k-2} F'_{k-2}). \tag{2.3}$$

This process can be continued until the matrix element produced is not expressed by "proper" quantum numbers, i.e., by the quantity $(\alpha J \| A^{(\lambda)}(J) \| \alpha' J')$.

Interactions Comparable in Magnitude. If the coupling energies of the individual nuclei are of the same order of magnitude, the eigenvalues of energy can be calculated approximately in the coupling scheme described by diagonalizing over all intermediate momenta F_k. However, this coupling scheme does not present any advantage compared to any other in which other momenta are given as intermediate quantum numbers. In a number of cases it is more convenient to use a more symmetrical coupling scheme, where the quantum numbers of the moments

$$\mathbf{j}_1 + \mathbf{j}_2 = \mathbf{I}_2; \ \mathbf{I}_2 + \mathbf{j}_3 = \mathbf{I}_3; \ \ldots \mathbf{I}_{n-1} + \mathbf{j}_n = \mathbf{I}; \ \mathbf{J} + \mathbf{I} = \mathbf{F} \tag{2.4}$$

are given. In the corresponding representation the matrix element of the interaction energy of the k-th nucleus is calculated from the formula (1.119):

$$(\alpha J j_1 j_2 I_2 \ldots I F \mid (A^{(\lambda)}(J) \cdot B^{(\lambda)}(j_k)) \mid \alpha' J' j_1 j_2 I'_2 \ldots I' F) =$$
$$= (-1)^{J'+I+F+\lambda} \begin{Bmatrix} \lambda & J & J' \\ F & I' & I \end{Bmatrix} (\alpha J \| A^{(\lambda)}(J) \| \alpha' J') \times$$
$$\times (I_{k-1} j_k I_k \ldots I \| B^{(\lambda)}(j_k) \| I'_{k-1} j_k I_k). \tag{2.5}$$

By repeated application of (1.121) the last matrix element is brought to the form

$$(I_{k-1} j_k I_k, \ldots, I \| B^{(\lambda)}(j_k) \| I_{k-1} j_k, \ldots, I) =$$
$$= f(I_{k-1}; j_{k+1}, j_{k+2}, \ldots, j_n; I_{k+1} I'_{k+1}, \ldots, II')(j_k \| B^{(\lambda)}(j_k) \| j_k). \tag{2.6}$$

The next stage in calculating the energy eigenvalues amounts to diagonalizing the interaction energy matrix for all the nuclei over the intermediate momenta I_k.

2. Construction of Symmetric Wave Functions

Statement of the Problem. The procedure described above for calculating the hyperfine interaction energy of several nuclei is extremely tedious if the interaction energies of the individual nuclei with the molecular field are of the same order of magnitude. However, very frequently cases are encountered in which several interacting nuclei are identical and take equivalent positions in a symmetrical molecule or in a symmetrical molecular complex in the crystalline lattice of a solid. In this case, the calculation procedure can be significantly simplified by choosing corresponding wave functions in the first-order perturbation theory.

Since the interaction Hamiltonian does not change when identical nuclei are permuted, it should be diagonal in the representation that is characteristic of the specific type of permutation symmetry of the states of identical nuclei.

Consequently, the problem amounts to discovering the matrix of the transformation from the representation given by the set of nuclear momenta $j_1, j_2, \ldots, j_n$ and the intermediate momenta $I_2, I_3, \ldots, I_n$ to the representation characterized by the specific type of symmetry with respect to the permutations of n identical nuclei. The latter representation can be denoted by $j^n \alpha_s(\beta) I$, where j is the spin of each of the identical nuclei, α_s signifies the type of irreducible representation of the permutation group, where s labels the row of this representation if it is not one-dimensional [22], I is the total spin moment of the identical nuclei, and β represents the other possible quantum numbers.

If the transformation to this representation is found, then the only quantum number with respect to which it is necessary to diagonalize is the total spin I, since the characters of the irreducible representations α_s are integrals of the motion [21]. An additional simplification results from the fact that the quantum number I can take on limited values for a given α_s.

Genealogical Coefficients. This problem is solved for one important and frequent case, namely, for the transition to an antisymmetric representation of the permutation group on n identical particles. The elements of the corresponding transformation are known as the genealogical coefficients (coefficients of fractional parentage) (see Sec. 2, Chap. 13, of [14]).

TABLE I

Representation	Group elements					
	E	(123)	(132)	(23)	(31)	(12)
A_1	1	1	1	1	1	1
A_2	1	1	1	—1	—1	—1
E	$\begin{pmatrix}1&0\\0&1\end{pmatrix}$	$\begin{pmatrix}-\frac{1}{2}&-\frac{\sqrt{3}}{2}\\ \frac{\sqrt{3}}{2}&-\frac{1}{2}\end{pmatrix}$	$\begin{pmatrix}-\frac{1}{2}&\frac{\sqrt{3}}{2}\\ -\frac{\sqrt{3}}{2}&-\frac{1}{2}\end{pmatrix}$	—1 $\begin{pmatrix}-1&0\\0&1\end{pmatrix}$	$\begin{pmatrix}\frac{1}{2}&\frac{\sqrt{3}}{2}\\ \frac{\sqrt{3}}{2}&-\frac{1}{2}\end{pmatrix}$	$\begin{pmatrix}\frac{1}{2}&-\frac{\sqrt{3}}{2}\\ -\frac{\sqrt{3}}{2}&-\frac{1}{2}\end{pmatrix}$

In our case we are interested not only in the transformation matrix to an antisymmetric representation, but also to all possible irreducible representations. The reason for this is that the spin wave function of the nuclei constitutes only part (entering as a factor) of the total wave function of the molecule, whereas the condition of complying with Bose or Fermi statistics with respect to permutations of identical particles must be applied to the total wave function of the molecule.

The physical reason for the fact that the spin wave functions of identical nuclei can obey arbitrary statistics consists, in the end, in the fact that the spatial part of nuclear wave functions in a molecule do not overlap. Having in mind the generalization to the case of arbitrary irreducible representations, which are not necessarily antisymmetric, we may call the elements of the matrix we are seeking generalized genealogical coefficients. The method of calculation of genealogical coefficients set forth in Chapter 13 of [14] can be generalized in an obvious way to the case of an arbitrary irreducible representation.

We also note that the calculation of the generalized genealogical coefficients amounts to the same thing as the classical problem of group theory – the discovery of the bases of irreducible representations of a group in a given vector space. In the present case the problem is one of finding the bases of the irreducible representations of the permutation group in the 2I + 1 dimensional space of the irreducible representation of a rotation group characterized by the total spin momentum I.

The Case of Three Identical Nuclei. We limit ourselves to a consideration of the simplest nontrivial case – three identical nuclei.* At the same time, this case is the most representative one for the molecules containing

*When the number of particles is increased, the complexities of the problem are rapidly intensified. If the generalized genealogical coefficients for three particles are expressed in terms of 6j-symbols, then for four particles

identical nuclei that are investigated in gaseous radiospectroscopy.

The permutation group on three identical nuclei is isomorphic to the corresponding point groups of molecular symmetry C_{3v} or D_3 [22]. These groups have three irreducible representations: totally symmetric A_1, antisymmetric A_2, and the two-dimensional representation E. The matrices representing the elements of this group can be chosen to be real (see Table I).

As a beginning, we choose the representation in which the intermediate momentum **L** is the sum of the spins of the nuclei taking the places 2 and 3. The resultant momentum **I** is obtained by coupling the spin of the nucleus at 1 with the momentum **L**, i.e., $\mathbf{I} = \mathbf{j} + \mathbf{L}$. We will denote the corresponding state vector by $|j, j^2(L)I)$, omitting the nonessential index of the projection of **I** on the z axis. The intermediate momentum takes on integral values lying in the interval

$$|I-j| \leqslant L \leqslant \min\{2j,\ I+j\}. \tag{2.6'}$$

The vectors $|j, j^2(L)I)$ are orthonormal, i.e.,

$$(j,\ j^2(L)I\,|\,j,\ j^2(L')I') = \delta_{LL'}\cdot\delta_{II'}. \tag{2.7}$$

We denote the representation sought by $j^3\alpha_s I$. The index α_s takes the values A_1, A_2, $E_{1,2}$, where E_1 and E_2 refer, respectively, to the first and second rows of the representation E. The transformation matrix will be denoted by $(j, j^2(L)I\,|\,j^3\alpha_s(L)I)$.

Permutation Operators. The action of the permutation operator on the state vector $|j, j^2(L)I)$ amounts to a change in the order of coupling the separate spins. For example, the effect of the permutation (123) is

$$\begin{aligned} P(123)\,|\,j_1,\ (j_2j_3)L_1I) &= |\,j_3,\ (j_1j_2)LI) = \\ &= \sum_{L'} |\,j_1, (j_2j_3)L'I)(j_1, (j_2j_3)L'\,|\,j_3, (j_1j_2)L)^{(I)}. \end{aligned} \tag{2.8}$$

Using the well-known formula from representation theory [21],

$$F = \sum_{a,b} |\,a)(a\,|\,F\,|\,b)(b\,|, \tag{2.9}$$

we can write

$$P_I(123) = \sum_{LL'} |\,j,\ j^2(L)I)(L\,|\,P(123)\,|\,L')^{(I)}(j,\ j^2(L')I\,|, \tag{2.10}$$

where

$$(L\,|\,P(123)\,|\,L')^{(I)} = (j_1,\ (j_2j_3)L\,|\,j_3,\ (j_1j_2)L)^{(I)}. \tag{2.11}$$

The last matrix is expressed in terms of 6j-symbols,

$$\begin{aligned} (j_1,\ (j_2j_3)L\,|\,j_3,\ (j_1j_2)L')^{(I)} &= (-1)^{j_3+L'-I}(j_1,\ (j_2j_3)L\,|\,(j_1j_3)L',\ j_3)^{(I)} = \\ &= (-1)^{j_1+j_2+2j_3+L'}\,[(2L'+1)(2L+1)]^{1/2}\begin{Bmatrix} j_1 & j_2 & L' \\ j_3 & I & L \end{Bmatrix}. \end{aligned} \tag{2.12}$$

Since all the values of the nuclear spins are the same, then, setting $j_1 = j_2 = j_3 = j$, we obtain, finally,

---

they must be expressed in terms of 9j-symbols, and for six particles by 15j-symbols. Generalization of the recurrence formulas of Redmond [24] to the case of multidimensional representations leads also to extremely cumbersome expressions.

$$(L \mid P(123) \mid L')^{(I)} = (-1)^{L'} [(2L+1)(2L'+1)]^{1/2} \begin{Bmatrix} j & j & L \\ j & I & L' \end{Bmatrix}. \quad (2.13)$$

Expressions for the other permutation operators are obtained in similar fashion.

Projection Operators. Table I gives the matrices of the irreducible representations $P^{(\alpha)}_{s's}(g)$ (g runs over the group elements), which possess the property of irreducibility. This is expressed by the relation [22]

$$\sum_g P^{(\alpha_1)*}_{s_1' s_1}(g) P^{(\alpha_2)}_{s_2' s_2}(g) = \frac{G}{f_{\alpha_1}} \delta_{\alpha_1\alpha_2} \delta_{s_1 s_2} \delta_{s_1' s_2'}, \quad (2.14)$$

where f_{α_1} is the dimension of the α-th irreducible representation, and the summation is carried out over all elements g of the group, the total number of which equals G. In our case, $G = 6$, $f_{A_1} = f_{A_2} = 1$, $f_E = 2$. With the aid of the matrices of the irreducible representation we construct the projection operators

$$P^{(\alpha)}_s(I) = \frac{f_\alpha}{G} \sum_g P^{(\alpha)}_{s's}(g) P_I(g), \quad (2.15)$$

where $P_I(g)$ is defined by equations like (2.10). The action of the projection operator $P^{(\alpha)}_s(I)$ on an arbitrary state vector $|j, j^2(L)I)$ results in either zero or leads to the formation of a vector (generally speaking, unnormalized) that transforms according to the s-th row of the irreducible representation $P^{(\alpha)}$, i.e., according to the law

$$P(g) \left| j^3\alpha_s I\right) = \sum_{s'} \left| j^3\alpha_{s'} I\right) P^{(\alpha)}_{s's}(g). \quad (2.16)$$

The operator $P^{(\alpha)}_s(I)$ can be represented in the form

$$P^{(\alpha)}_s(I) = \sum_{L,L'} |j, j^2(L)I)(L \mid P^{(\alpha)}_s(I) \mid L')(j, j^2(L')I|, \quad (2.17)$$

where

$$(L \mid P^{(\alpha)}_s(I) \mid L) = \frac{f_\alpha}{G} \sum_g P^{(\alpha)*}_{ss}(g) (L \mid P(g) \mid L)^{(I)}. \quad (2.18)$$

By virtue of (2.14) a second application of the projection operator to the vector $|j, j^2(L)I)$ does not give a new result. In this sense we can write

$$[P^{(\alpha)}_s(I)]^2 = P^{(\alpha)}_s(I). \quad (2.19)$$

Vectors of Symmetric States. The desired normalized vectors have the form

$$|j^3\alpha_s(L_0)I) = [N(\alpha_s I;\ L_0)]^{-1/2} P^{(\alpha)}_s(I) \mid j, j^2(L_0)I). \quad (2.20)$$

Since the states $|j, j^2(L_0)I)$ are orthonormal, the normalization factors are easily calculated from (2.19):

$$N(\alpha_s I;\ L_0) = \sum_L (L_0 \mid P^{(\alpha)}_s(I) \mid L)(L \mid P^{(\alpha)}_s(I) \mid L_0) = (L_0 \mid P^{(\alpha)}_s \mid L_0). \quad (2.21)$$

Consequently, the unitary matrix that transforms to the representation $j^3\alpha_s I$ (genealogical coefficients) is

$$(j, j^2(L)I \mid j^3\alpha_s(L_0)I) = \frac{(L \mid P^{(\alpha)}_s(I) \mid L_0)}{[(L_0 \mid P^{(\alpha)}_s(I) \mid L_0)]^{1/2}}, \quad (2.22)$$

and we can write

$$\left| j^3\alpha_s(L_0)I\right) = \sum_L \left| j, j^2(L)I\right) \frac{(L \mid P^{(\alpha)}_s(I) \mid L_0)}{[(L_0 \mid P^{(\alpha)}_s(I) \mid L_0)]^{1/2}}. \quad (2.23)$$

Multiplicity of Representations. As follows from (2.23), the vectors obtained, generally speaking, depend on the value of the initial intermediate momentum L_0. The number of such different linearly independent systems of vectors which can be obtained from different L_0 equals the multiplicity of the irreducible representation α in the vector space with a given value of I. This number equals [22]

$$m_I^{(\alpha)} = \frac{1}{G}\sum_g P^{(\alpha)}_{s's}(g)\sum_L (L\,|\,P(g)\,|\,L). \tag{2.24}$$

If $m_I^{(\alpha)} = 0$, then for a given value of I it is impossible to construct a vector that transforms according to the representation α. If the multiplicity equals 1, the systems of vectors $|\,j^3\alpha_s(L_0)\,I)$ constructed from different L_0 can be differentiated from each other only by a common phase factor, which is nonessential in the calculation of the diagonal (in α_s) matrix elements. These vector systems can be constructed only from such L_0 for which $(L_0\,|\,P_s^{(\alpha)}(I)|\,L_0) \neq 0$.

Multiplicities Greater Than Unity. If the multiplicity of the representation is greater than one, then, as a rule, it is possible to select a set of initial values L_0 such that the systems of vectors $|j^3\alpha_s(L_0)\,I)$ obtained turn out to be orthogonal for different L_0. And if this condition is not fulfilled, then the vector systems produced by different L_0 can always be orthogonalized by the Schmidt process. The scalar products necessary for this are calculated from the formula

$$\begin{aligned}(j^3\alpha_s(L_0)\,I\,|\,j^3\alpha_s(L_0')\,I) &= \sum_L (j^3\alpha_s(L_0)\,I\,|\,j,\,j^2(L)\,I)\cdot(j,\,j^2(L)\,I\,|\,j^3\alpha_s(L_0')\,I) = \\ &= [(L_0\,|\,P_s^{(\alpha)}(I)\,|\,L_0)(L_0'\,|\,P_s^{(\alpha)}(I)\,|\,L_0')]^{-1/2}(L_0\,|\,P_s^{(\alpha)}(I)\,|\,L_0'),\end{aligned} \tag{2.25}$$

which follows directly from (2.23) and (2.19). The different eigenvectors so obtained can be enumerated by additional index β, which takes the values 1, 2, . . . , $m_I^{(\alpha)}$. These vectors will be orthonormal, i.e.,

$$\begin{aligned}&(j^3\alpha_s(\beta)\,I\,|\,j^3\alpha_3'(\beta')\,I) = \\ &= \sum_L (j^3\alpha_s(\beta)\,I\,|\,j,\,j^2(L)\,I)\;(j,\,j^2(L)\;I\,|\,j^3\alpha_{s'}'(\beta')\,I) = \delta_{\alpha\alpha'}\delta_{\beta\beta'}\delta_{ss'}.\end{aligned} \tag{2.26}$$

In some cases a simpler way can be found to solve the system of equations

$$\sum_{L'} (L\,|\,P_s^{(\alpha)}\,|\,L')(j,\,j^2(L')\,I\,|\,j^3\alpha_s(\beta)\,I) = (j,\,j^2(L)\,I\,|\,j^3\alpha_s = (\beta)\,I). \tag{2.27}$$

Then the $m_I^{(\alpha)}$ of the linearly independent solutions of this system, normalized in accordance with (2.26), give the desired genealogical coefficients.

Symmetrical Spin Functions of Three Identical Nuclei. We write below the spin functions for three identical nuclei which transform according to the irreducible representations A_1, A_2, and E of the permutation group, in the representation j, j^2(L) I. For sufficiently small values of j, e.g., 1, $^3/_2$, the multiplicity of all representations is one. Consequently, these wave functions can be calculated from Eq. (2.22), and after normalization only the common phase factor for a given α depends on the starting intermediate momentum L_0. The phase factors can always be chosen real, since the matrices of the irreducible representations $P^{(\alpha)}_{s's}(g)$ and the matrices of the recoupling coefficients are real. The result can be presented as follows:

$$(j,\,j^2(L)\,I\,|\,j^3A_1I) = D(2j,\,L)\,D(L,\,L_0)\,\frac{\delta(L,\,L_0) + 2C\,(j^3I;\,LL_0)}{[3\,(1 + 2C\,(j^3I;\,L_0L_0))]^{1/2}}; \tag{2.28}$$

$$(j,\,j^2(L)\,I\,|\,j^3A_2I) = D(2j+1,\,L)\,D(L,\,L_0)\,\frac{\delta(L,\,L_0) + 2C\,(j^3I;\,LL_0)}{[3\,(1 + 2C\,(j^3I;\,L_0L_0))]^{1/2}}; \tag{2.29}$$

$$(j,\,j^2(L)\,I\,|\,j^3E_1I) = D(2j,\,L)\,D(L,\,L_0)\sqrt{\frac{2}{3}}\cdot\frac{\delta(L,\,L_0) - C\,(j^3I;\,LL_0)}{[1 - C\,(j^3I;\,LL_0)]^{1/2}}; \tag{2.30}$$

$$(j, j^2(L)I \mid j^3E_2I) = D(2j+1, L)\,D(L, L_0)\sqrt{\frac{2}{3}}\,\frac{\delta(L, L_0) - C(j^3I;\ LL_0)}{[1 - C(j^3I;\ LL_0)]^{1/2}}. \tag{2.31}$$

Besides the Kronecker $\delta(n, n')$ we have introduced the symbols

$$D(n, n') = {}^1\!/_2\,[1 + (-1)^{n-n'}] = \begin{cases} 1, \text{ when } & (-1)^n = (-1)^{n'}, \\ 0, \text{ when } & (-1)^n = -(-1)^{n'}, \end{cases} \tag{2.32}$$

and

$$C(j^3I;\ LL') = (-1)^{L'}[(2L+1)(2L'+1)\,J^{1/2}\begin{Bmatrix} jjL \\ jIL' \end{Bmatrix}. \tag{2.33}$$

We have calculated the functions (2.28) through (2.31) for $j = 1, {}^3\!/_2$. The results are tabulated in Appendix B.

3. Calculation of Matrix Elements

Reduction Theorem for the Matrix Elements. The calculation of the matrix elements of the interactions of a group of identical nuclei can be reduced to the calculation of the matrix elements of one (or two, if the interaction is pairwise) of the identical nuclei on the basis of a theorem analogous to the Wigner-Eckart theorem (1.112).

Let us consider the action of the operations of the symmetry group P(g) of the molecule on the wave functions and Hamiltonian of the interaction of identical nuclei. These operations include, in particular, the permutation of the coordinates and spin variables of the identical nuclei. If we represent the interaction Hamiltonian of the first nucleus with the molecular field by $\mathcal{H}_1$ and of the second by $\mathcal{H}_2$, then, to be specific, we can write

$$\mathcal{H}_2 = P(12)\,\mathcal{H}_1 P^{-1}(12), \tag{2.34}$$

where P(12) is the symmetry operation that interchanges nuclei 1 and 2. The Hamiltonian of the interaction of N identical nuclei can be represented in the form

$$\mathcal{H} = \frac{N}{N!}\sum_g P(g)\,\mathcal{H}_n P(g), \tag{2.35}$$

where the summation is over all the elements of the group. As a starting (generating) element $\mathcal{H}_n$ can be chosen the Hamiltonian of any of the identical nuclei, for example, the first $\mathcal{H}_1$. The Hamiltonian (2.35) is invariant under any transformation P(g') of the symmetry group of the molecule:

$$P(g')\,\mathcal{H}\,P^{-1}(g') = \frac{N}{N!}\sum_g P(g'g)\,\mathcal{H}_1 P^{-1}(g'g) = \frac{N}{N!}\sum_{g''} P(g'')\,\mathcal{H}_1 P^{-1}(g''), \tag{2.36}$$

where P(g") = P(g'g) = P(g')P(g). In other words, it transforms according to the fully symmetrical representation of the symmetry group A_1, i.e., it is a scalar of the symmetry group. The matrix elements of the scalar $\mathcal{H}$ differ from zero only between states that transform according to the same representation of the symmetry group of the molecule [22]. We denote the vectors of the states that transform according to the s-th row of the irreducible representation α by $|\psi_s^{(\alpha)}(\gamma)>$, where γ represents the other quantum numbers. Then, to be specific,

$$\begin{aligned} P(g)\,|\,\psi_s^{(\alpha)}(\gamma)> &= \sum_{s'}|\,\psi_{s'}^{(\alpha)}(\gamma)> P_{s's}^{(\alpha)}(g), \\ <\psi_s^{(\alpha)}(\gamma)\,|\,P^{-1}(g) &= \sum_{s'} P_{s's}^{(\alpha)*}(g) <\psi_{s'}^{(\alpha)}(\gamma)\,|, \end{aligned} \tag{2.37}$$

where $P_{s's}^{(\alpha)}(g)$ are the matrices of the irreducible representations of the molecular symmetry group. These matrices satisfy (2.14), in which it is necessary to set G = N!. Using (2.37) and (2.14), we obtain for the matrix elements of the Hamiltonian (2.35)

$$<\psi_s^{(\alpha)}(\gamma)\,|\,\mathcal{H}\,|\,\psi_{s'}^{(\alpha')}(\gamma')> = N\sum_{s_1 s_2} <\psi_{s_1}^{(\alpha)}(\gamma)\,|\,\mathcal{H}_1\,|\,\psi_{s_2}^{(\alpha')}(\gamma')> \times$$

$$\times \frac{1}{N!}\sum_g P_{s_1 s}^{(\alpha)*}(g)\,P_{s_2 s'}^{(\alpha')}(g) = \frac{N}{f_\alpha}\delta_{\alpha\alpha'}\delta_{ss'}\sum_{s'} <\psi_{s_1}^{(\alpha)}(\gamma)\,|\,\mathcal{H}_1\,|\,\psi_{s_2}^{(\alpha')}(\gamma')>. \tag{2.38}$$

In the case of free molecules, the total molecular wave function must transform according to the representation A_1 if the identical nuclei are bosons, or according to the representation A_2 if the nuclei are fermions. Both these representations are one-dimensional ($f_\alpha = 1$) and, consequently, it is not necessary to specify the indices α and s for the total wave function. Hence, for free molecules we may write

$$<\gamma\,|\,\mathcal{H}\,|\,\gamma'> = N <\gamma\,|\,\mathcal{H}_2\,|\,\gamma'>. \tag{2.39}$$

If the molecule (or molecular complex) is bound in a crystalline solid, then, generally speaking, the space parts of the wave functions overlap.

In this case, the molecular state can transform according to an arbitrary representation α and it is necessary to use the general result (2.38) to reduce the matrix elements.

Reduction Coefficients of Single Interactions. We return now to the calculation of the matrix elements of the separate interactions of the identical nuclei, the typical term having the form

$$\mathcal{H}_1^\lambda = (A^{(\lambda)} \cdot B^{(\lambda)}), \tag{2.40}$$

where $B^{(\lambda)}$ is an operator acting only on the quantum numbers of one of the identical nuclei, and $A^{(\lambda)}$ is the molecular field operator acting on the nucleus. After application of the reduction theorem (2.38), the problem comes down to calculation of a matrix element of type (2.7).

In a representation with a given type of spin-state symmetry α_s this matrix element can be reduced to the form $\Delta(\lambda, \alpha_s)(j\|B^{(\lambda)}\|j)$, where the factor $\Delta(\lambda, \alpha_s)$ will be called a reduction coefficient.

Using the wave functions (2.28)-(2.31), we calculate the reduction coefficients for the case of three identical nuclei, which we define in the following way:

$$\Delta(\lambda\alpha_s j;\ II') = (-1)^{I-j}\,3\,\frac{(j^3\alpha_s I\,\|\,B^{(\lambda)}\,\|\,j^3\alpha_s I')}{(j\,\|\,B^{(\lambda)}\,\|\,j)}. \tag{2.41}$$

To calculate the matrix element $(j^3\alpha_3 I\|B^{(\lambda)}\|j^3\alpha_3 I')$, we transform to the representation j, $j^2(L)I$, considering the nucleus to be in the place numbered 1, and using (1.93). Substituting the result in (2.41), we obtain

$$\Delta(\lambda\alpha_3 j;\ II') = 3\cdot(-1)^{\lambda+I+I'}[(2I+1)(2I'+1)]^{1/2}\times$$

$$\times\sum_L(-1)^L(j^3\alpha_s I\,|\,j,\,j^2(L)\,I)\begin{Bmatrix}\lambda I I'\\ L j j\end{Bmatrix}(j,\,j^2(L|\,I\,|\,j^3\alpha_s I'). \tag{2.42}$$

For vector interactions ($\lambda = 1$) the reduction coefficients can be calculated starting from elementary considerations, since the Hamiltonian of vector interactions is diagonal in the total spin I.

It is easy to show that

$$\Delta(1\alpha_s j;\ II') = \delta_{I,I'}\left(\frac{I(I+1)(2I+1)}{j(j+1)(2j+1)}\right]^{1/2}(-1)^{I-j}. \tag{2.43}$$

For interactions of the quadrupole type ($\lambda = 2$) the coefficients $\Delta(2\alpha_s j;\ II')$ have been calculated for the cases j = 1 and $^3/_2$ and can be found in Appendix B.

Reduction Coefficients for Pair Interactions. A somewhat more complicated expression is obtained for the reduction coefficients of pair or spin-spin interactions, containing operators of the form $[j_1^{(1)} \times j_2^{(1)}]_\mu^{(2)}$. We define them as follows:

$$3\,(j^3\alpha_s I \,\|\, [j^{(1)} \times j^{(1)}]^{(2)} \,\|\, j^3\alpha_s I') =$$
$$= (-1)^{j+I+1}\sqrt{5}\,j\,(j+1)\,(2j+1)\,\sigma\,(\alpha_s j;\ II'). \qquad (2.44)$$

This matrix element is calculated, transforming to the j, j^2(L) I representation, using (1.121) and (1.118). First we obtain

$$(j,\, j^2(L)\, I \,\|\, [j^{(1)} \times j^{(1)}]^{(2)} \,\|\, j,\, j^2(L')\, I = (-1)^{j+L'+I+1}\cdot j\,(j+1)\,(2j+1)\times$$
$$\times\, [(2I+1)(2I'+1)(2L+1)(2L'+1)\cdot 5]^{1/2} \begin{Bmatrix} 2LL' \\ jI'I \end{Bmatrix} \cdot \begin{Bmatrix} jjL \\ jjL' \\ 112 \end{Bmatrix}, \qquad (2.45)$$

and then for the coefficient σ in expression (2.44),

$$\sigma\,(\alpha_s j;\ II') = [(2I+1)(2I'+1)]^{1/2} \sum_{LL'} (-1)^L D(L,L')\,[(2L+1)(2L'+1)]^{1/2} \times$$
$$\times\, (j^3\alpha_s I \,|\, j,\, j^2(L)\, I) \begin{Bmatrix} 2II' \\ jL'L \end{Bmatrix} \frac{3\,(2L+1) \begin{Bmatrix} jjL' \\ 1\lambda j \end{Bmatrix} \begin{Bmatrix} jjL \\ 1\lambda j \end{Bmatrix} + (-1)^{2j+L+1} \begin{Bmatrix} Ljj \\ 1jj \end{Bmatrix}}{5(2L+1) \begin{Bmatrix} 2LL' \\ \lambda 11 \end{Bmatrix}} \times$$
$$\times\, (j,\, j^2(L')\, I \,|\, j^3\alpha_s I'). \qquad (2.46)$$

To calculate this expression we used the expression for the 9j-symbol given in Appendix A. The symbol D(L, L'), defined according to (2.32), is introduced in order to simplify the calculation of the selection rule for $\Delta L = L - L'$. The quantity λ entering into (2.46) should be set equal to $\frac{1}{2}$ (L + L'). The numerical values of the coefficients $\sigma(\alpha_s j;\ II')$ for j = 1 and $^3/_2$ are also presented in Appendix B. The practical convenience of the reduction coefficients $\Delta(\lambda\,\alpha_s j;\ II')$ and $\sigma(\alpha_s j;\ II')$ resides, of course, in the fact that they do not depend on the magnitude of the interaction and are in this sense universal.

Section 3

PROPERTIES OF MOLECULAR WAVE FUNCTIONS

The magnitudes of the intramolecular electric and magnetic fields and consequently the hyperfine interactions depend strongly on the electronic structure of the molecule. Although contemporary notions about the electronic state of molecules provides an adequate basis for the theory of the chemical bond, they cannot pretend to be more than a crude qualitative description of the intramolecular fields [25]. Some characteristics of these fields, e.g., the magnitude of the electric field gradient, can be fairly well evaluated on the basis of the semiqualitative theory of Daily-Townes [26]. However, estimates of this nature represent a special problem lying outside the scope of our treatment.

The Hamiltonian of the molecule has a whole series of symmetry properties which specifically limit the wave functions and matrix elements of the hfs operators.

In the case to which we limit our discussion, namely when the electronic state of the molecule is nondegenerate, the symmetry properties lead to a compensation of the molecular magnetic fields in the first order of perturbation theory. This determines the character of the hfs spectrum of such molecules where, in contrast to atoms, the magnetic interactions are two or three orders of magnitude smaller than electric quadrupole interactions. The use of these symmetry properties allows us to consider hyperfine interactions without concrete knowledge about the properties of the electronic structure of the molecule.

Use of the symmetry properties of the rotational states of the molecule turns out to be more successful. In this case, the theory of rotation groups permits the complete calculation of the matrix elements of the hfs operators in the symmetric top representation.

In what follows, we shall analyze the symmetry properties of molecular wave functions from the viewpoint of their influence on the matrix elements of the operators of the intramolecular fields.

1. Electronic States

Adiabatic Approximation. Usually, the different types of molecular motion can be separated in the adiabatic approximation. The total molecular function is represented as a product of wave functions, each describing separately the electronic, vibrational, and rotational state of the molecule and the spin state of the nuclei:

$$\psi = \psi_e \psi_v \psi_r \chi_n. \tag{3.1}$$

For investigating hfs the molecules of most interest are those in the electronic and vibrational ground state, since the population of the excited electronic and vibrational levels is extremely small at the temperatures usually used to obtain sufficiently intense spectra.

The vibrational state of the molecule does not significantly affect the hfs. The ground vibrational state of a molecule usually transforms according to the totally symmetric representation of the symmetry point group of the molecule. Hence, it is of no consequence in the consideration of the symmetry properties of the total wave function of the molecule.

The rotational state will be considered specially. As for the nuclear spin states, they were treated in the preceding chapter. We turn to a more thorough consideration of the electronic state.

LS-Coupling. The principal contribution to the energy of the electronic state comes from the Coulomb interaction of the electrons with the nuclei and with each other, whereas the magnetic interactions of the electrons play an exceedingly small part. Thus, for molecules, especially those containing light and medium atoms, the Russel-Saunders approximation [11] is valid to a high degree; according to this the electron spins couple independently of their orbital state. This is the basis for saying that operators not involving electron spins act only on the quantum numbers of the orbital states, and the spin operators act only on electron spin functions.

Usually, complete compensation of all electronic spin moments is required for the formation of a stable chemical bond. In the LS-coupling approximation the ground state of such a molecule is a singlet, i.e., the total spin of the electrons **S** equals zero. Henceforth we consider only singlet states. Any interaction depending on electron spins is zero in a singlet state, since the Russel-Saunders approximation is fulfilled.

Parity. It is known [27] that nuclei in the ground state have a definite parity, and this imposes a limitation on the expectation values of the nuclear multipole moment operators. That is to say, in the ground state the electric multipoles only of even parity (charge, quadrupole moment) and the magnetic multipoles only of odd parity (dipole, octopole moments) can be different from zero. Since, in the Hamiltonian, the hfs interactions enter as products of molecular multipoles and nuclear multipoles of the same parity, we cannot consider molecular multipoles (e.g., molecular electric dipoles or magnetic quadrupoles) for which the corresponding nuclear multipoles are always zero.

At the same time, it is interesting to note that the Hamiltonian of the molecule is invariant with respect to the parity operation, as is the Hamiltonian of nuclei and atoms. However, in contrast with the latter, the overwhelming majority of molecules have dipole electric moments – a consequence of extreme importance for molecular spectroscopy.

The theorem according to which the mean values of the odd electric and the even magnetic multipoles of a quantum mechanical system should be zero [27] is true only for nondegenerate levels (and in a coordinate system having an inversion center). This means that for molecules having average electric dipole moments the conditions of the theorem are not fulfilled, and in fact all the energy levels of such molecules are doubly degenerate. Usually the inversion doubling in molecules is of no consequence except for one case – the ammonia molecule. Since this molecule is of direct interest, we shall consider this question more thoroughly.

The invariance of the total Hamiltonian of the molecule relative to the inversion operation I means, as is well known, the equivalence of the description of the molecule in left- and right-handed coordinate systems. Strictly speaking, the effect of the operation I on the molecular wave function involves the simultaneous change of the signs of the coordinates of all particles, i.e., the nuclei and electrons, as well as a corresponding transformation of their spin wave functions. We see at once that the internal parity of the nuclei and electrons is immaterial in considering the inversion state of the molecule. This is because the physical results depend only on the Hermitian products of the spin wave functions $\chi_i^* \chi_i$, since the nuclei and electrons can be considered as point particles. In virtue of this we can state formally that the parity of the spin states equals +1.

Further, the inversion operation, defined as a simultaneous change of the signs of all coordinates $\mathbf{r}_i \rightarrow -\mathbf{r}_i$, commutes with the rotations and, since the rotational and vibrational states can be separated, we can consider that the inversion operation does not change the wave functions of the molecular rotation states. Thus, it makes sense to consider the inversion only in relation to the electronic-vibrational states of the molecule.

The invariance of the electron-vibrational Hamiltonian relative to inversion obliges us to consider together the usual configuration space of the molecule and the inverted configuration space obtained by inversion relative to an inversion center lying in the molecule. Both these spaces form only one configuration space, separated, as a rule, by an infinitely high potential barrier that precludes a spontaneous transition of the molecule from the left-handed configuration to the right-handed one, and vice-versa. In this case, the two electron-vibrational eigenfunctions of the Hamiltonian and the inversion operator that are considered in the total configuration space – symmetric and antisymmetric – correspond to the same energy level, and we have double inversion degeneracy, rendering inapplicable the aforementioned theory about the zero average electric dipole moment of a molecule.

There are two exceptions to the situation described. The first of these are molecules having a center of symmetry, e.g., H_2 or CH_4. In such molecules the inverted configuration is the same as the initial one, and inversion doubling is absent. But in this case the theorem is still applicable, and the mean value of the electric dipole moment is zero.

The second exception is the case of NH_3 or ND_3 molecules. Here the potential barrier separating the mutually inverted halves of the configuration space is finite, and the probability of a spontaneous transition of the molecule from the left-handed configuration to the right-handed one, and vice versa, is nonzero. As a result, the inversion degeneracy is lifted, and the well-known splitting occurs [1]. The two inversion states (symmetric and antisymmetric) become nondegenerate, and in each of them the mean value of the molecular electric dipole moment equals zero because of the applicability of the aforementioned theorem. We note that any internal interaction that is invariant under inversion, among them the hyperfine interaction, should be diagonal with respect to the parity of the inversion states. Hence, the hfs must be considered separately for even and odd inversion states. Since the electronic-vibrational wave functions of these states are different, then also the hfs constants for even and odd states can, in general, be different, as is observed in the case of NH_3 (see Section 5). On the other hand, interactions with the external electric fields proceed only with a change in the parity of the inversion state (accompanied by a change in parity of the photon), and this leads to the appearance of a rotational-inversion spectrum in ammonia [1]. The specific parity of the electronic-vibrational states of ammonia should be kept in mind in considering the symmetry properties of the total molecular wave function. This question is considered in the last part of this section.

Time Reversal and the Symmetry of Molecular Terms. In the absence of a magnetic field, the Hamiltonian of the molecule is invariant under time reversal. This means that in a nondegenerate state the mean value of the internal magnetic field created by motion of the electrons vanishes.

Let us consider the operation of time reversal T applied to a molecule more thoroughly. The application of T to a state vector $|\psi\rangle$ amounts to Hermitian conjugation and a certain unitary transformation U, so that [23]

$$T|\psi\rangle = \langle\psi'| = \langle\psi|U^{-1}, \tag{3.2}$$

*This theorem also applies to internal multipole moments of the molecule, but only at the point relative to which inversion is produced, viz., the inversion center of the molecule. The consequences of this are trivial. For example, the dipole electric derivative (electrical field gradient) is zero at the inversion center, and upon inversion of the molecule the nitrogen atom occupies a position of unstable equilibrium in the plane of the hydrogens.

$$<\psi|T^{-1} = <\psi|T^{+} = |\psi'> = U|\psi>. \tag{3.3}$$

Since the operator T involves Hermitian conjugation, it is antilinear, i.e., if C is some number, then

$$TC|\psi> = C^{*}T|\psi>. \tag{3.4}$$

The matrices of the irreducible representation of the symmetry group of the Hamiltonian $P^{(\alpha)}_{s's}(g)$, according to which the eigenvectors $|\psi^{(\alpha)}_{s}>$ transform,

$$P(g)|\psi^{(\alpha)}_{s}> = \sum_{s'} |\psi^{(\alpha)}_{s'}> P^{(\alpha)}_{s's}(g), \tag{3.5}$$

transform under time reversal to their complex conjugates by (3.2):

$$TP(g)T^{-1}(T|\psi^{(\alpha)}_{s}>) = \sum_{s'} (T|\psi^{(\alpha)}_{s'}>) P^{(\alpha)*}_{s's}(g). \tag{3.6}$$

Following Wigner [28], we distinguish three cases:

$$\sum_{g}\sum_{s} P^{(\alpha)}_{ss}(g^{2}) = \begin{cases} N & \text{case (a)} \\ 0 & \text{" (b)} \\ -N & \text{" (c)} \end{cases} \tag{3.7}$$

In case (b) the representations acting in the initial and the time-reversed schemes, i.e., $P^{(\alpha)}_{s's}(g)$ and $P^{(\alpha)*}_{s's}(g)$, are inequivalent. Because of this, the bases of the representations $|\psi^{(\alpha)}_{s}>$ and $T|\psi^{(\alpha)}_{s}>$ are linearly independent, which leads to degeneracy of the corresponding term E_{α}. In cases (a) and (c) the matrices $P^{(\alpha)}_{s's}$ and $P^{(\alpha)*}_{s's}$ are equivalent, but the difference consists in that, in case (a), the matrices $P^{(\alpha)}_{s's}$ can be brought into real form, and in case (c) they cannot.

Actually, for symmetry point groups, case (c) for molecules with an even number of electrons and case (a) for molecules with an odd number of electrons are not of physical interest. For systems with an even number of electrons the state vectors $|\psi>$ correspond to integral spin and have the property $T^{2}|\psi> = |\psi>$. According to [14] they transform in this case under rotations according to the representations D_{j}, where j can only be integral. But any representation can be made real [14], so that case (c) is not realized for symmetry point groups in the space of the eigenvectors of molecules with an even number of electrons.

For systems containing an odd number of electrons, we have the property $T^{2}|\psi> = -|\psi>$, demonstrating Kramers degeneracy. However, in this case, the situation is reversed; the vectors transform according to the double-valued representations corresponding to half-integral values of j which, according to [14], cannot be brought into a real form. This eliminates case (a) for systems with an odd number of electrons. Of course, these assertions are valid only for those symmetry groups that contain elements leading to the existence of twofold symmetry axes (to such elements also belong reflections in the plane of symmetry, since this amounts to a pure inversion plus a rotation by π about an axis perpendicular to the plane of symmetry). Consequently, the exceptions are only the groups C_{n}, where n is odd [22], which are of slight interest.

In case (a) for systems with an even number of electrons and case (b) for systems with an odd number of electrons, the vectors $|\psi^{(\alpha)}_{s}>$ and $T|\psi^{(\alpha)}_{s}>$ transform according to the same irreducible representation. However, in systems with an odd number of electrons $|\psi^{(\alpha)}_{s}>$ and $T|\psi^{(\alpha)}_{s}>$ are linearly independent, leading to Kramers degeneracy.

Molecular Magnetic Multipoles. We consider a molecule in a nondegenerate electronic state. It follows from what was said in the preceding part that the eigenfunctions transform according to a one-dimensional representation belonging to the case (a) for a system with integral spin. A typical case where this situation is realized is the ground state of a molecule with an even number of electrons, which transforms according to the totally symmetric representation of the point group of molecular symmetry. This nondegenerate level must be a singlet, since, in this state, the electron spin, which changes sign under time reversal, must be zero. For the same reason the orbital moment of the electrons reverts to zero in a nondegenerate state. (Note that the compensation of the spin and orbital moments separately comes from the condition of fulfillment of the Russel-Saunders approximation.)

If the vector of the nondegenerate state is symbolized by $|n>$, then

$$\left.\begin{array}{l} T\,|\,n> = \alpha_n < n\,| \\ < n\,|\,T^{-1} = \alpha_n^*\,|\,n> \end{array}\right\}, \text{ where } |\,\alpha_n\,|^2 = 1. \tag{3.8}$$

In accordance with (1.111) and (3.6), we obtain for the magnetic multipole matrix element

$$\begin{aligned} < n\,|\,m_\nu^{(l,\lambda)}\,|\,n> &= < n\,|\,T^{-1}\,(T m_\nu^{(l,\lambda)} T^{-1})\,T\,|\,n> = \\ &= -\,|\,\alpha_n\,|^2 < n\,|\,m_\nu^{(l,\lambda)}\,|\,n> = 0. \end{aligned} \tag{3.9}$$

We have come to an important conclusion: the average value of the molecular magnetic multipole created by the electrons in the molecule is zero in a nondegenerate electronic state. The magnetic interactions of the electrons with the nuclei can arise only in second order. In fact, this interaction arises from the excitation of electronic states by the rotation of the molecule.

Rotational-Electronic Interaction. In rotating molecules a magnetic interaction between the electrons and nuclei arises in the second order of perturbation theory as a consequence of the interaction between the electronic and rotational motions [1]. The energy associated with the orbital motion of electrons in a rotating molecule equals

$$-\,\boldsymbol{\omega}\mathbf{L} = -\,B_{ik}J_iL_k, \tag{3.10}$$

where ω is the angular velocity of molecular rotation, J_i are the components of the total angular momentum of the molecule (without taking nuclear spins into account), B_{ik} is a tensor involving the rotational constants of the molecules, so that

$$\omega_k = J_iB_{ik}. \tag{3.11}$$

The operators J_i do not act on the eigenvectors of the electronic state and in relation to them can be considered simply as numbers. The operators L_x and L_y in a symmetric top molecule can cause transitions from the ground $^1\Sigma$-state to a singlet excited $^1\Pi$-state, in correspondence with the selection rule $\Lambda \rightarrow \Lambda' \pm 1$, where Λ is the projection on the axis of symmetry of the molecule. The operator L_z is diagonal in a representation with a given Λ, and, since

$$L_z\,|^1\Sigma> = 0, \tag{3.12}$$

it does not cause transitions to excited states.

Only the first-order correction to the basis vector of the ground $^1\Sigma$-state need be calculated, since the ratio of rotational energy to the electronic energy is 10^{-4} in order of magnitude. The magnetic field created by the electrons is reduced by this ratio in comparison with that ordinarily found in an atom. It is of the same order of magnitude as that generated by the rotation of nuclear charges [1]. If the eigenvectors of the excited $^1\Pi$-states are symbolized by $|n\Pi>$, then in the first order of the electronic-rotational interaction we obtain for the basis vector of the ground state

$$|\,0> = |\,\Sigma> = J_iB_{ik}\Sigma\,\frac{|\,n\Pi><n\Pi\,|\,L_k\,|\,\Sigma>}{E_{n\Pi} - E_\Sigma}, \tag{3.13}$$

where for simplicity we omit the multiplicity indices for the electronic states, since they are all singlets.

With the excitation taken into account, the average value of the multipole moment operator $F_\nu^{(\lambda)}$ is

$$\begin{aligned} \langle F_\nu^{(\lambda)}\rangle &= \langle 0\,|\,F_\nu^{(\lambda)}\,|\,0\rangle = \langle\Sigma\,|\,F_\nu^{(\lambda)}\,|\,\Sigma\rangle + \\ &+ J_iB_{ik}\sum_n (E_{n\Pi} - E_\Sigma)^{-1}\,[\langle\Sigma\,|\,F_\nu^{(\lambda)}\,|\,n\Pi\rangle\langle n\Pi\,|\,L_k\,|\,\Sigma\rangle + \langle\Sigma\,|\,L_k\,|\,n\Pi\rangle\langle n\Pi\,|\,F_\nu^{(\lambda)}\,|\,\Sigma\rangle]. \end{aligned} \tag{3.14}$$

According to (3.9) for the magnetic multipole the first term of this expression becomes zero; as a result, the average value $\langle m_\nu^{(l,\lambda)}\rangle$ turns out to be proportional to the vector **J**.

We introduce the irreducible tensor $(BL)_\mu^{(1)}$ formed from the components of the vector $B_{ik}L_k$,

$$(BL)^{(1)}_{\pm 1} = \frac{1}{\sqrt{2}}(B_{yk}L_k \mp iB_{xk}L_k); \quad (BL)^{(1)}_0 = iB_{zk}L_k, \tag{3.15}$$

and we define the irreducible tensor operators (g-factors) of the molecule

$$G_q^{(\lambda,k)} = \sum_{\mu\nu}(\lambda 1 k q\,|\,\lambda\nu 1\mu)\sum_n (E_{n\Pi} - E_\Sigma)_\lambda^{-1} \times$$
$$\times\,[\langle\Sigma\,|\,m_\nu^{(\lambda,\lambda)}\,|\,n\Pi\rangle\langle n\Pi\,|\,(BL)_\mu^{(2)}\,|\,\Sigma\rangle + \langle\Sigma\,|\,(BL)_\mu^{(2)}\,|\,n\Pi\rangle\langle n\Pi\,|\,m_\nu^{(\lambda,\lambda)}\,|\,\Sigma\rangle]. \tag{3.16}$$

The tensor $G_q^{(\lambda,k)}$ does not change sign on inversion of coordinates, since it is expressed in terms of matrix elements between states with a specified parity. It represents the product of two axial tensors $m_\nu^{(\lambda,\lambda)}$ and $(BL)_\mu^{(1)}$, and hence is a real tensor, whence the rank of the tensor $G_\nu^{(\lambda,k)}$ can take only even values $k = \lambda \pm 1$. $G_\nu^{(\lambda,\lambda-1)}$ can be called a purely isotropic "g-factor," while $G_\nu^{(\lambda,\lambda+1)}$ is purely anisotropic. The average value of the magnetic multipole can be represented in the form of a product of the irreducible tensors $J_\mu^{(1)}$ and $G_q^{(\lambda,\lambda\pm 1)}$:

$$\langle m_\nu^{(\lambda,\lambda)}\rangle = [J^{(1)} \times G^{(\lambda,\lambda+1)}]_\nu^{(\lambda)} + [J^{(1)} \times G^{(\lambda,\lambda-1)}]_\nu^{(\lambda)}. \tag{3.17}$$

The quantities $G_q^{(\lambda,\lambda\pm 1)}$ in the approximation considereddo not depend on the rotational state and, hence, are rotational constants. The total number of independent constants characterizing the 2^λ-pole magnetic interaction is in general $4\lambda + 2$.

In averaging the electric multipole operator, the first term of (3.14) for $\langle q_\nu^{(\lambda)}\rangle$ represents the principal contribution. The second-order correction, which contains matrix elements of the type $\langle\Sigma\,|\,q_\nu^{(\lambda)}\,|\,n\Pi\rangle \cdot \langle n\Pi\,|\,L_k\,|\,\Sigma\rangle$, is zero for the same reason applied in the first-order correction for the magnetic interactions. In fact, the expression

$$\sum_n[\langle\Sigma\,|\,q_\nu^{(\lambda)}\,|\,n\Pi\rangle\langle n\Pi\,|\,L_k\,|\,\Sigma\rangle + \langle\Sigma\,|\,L_k\,|\,n\Pi\rangle\langle n\Pi\,|\,q_\nu^{(\lambda)}\,|\,\Sigma\rangle], \tag{3.18}$$

where the summation is over all degenerate intermediate states if they exist, changes sign upon time reversal, from which it follows that it becomes zero. The consequent difference from zero of the third-order correction is negligibly small. Thus, we can write for the average value of the electric molecular multipole,

$$\langle q_\nu^{(\lambda)}\rangle = \langle\Sigma\,|\,q_\nu^{(\lambda)}\,|\,\Sigma\rangle. \tag{3.19}$$

Similar expressions for the average values of the molecular multipole operators can also be obtained for asymmetric top molecules. We shall not write them out here, since they differ from those presented above only in the labeling of the electronic states.

2. The Rotational State

Rotational Wave Functions. The problem of quantization of rotational motion of molecules has been well studied [1]. However, for the purposes of computation of the matrix elements of the molecular multipole operators and determination of the independent hfs constants, it is desirable to turn once again to a consideration of the transformation properties of rotational wave functions.

The vector of the electronic vibrational and rotational states of the molecule are written $|\,nvE_{J\tau}M\rangle$, where n and v are indices for the electronic and vibrational states, and $E_{J\tau}$ is the rotational energy of the molecule having a total angular momentum **J**, and M is the projection of **J** on the laboratory axis. It is unnecessary for us to consider the deviation from adiabaticity of the electronic and rotational motions, since the dependence of the electronic state on the rotational state, which is represented by equations of the type (3.13), is already taken into account in averaging the molecular field operators and is contained in Eq. (3.14). Therefore, we can consider that the eigenvector of the rotational state $|E_{J\tau}M)$ is independent of the electronic and vibrational states.

In the coordinates of the representation given by the Euler angles φ, θ, ψ, the vector $|E_{J\tau}M)$ can be expressed in the form

$$(\varphi, \theta, \psi \mid E_{J\tau}M) = \sum_{k=-J}^{J} (\varphi\theta\psi \mid JKM)(JK \mid E_{J\tau}), \tag{3.20}$$

where $(\varphi\theta\psi|JKM)$ is the matrix transforming from the coordinate representation to the symmetric top representation JKM (1.43), which was considered in detail in Section 1.

The wave functions $(JK|E_{J\tau})$ must diagonalize the Hamiltonian for the rotational motion of the molecule,

$$\mathcal{H}_r = {}^1/_2 [B_{x'x}J_{x'}^2 + B_{y'y'}J_{y'}^2 + B_{z'z'}J_z^2], \tag{3.21}$$

which is written in a coordinate system fixed to the nuclear framework of the molecule, and the axes x', y', and z' are directed along the principal axes of the moment of inertia tensor of the molecule (i.e., they coincide with the principal axes of the tensor of rotational constants B_{ik}).

In this way, the symmetry of the rotational wave functions associated with the free rotation of the molecule in space is taken into account by the functions $(\varphi\theta\psi|JKM)$ of (1.43). The symmetry of the Hamiltonian (3.21) must be reflected in the wave functions $(JK|E_{J\tau})$. Before discussing this, it is convenient to consider first the matrix elements of the molecular multipole operators in the JKM- and $E_{J\tau}M$-representations.

Matrix Elements. In the preceding part operators averaged over electronic and vibrational states were considered. In what follows we shall omit the averaging symbol over these states, since this will not cause misunderstanding. The averaged operators can be divided into two types. The first of these, like (3.17), does not contain components of the operator J and therefore does not act directly on the eigenvector of a rotational state if it is written in the coordinate system fixed in the molecule.

To calculate the matrix elements of this operator $A_\nu^{(\lambda)}$ we shall project it into a motionless coordinate system, where its components will be denoted by $A_\mu^{(\lambda)}$. Then according to (1.50) we have

$$A_\mu^{(\lambda)} = \sum_\nu A_\nu^{(\lambda)} D_{\nu\mu}^{(\lambda)}(\psi, \theta, \varphi). \tag{3.22}$$

The matrix elements of $A_\mu^{(\lambda)}$ are now specified in terms of the matrix elements of the function $D_{\nu\mu}^{(\lambda)}$ and the "molecular constants" $A_\nu^{(\lambda)}$. Using (1.45) and (1.67), we obtain

$$(JKM \mid A_\mu^{(\lambda)} \mid J'K'M') = (-1)^{J-M} \begin{pmatrix} J' & J & \lambda \\ M & -M & \mu \end{pmatrix} \times$$
$$\times [(2J+1)(2J'+1)]^{1/2} \sum_\nu (-1)^{J-K} \begin{pmatrix} J' & J & \lambda \\ K' & -K & \nu \end{pmatrix} A_\nu^{(\lambda)}. \tag{3.23}$$

Since the values K' and K are given, the value of $\nu = K - K'$ is specified and the summation symbol over ν can be omitted.

In accordance with the Wigner-Eckart theorem (1.112) and on the basis of (3.23), we can write

$$(JK \| A^{(\lambda)} \| J'K') = [(2J+1)(2J'+1)]^{1/2} (-1)^{J-K} \begin{pmatrix} J' & J & \lambda \\ K' & -K & \nu \end{pmatrix} A_\nu^{(\lambda)}. \tag{3.24}$$

We note that Eq. (3.24) is basic for the application of the theory of irreducible tensor operators to the calculation of the hfs of the rotation spectra of molecules. If the rotational wave functions are known in the JKM-representation $(JKM|E_{J\tau}M) = (JK|E_{J\tau})$, then the matrix element $(E_{J\tau}\|A^{(\lambda)}\|E_{J'\tau'})$ is calculated according to the usual rules of representation theory [21]:

$$(E_{J\tau} \| A^{(\lambda)} \| E_{J'\tau'}) = \sum_{K,K'} (E_{J\tau} | JK)(JK \| A^{(\lambda)} \| J'JL')(J'K' | E_{J'\tau'}). \tag{3.25}$$

We now consider the averaged operators of the second type, which are proportional to the vector **J**. In calculating their matrix elements it is necessary to take certain precautions, since the matrix elements of the operator **J** have an interesting peculiarity. In the laboratory coordinate system the components of the operator $J^{(1)}_{\mu}$, where

$$J^{(1)}_{\pm 1} = \frac{1}{\sqrt{2}}(J_y \mp iJ_x); \quad J^{(1)}_0 = iJ_z, \tag{3.26}$$

have matrix elements according to (1.112),

$$(JKM | J^{(1)}_{\mu} | JK'M') = \delta_{K,K'} (-1)^{J-M} \begin{pmatrix} J & J & 1 \\ M' & -M & \mu \end{pmatrix} (J \| J^{(1)} \| J). \tag{3.27}$$

It follows from a comparison of these equations with (1.2) that

$$(J \| J^{(1)} \| J) = i[J(J+1)(2J+1)]^{1/2}. \tag{3.28}$$

However, the matrix elements of the operator $J^{(1)}_{\nu}$ in the rotated system, which depend on K, must be transformed with respect to (3.27) in virtue of the properties of the vectors $|JK)$ which were discussed in Part 1 of Section 1. Hence,

$$(JKM | J^{(1)}_{\nu} | JK'M) = \delta_{M,M'} (-1)^{J-K'} \begin{pmatrix} J & J & 1 \\ K & -K' & \nu \end{pmatrix} (J \| J^{(1)} \| J), \tag{3.29}$$

where the reduced matrix element is expressed by the same relation (3.28). On the other hand, we can make use of the relation

$$J^{(1)}_{\mu} = \sum_{\nu} J^{(1)}_{\nu} D^{(1)}_{\nu\mu}(\psi, \theta, \varphi) \tag{3.30}$$

and calculate the matrix element on the basis of (1.45) taking (3.29) into account. As a result we obtain

$$(-1)^{J-M} \begin{pmatrix} J & J & 1 \\ M & -M & \mu \end{pmatrix} (J \| J^{(1)} \| J) = \sum_{\nu, K''} (-1)^{J-K'} \begin{pmatrix} J & J & 1 \\ K & -K'' & \nu \end{pmatrix} \times$$
$$\times (2J+1)(-1)^{J-K''} \begin{pmatrix} J & J & 1 \\ K' & -K'' & \nu \end{pmatrix} (-1)^{J-M} \begin{pmatrix} J & J & 1 \\ M & -M' & \mu \end{pmatrix} (J \| J^{(1)} \| J). \tag{3.31}$$

This equality is satisfied identically if account is taken of the orthogonality property of the 3j-symbols (Appendix A) in summing over ν and K". On this basis we can formally write

$$(JK \| J^{(1)} \| JK') = \delta_{KK'} \cdot i[J(J+1)(2J+1)]^{1/2} \tag{3.32}$$

for the reduced matrix element and, in general, because of the orthonormality of the wave functions $(JK|E_{J\tau})$,

$$(E_{J\tau} \| J^{(1)} \| E_{J'\tau'}) = \delta_{J,J'} \cdot \delta_{\tau,\tau'} \cdot i[J(J+1)(2J+1)]^{1/2}. \tag{3.33}$$

Now it is easy to calculate the reduced matrix element of an operator of the form (3.15) containing **J**. For this it is necessary to use Eqs. (1.123) and (3.33), resulting in

$$(E_{J\tau} \| [J^{(1)} \times G^{(\lambda,k)}]^{(\lambda)} \| E_{J'\tau'}) =$$
$$= (-1)^{J+J'+\lambda} (2\lambda+1)^{1/2} (J \| J^{(1)} \| J) \begin{Bmatrix} \lambda & J & J' \\ J & K & 1 \end{Bmatrix} (E_{J\tau} \| G^{(\lambda,K)} \| E_{J'\tau'}). \tag{3.34}$$

In particular, for the magnetic multipole operator (3.15) in the JKM-representation we have

$$(JKM \| m^{(\lambda,\lambda)} \| J'K') = (-1)^{\lambda+J'-K} (J \| J^{(1)} \| J) [(2\lambda+1)(2\lambda+1)(2J'+1)]^{1/2} \times$$
$$\times \left[\begin{Bmatrix} \lambda & J & J' \\ J\lambda+1 & 1 & \end{Bmatrix} \begin{pmatrix} J' & J & \lambda+1 \\ K & -K & \nu \end{pmatrix} G_{\nu}^{(\lambda,\lambda+1)} + \begin{Bmatrix} \lambda & J & J' \\ J & \lambda-1 & 1 \end{Bmatrix} \begin{pmatrix} J' & J & \lambda-1 \\ K' & -K & \nu \end{pmatrix} G_{\nu}^{(\lambda,\lambda-1)} \right]. \tag{3.35}$$

We introduce the coefficients

$$b_{\nu}^{\lambda}(J_{\tau}, J'\tau') = [(2J+1)(2J'+1)]^{1/2} \sum_{K} (E_{J\tau} \mid JK) (-1)^{J-K} \begin{pmatrix} J' & J & \lambda \\ K' & -K & \nu \end{pmatrix} (J'K-\nu \mid E_{J'\tau'}). \tag{3.36}$$

Then the reduced matrix elements are expressed in terms of rotational constants of type $q_{\nu}^{(\lambda)}$ and $G_{q}^{(\lambda,K)}$ and the coefficients (3.36). For operators of the first type we have

$$(E_{J\tau} \| A^{(\lambda)} \| E_{J'\tau'}) = \sum_{\nu} A_{\nu}^{(\lambda)} b_{\nu}^{\lambda} (J\tau; J'\tau'). \tag{3.37}$$

For operators of the second type,

$$(E_{J\tau} \| m^{(\lambda,\lambda)} \| E_{J'\tau'}) = (-1)^{J'+J+\lambda} (2\lambda+1)^{1/2} \times$$
$$\times (J \| J^{(1)} \| J) \left[\begin{Bmatrix} \lambda & J & J' \\ J\lambda+1 & 1 & \end{Bmatrix} \sum_{\nu} G_{\nu}^{(\lambda,\lambda+1)} b_{\nu}^{(\lambda+1)} (J\tau; J'\tau') + \right.$$
$$\left. + \begin{Bmatrix} \lambda & J & J' \\ J & \lambda-1 & 1 \end{Bmatrix} \sum_{\nu} G^{(\lambda,\lambda-1)} b_{\nu}^{\lambda-1} (J\tau; J'\tau') \right]. \tag{3.38}$$

The constants $A^{(\lambda)}$ and $G_{\nu}^{(\lambda,\lambda\pm1)}$ entering into these expressions do not depend on the rotational state. This dependence is contained in the coefficients in $b_{\nu}^{\lambda}(J\tau; J'\tau')$.

It remains for us to investigate the general properties of the coefficients (3.36) so that we can determine the number of independent constants entering in (3.37) and (3.38) for the diagonal reduced matrix elements. These constants determine the number of independent parameters of the hyperfine splitting of a given rotational level $E_{J\tau}$. We shall consider the cases of the symmetric and asymmetric tops separately.

Symmetric Top. We consider the consequences of the symmetry of the rotational Hamiltonian (3.21) in the case of the symmetric top, where $B_{x'x'} = B_{y'y'}$. Since the operation of inversion commutes with rotations, only rotations are of importance in the symmetry operations of the rotational Hamiltonian. The rotational Hamiltonian of the symmetric top has the symmetry group D_{∞}, in which there are arbitrary rotations about the symmetry axis z' of the molecule and a rotation by the angle π about a perpendicular axis. The wave functions that transform according to the irreducible representations of this group are characterized by specific absolute values $|K|$ and can be written down at once:

$$\left. \begin{aligned} (JK' \mid E_{JK}) & \frac{1}{\sqrt{2}} (\delta_{K',K} \pm \delta_{K',-K}), \\ (JK' \mid E_{J0}) & = \delta_{K',0}. \end{aligned} \right\} \tag{3.39}$$

The energy levels E_{JK} are degenerate with respect to the sign of the projection of K. In the expression for the diagonal matrix element $(E_{JK} \| A^{(\lambda)} \| E_{JK})$ enter only those components of the tensor $A_{\nu}^{(\lambda)}$ that transform according to the identity representation of D_{∞}. Such properties are possessed by linear combinations $\frac{1}{2}(A_{\nu}^{(\lambda)} + A_{-\nu}^{(\lambda)})$ with even values of λ and ν. For example, out of the components of the quadrupole derivative these properties are possessed by two linear combinations $q_0^{(2)} = -q_{zz}$, $\frac{1}{2}(q_2^{(2)} + q_{-2}^{(2)}) = \frac{1}{4}(q_{yy} - q_{xx})$. If the nucleus is situated on the symmetry axis of the molecule, the only component differing from zero is $q_0^{(2)} = -q_{zz}$. If the nucleus is not on the symmetry axis, the second component $\frac{1}{4}(q_{yy} - q_{xx})$ is also different from zero. Its matrix elements differ for states with $|K| = 1$, according to the plus or minus sign in Eq. (3.39), resulting in a removal of degeneracy in sign of $K = 1$. The degeneracies of levels with $|K| = 2$ can be removed only by octupole or hexadecapole interactions containing constants of the type $\frac{1}{2}(A_4^{(4)} + A_{-4}^{(4)})$.

Asymmetric Top. The rotational Hamiltonian of the asymmetric top has the symmetry of the rhombic group D_2. Each energy level $E_{J\tau}$, where $\tau = -J, -J+1, \ldots, J$, belongs to one of the irreducible representations of the group D_2. Linear combinations of the components $A_\nu^{(\lambda)}$ that transform according to the identity representation A of the group D_2 enter into the expressions for $(E_{J\tau} \| A^{(\lambda)} \| E_{J\tau})$.

From the components of the second-rank tensor $A_\nu^{(2)}$, two such combinations $A_0^{(2)} = -A_{zz}$ and $\frac{1}{2}(A_2^{(2)} + A_{-2}^{(2)}) = \frac{1}{2}(A_{yy} - A_{xx})$ can be formed.

In the general case, the linear combinations $\frac{1}{2}[A_\nu^{(\lambda)} + (-1)^\lambda A_{-\nu}^{(\lambda)}]$, with even values of ν, transform according to the identity representation. The number of such linear combinations that differ from zero which can be made up from the components of the irreducible tensors of the molecular multipoles equals the number of independent molecular parameters involved in the hfs of the rotational levels.

In higher orders of approximation the hfs depends on off-diagonal elements of the type $(E_{J\tau} \| A^{(\lambda)} \| E_{J'\tau'})$, in which all possible components of $A_\nu^{(\lambda)}$ usually enter.

3. Calculation of the Nuclear Spin States

The total state vector of a molecule including nuclear spin is constructed with the aid of the Clebsch-Gordan coefficients. If the state vectors of a nucleus with spin I and projection $I_z = m$ are denoted by $|Im)$, the coupling scheme $\mathbf{J} + \mathbf{I} = \mathbf{F}$ corresponds to the total state vector,

$$|nvE_{J\tau}FM_F) = \sum_{mM} |nvE_{J\tau}M)\,|Im)\,(JMIm\,|\,JIFM), \tag{3.40}$$

where $M_F = M + m$ is the projection of $\mathbf{F}$ on the laboratory z axis. If the molecule contains identical nuclei, then this vector is subject to the requirements of Fermi or Bose statistics. This case merits more complete consideration.

For concreteness, we limit ourselves to the case of three identical nuclei, for which the spin functions of the nuclei having a given permutation symmetry were already obtained in Section 2. Suppose we have state vectors of the type (3.40), $|nvE_{J\tau}\ldots F_1M_F)$, where $\mathbf{F}_1$ is some resultant angular momentum, M_{F_1} is its projection on the z axis, and the spin states of the identical nuclei are $|j^3\alpha_s Im)$. It is required to construct vectors $|nvE_{J\tau}\ldots F_1)j^3\alpha_s IFM_F)$ which correspond to the coupling scheme $\mathbf{F}_1 + \mathbf{I} = \mathbf{F}$ and which transform according to the irreducible representation of the molecular symmetry group A_1 or A_2, depending on whether the spin of the nuclei j is integral or half-integral.

Since the transformation law for the nuclear spin vectors is known,

$$P(g)\,|j^3\alpha_s Im) = \sum_{s'} |j^3\alpha_{s'} Im)\,P_{s's}^{(\alpha)}(g), \tag{3.41}$$

where $P(g)$ is the permutation operator, $P_{s's}^{(\alpha)}(g)$ is the matrix of the irreducible representation α corresponding to the permutation $P(g)$, then the desired vector can be easily constructed by acting with the projection operator on the vector $|nvE_{J\tau}\ldots F_1j^3FM_F)$. For this it is necessary to know how the vector $|nvE_{JK}M)$ (without nuclear spin) transforms under the operations of the molecular symmetry group. These transformation properties are easily established if it is realized that under rotation by φ about the symmetry axis the vector $|JKM)$ is multiplied by $e^{i\varphi K}$, and under rotation by π about the y' axis it is transformed according to the law

$$C_{y'}(\pi)\,|JKM) = (-1)^{J-K}\,|J\,-KM). \tag{3.42}$$

Having in mind later application to the ammonia molecule (Section 5), we shall also take into account the symmetry of the inversion state, which is described by vibrational quantum number v. That is to say, the lower (symmetric) inversion state corresponds to v = 0, and the upper (antisymmetric) to v = 1. The inversion of the ammonia molecule [1] can be represented as a product of a pure inversion ($\mathbf{r} \to -\mathbf{r}$) relative to the center of mass of the molecule and a rotation by π about the z' axis. The first of these commutes with rotation and leads to multiplication of the vector $|nv)$ by $(-1)^v$ (we consider the vector of the ground electronic state as totally symmetric); the second operation leads to multiplication of the vector $|JKM)$ by $(-1)^k$.

We set the y' axis through nucleus 1 and the positive direction of the z' axis in the direction of motion of a right-handed screw in a rotation that carries nucleus 1 into nucleus 2, nucleus 2 into nucleus 3, and nucleus 3 into nucleus 1. Then the symmetry operation that interchanges nuclei 2 and 3 is the same as a rotation by π about the y' axis followed by inversion of the molecule, since in this the cyclic order of 1, 2, 3 is changed. As a result, this operation gives

$$P(23)\,|\,nvJKM) = (-1)^{J+v}\,|\,nvJ - KM). \tag{3.43}$$

The operation of interchanging nuclei 1 and 3 and 1 and 2 amounts to (3.41) and an additional rotation about z' by $2\pi/3$ or $4\pi/3$, respectively.

The total state vector is unchanged under operations exchanging two pairs of identical nuclei and is multiplied by (-1) under operations exchanging only one pair.

Taking all this into account, it is a simple matter to find the desired state vectors. We write them out here, eliminating the nonessential projection M_F, for all cases for which the symmetry index of the nuclear spin states takes on the values $\alpha_s = A_1, A_2$, and $E_{1,2}$:

$$|\,nvE_{JK}\ldots F_1 j^3 A_1 IF) = \frac{1}{\sqrt{2}}\{|\,nvJK\ldots F_1 j^3 A_1 IF) + \\ + (-1)^{J+v+2j}\,|\,nvJ, -K\ldots F_1 j^3 A_1 IF)\}. \tag{3.44}$$

$$|\,nvE_{JK}\ldots F_1 j^3 A_2 IF) = \frac{1}{\sqrt{2}}\{|\,nvJK\ldots F_1 j^3 A_2 IF) - \\ - (-1)^{J+v+2j}\,|\,nvJ, -K\ldots F_1 j^3 A_2 IF)\}. \tag{3.45}$$

$$|\,nvE_{JK}\ldots F_1 j^3 EIF) = \frac{1}{2}\{|\,nvJK\ldots F_1 j^3 E_1 IF) - |\,nvJK\ldots F_1 j^3 E_2 IF)\} + \\ + \frac{1}{2}(-1)^{J+v+2j}\{|\,nvJ, -K\ldots F_1 j^3 E_1 IF) + |\,nvJ, -K\ldots F_1 j^3 E_2 IF)\}. \tag{3.46}$$

Section 4

CALCULATION OF THE HYPERFINE STRUCTURE OF MOLECULAR ROTATIONAL LEVELS

1. Hyperfine Interaction Hamiltonian

Expansion in Multipole Moments. The energy of the hyperfine interaction of a nucleus in a molecule can be written in the form

$$\mathcal{H} = \int \rho_n(\mathbf{r})\,\varphi(\mathbf{r})\,dv - \frac{1}{c}\int \mathbf{j}_n(\mathbf{r})\cdot A(\mathbf{r})\,dv, \tag{4.1}$$

where $\rho_n(\mathbf{r})$ and $\mathbf{j}_n(\mathbf{r})$ are charge density and current density operators in the nucleus, and

$$\varphi(\mathbf{r}) = \int \frac{\rho_m(\mathbf{r})}{|\mathbf{r}-\mathbf{r}'|}\,dv' - \int \frac{\rho_m(\mathbf{r}')}{|\mathbf{r}|}\,dv' \tag{4.2}$$

and

$$A(\mathbf{r}') = \frac{1}{c}\int \frac{j_m(\mathbf{r}')}{|\mathbf{r}-\mathbf{r}'|}dv' \tag{4.3}$$

are the scalar and vector potentials created by the distribution of charge $\rho_\mu(\mathbf{r})$ and current $\mathbf{j}_m(\mathbf{r})$ in the molecule. In the scalar potential (4.2) the part which gives a purely isotropic energy of the coupling between nucleus and field and does not enter into the hyperfine interaction energy has been left out. It is convenient to apply the Coulomb gauge condition to the vector potential of the molecule:

$$\operatorname{div} \mathbf{A}(\mathbf{r}) = 0; \tag{4.4}$$

then the first term on the right-hand side of (4.1) describes the instantaneous Coulomb interaction. The wavelength of the hfs quanta exceeds the dimensions of the molecule more than 10^8 times, and the accuracy with which we compute the hyperfine interactions is tens of cycles per second, which is 10^{-6} of the maximum hyperfine energy. Because of this, we can neglect retardation effects and consider the potentials $\varphi(\mathbf{r})$ and $\mathbf{A}(\mathbf{r})$ as stationary.

In calculating the matrix elements of the Hamiltonian (4.1) it is convenient to expand it in terms of multipole moments. For this we use the expansions (1.98) and (1.99) in which the second terms in square brackets are omitted. Thus we do not consider effects arising from the penetration of electrons into the nucleus, which can be neglected for molecules in singlet electronic states [1]. Because of the condition (4.4), we need retain in the vector potential expansion only those terms which contain multipole moments $m^{(l,\lambda)}$ of magnetic character, for which $l = \lambda$. Consequently, we need to consider electrical and magnetic molecular multipoles,

$$q_\mu^{(\lambda)} = \int r^{-(\lambda+1)} Y_\mu^{(\lambda)}(\theta', \varphi')\, \rho_m(\mathbf{r}')\, dv; \tag{4.5}$$

$$m_\mu^{(\lambda,\lambda)} \equiv m_\mu^{(\lambda)} = \frac{1}{c}\int r^{-(\lambda+1)} \mathbf{Y}_\mu^{(\lambda,\lambda)}(\theta', \varphi')\, \mathbf{j}_m(\mathbf{r}')\, dv', \tag{4.6}$$

specified in the laboratory coordinate system, the origin of which coincides with the center of mass of the nucleus under consideration; the integration is carried out over the entire volume of the molecule, and the charge and current densities ρ_m and $\mathbf{j}_m$ are created by all particles in the molecule that do not enter into the constitution of the nucleus under consideration.

The corresponding nuclear multipole moments

$$Q_\mu^{(\lambda)} = \int r^\lambda Y_\mu^{(\lambda)}(\theta, \varphi)\, \rho_n(\mathbf{r})\, dv \tag{4.7}$$

and

$$M_\mu^{(\lambda,\lambda)} \equiv M_\mu^{(\lambda)} = \frac{1}{c}\int r^\lambda\, \mathbf{Y}_\mu^{(\lambda,\lambda)}(\theta, \varphi) \cdot \mathbf{j}_n(\mathbf{r})\, dv \tag{4.8}$$

are determined from the charge and current densities ρ_n and $\mathbf{j}_n$ created by the nucleons constituting the nucleus.

The multipole moments (4.5)-(4.8) defined by spherical harmonics (1.76) and (1.86) are self-conjugate irreducible tensor operators. As already mentioned in Section 3, the average values of the electrical multipoles $q_\mu^{(\lambda)}$ and $Q_\mu^{(\lambda)}$ differ from zero only for even values of λ, and the magnetic multipoles $m_\mu^{(\lambda)}$ and $M_\mu^{(\lambda)}$ only for odd values of λ.

The Hamiltonian (4.1) represented as an expansion in multipole moments is expressed in the form of scalar products of molecular and nuclear multipoles:

$$\mathcal{H} = \sum_{\lambda=2n} (q^{(\lambda)} \cdot Q^{(\lambda)}) - \sum_{\lambda=2n+1} (m^{(\lambda)} \cdot M^{(\lambda)}), \tag{4.9}$$

where the first summation is carried out over even values of λ and the second over odd values. The magnitude of the scalar product of multipoles of order λ is proportional to $(r_n/r_a)^\lambda$, where r_n and r_a are mean distances characterizing the distribution of charges and currents in the nucleus and atom, respectively. Since $r_n/r_a \approx Z^{\frac{1}{2}} \cdot 10^{-4}$, octupole ($\lambda$ = 3) and hexadecapole (λ = 4) interactions can also be significant in molecules containing heavy nuclei, besides the dipole (λ = 1) and quadrupole (λ = 2) interactions. If we consider only these types of interactions, the Hamiltonian (4.9) takes the form

$$\mathcal{H} = (q^{(2)} \cdot Q^{(2)}) - (m^{(1)} \cdot M^{(1)}) + (q^{(4)} \cdot Q^{(4)}) - (m^{(3)} \cdot M^{(3)}). \tag{4.10}$$

Transformation of Magnetic Molecular Multipoles. Molecular magnetic multipoles are created by the complex distribution of currents and magnetic moments of nuclei and electrons in the molecule, which rotates as a whole relative to the considered nucleus, whereas the nucleus itself maintains its orientation in space, to a first approximation. Thus, it is necessary to consider further that the coordinate system fixed in the subject nucleus is not an inertial one if the nucleus is not situated at the center of mass of the molecule.

The total molecular current $\mathbf{j}_m$ is made up of the convection current from the molecular rotation and "spin currents" caused by the magnetic moments of the electrons and nuclei. Consequently, the total current is a sum of four terms,

$$\mathbf{j}_m = \mathbf{j}_v + \mathbf{j}_\omega + \mathbf{j}_s + \mathbf{j}_I. \tag{4.11}$$

These current density operators can be expressed in terms of the wave function Ψ of the electronic and vibrational states of the molecule:

$$\mathbf{j}_v = -e\Psi^*\left(\sum_i \mathbf{v}_i\right)\Psi + e\Psi^*\left(\sum_k Z_k \mathbf{v}_k\right)\Psi; \tag{4.12}$$

$$\mathbf{j}_\omega = e\boldsymbol{\omega}\times\Psi^*\left(\sum_k Z_k \mathbf{r}_k\right)\Psi - e\boldsymbol{\omega}\times\Psi^*\left(\sum_i \mathbf{r}_i\right)\Psi. \tag{4.13}$$

$$\mathbf{j}_s = -2\mu_e c\nabla\times\Psi^*\left(\sum_i \mathbf{S}_i\right)\Psi; \tag{4.14}$$

$$\mathbf{j}_I = \mu_n c\nabla\times\Psi^*\left(\sum_k q_k \mathbf{I}_k\right)\Psi. \tag{4.15}$$

The following symbols have been introduced here: the electrons are labeled by the index i, the nuclei by k; $\mathbf{r}_i$ and $\mathbf{v}_i$ are the coordinate and velocity of the i-th electron, and $\mathbf{r}_k$ and $\mathbf{v}_k$ are the corresponding quantities for the k-th nucleus in a coordinate system fixed in the nuclear framework of the molecule; $\mathbf{S}_i$ is the spin of the i-th electron, $\mathbf{I}_k$ and q_k the spin and g-factor of the k-th nucleus; μ_e and μ_n are the Bohr and nuclear magnetons, respectively; e is the charge on a proton.

We consider now the contributions to the multipole moments of each of these terms.

Only the first term in Eq. (4.12) is nonzero, since the velocities of all nuclei $\mathbf{v}_k$ are zero in the molecular coordinate system.

The current $\mathbf{j}_v$ gives a contribution

$$m_v^{(\lambda)}(L) = -\frac{e}{c}\int r^{-(\lambda+1)}\mathbf{Y}_v^{(\lambda,\lambda)}(\theta,\varphi)\Psi^*\sum_i \mathbf{v}_i\psi\,dv, \tag{4.16}$$

which we transform, using (1.91) and the relation

$$\mathbf{v}[\mathbf{r}\times\nabla f(\mathbf{r})] = -(\mathbf{r}\times\mathbf{v})\cdot\nabla f(\mathbf{r}). \tag{4.17}$$

As a result, we obtain

$$m_\nu^{(\lambda)}(L) = \frac{2\mu_e}{\sqrt{\lambda(\lambda+1)}} \int \nabla (r^{-(\lambda+1)} Y_\nu^{(\lambda)}(\theta, \varphi)\, \Psi^* \mathbf{L}' \Psi \, dv), \tag{4.18}$$

where

$$\mathbf{L}' = \sum_i m\, \mathbf{r}_i \times \mathbf{v}_i \tag{4.19}$$

is the orbital momentum of the electrons relative to the considered nucleus. Equation (4.17) can also be transformed to the form

$$m_\nu^{(\lambda)}(L) = -2\mu_e \left(\frac{2\lambda-1}{\lambda+1}\right)^{1/2} \int r^{-(\lambda+2)}\, \mathbf{Y}_\nu^{(\lambda-1,\,\lambda)}(\theta, \varphi)\, \Psi^* \mathbf{L}' \Psi \, dv, \tag{4.20}$$

if use is made of (1.92) and the relation

$$(\mathbf{r} \times \mathbf{v})\, \mathbf{Y}_\nu^{(\lambda+1,\,\lambda)}(\theta, \varphi) = -\left[\frac{\lambda(2\lambda-1)}{(\lambda+1)(2\lambda+3)}\right]^{1/2} (\mathbf{r} \times \mathbf{v})\, \mathbf{Y}_\nu^{(\lambda-1,\,\lambda)}(\theta, \varphi), \tag{4.21}$$

which follows from (1.90).

The contribution of the nuclear magnetic moments

$$\begin{aligned} m_\nu^{(\lambda)}(I) &= \frac{1}{c} \int r^{-(\lambda+1)}\, \mathbf{Y}_\nu^{(\lambda,\,\lambda)}(\theta, \varphi) \cdot \mathbf{j}_I \, dv = \\ &= \mu_n \int r^{-(\lambda+1)}\, \mathbf{Y}_\nu^{(\lambda,\,\lambda)}(\theta, \varphi) \cdot \nabla \times \Psi^* \sum_k g_k \mathbf{I}_k \Psi \, dv \end{aligned} \tag{4.22}$$

is transformed using

$$r^{-(\lambda+1)} \mathbf{Y}_\nu^{(\lambda,\,\lambda)} (\nabla \times \boldsymbol{\mu}) = \boldsymbol{\mu} \cdot (\nabla \times r^{-(\lambda+1)}\, \mathbf{Y}^{(\lambda,\,\lambda)}) + \nabla (r^{-(\lambda+1)}\, \mathbf{Y}^{(\lambda,\,\lambda)} \times \boldsymbol{\mu}) \tag{4.23}$$

and

$$\nabla \times r^{-(\lambda+1)}\, \mathbf{Y}_\nu^{(\lambda,\,\lambda)}(\theta, \varphi) = -\sqrt{\lambda(2\lambda+3)}\; r^{-(\lambda+2)}\, \mathbf{Y}_\nu^{(\lambda+1,\,\lambda)}(\theta, \varphi), \tag{4.24}$$

which follows from (1.94).

The first term on the right of (4.23) gives

$$m_\nu^{(\lambda)}(I) = -\mu_n [\lambda(2\lambda+3)]^{1/2} \sum_k g_k \mathbf{I}_k r_k^{-(\lambda+2)}\, \mathbf{Y}_\nu^{(\lambda+1,\,\lambda)}(\theta, \varphi), \tag{4.25}$$

where r_k, θ_k, and φ_k are the spherical coordinates of the k-th nucleus. The second term of (4.23) leads to the expression

$$\mu_n \int \nabla \left\{ r^{-(\lambda+1)}\, \mathbf{Y}_\nu^{(\lambda,\,\lambda)}(\theta, \varphi) \times \Psi^* \sum_k g_k \mathbf{I} \Psi \right\} dv, \tag{4.26}$$

which is easily seen to go to zero by transforming it to a surface integral,

$$\mu_n \int r^{-(\lambda+1)} \left(\mathbf{Y}_\nu^{(\lambda,\,\lambda)}(\theta, \varphi) \times \Psi^* \sum_k g_k \mathbf{I} \Psi \right) d\mathbf{s}. \tag{4.27}$$

We turn now to a consideration of the contribution of the rotational motion of the molecule,

$$m_\nu^{(\lambda)}(\omega) = \frac{1}{c}\int r^{-(\lambda+1)}\,\mathbf{Y}_\nu^{(\lambda,\lambda)}(\theta,\varphi)\cdot\mathbf{j}_\omega\,dv, \tag{4.28}$$

where $\mathbf{j}_\omega$ is given by the expression (4.13). This can be transformed using the relation

$$\frac{\mathbf{r}}{r}\times\mathbf{Y}_\nu^{(\lambda,\lambda)}(\theta,\varphi) = \frac{[\lambda(2\lambda+3)]^{1/2}}{2\lambda+1}\mathbf{Y}_\nu^{(\lambda+1,\lambda)}(\theta,\varphi) - \\ - \frac{[(\lambda+1)(2\lambda-1)]^{1/2}}{2\lambda+1}\mathbf{Y}_\nu^{(\lambda-1,\lambda)}(\theta,\varphi). \tag{4.29}$$

As a result, we obtain

$$m_\nu^{(\lambda)}(\omega) = \omega\frac{1}{2\lambda+1}\frac{e}{c}\int r^{-\lambda}\{\sqrt{\lambda(2\lambda+3)}\,\mathbf{Y}_\nu^{(\lambda+1,\lambda)}(\theta,\varphi) - \\ - [(\lambda+1)(2\lambda-1)]^{1/2}\mathbf{Y}_\nu^{(\lambda-1,\lambda)}(\theta,\varphi)\,\Psi^*\left[\sum_k Z_k\delta(\mathbf{r}-\mathbf{r}_k) - \sum_i \delta(\mathbf{r}-\mathbf{r}_i')\right]\Psi\,dv. \tag{4.30}$$

It remains to consider that part of the magnetic dipole

$$m_\nu^{(\lambda)}(S) = -2\mu e\int r^{-(\lambda+1)}\,\mathbf{Y}_\nu^{(\lambda,\lambda)}(\theta,\varphi|\cdot\nabla\times\Psi^*\mathbf{S}\Psi\,dv, \tag{4.31}$$

which comes from electron spins. Here we have introduced a symbol $\mathbf{S} = \sum_i \mathbf{S}_i$ for the total electronic spin.

Just as in (4.22), the contribution of the "spin" currents of the electrons is conveniently transformed to a form analogous to (4.25). The result is

$$m_\nu^{(\lambda)}(\mathrm{S}) = 2\mu_e[\lambda(2\lambda+3)]^{1/2}\int r^{-(\lambda+2)}\mathbf{Y}_\nu^{(\lambda+1,\lambda)}(\theta,\varphi)\,\Psi^*\,\mathbf{S}\,\Psi\,dv, \tag{4.32}$$

to which should be added a term arising from the penetration of electrons in the nucleus and leading to an interaction of the Fermi type [22]:

$$\mathcal{H}_F = \frac{16\pi\mu_e}{3c}g_0\,\mathbf{I}_0\mu_n\int\Psi^*\,\mathbf{S}\,\delta(\mathbf{r})\,\Psi\,dv, \tag{4.33}$$

where $\mu_n g_0 I_0$ is the magnetic moment of the nucleus. We shall omit this term from further consideration since, as has been mentioned above, effects associated with electron penetration are not of importance to us.

Summing up, we see that in our case (singlet electronic state), the total magnetic molecular multipole is made up of the four terms

$$m_\nu^{(\lambda)} = m_\nu^{(\lambda)}(L) + m_\nu^{(\lambda)}(I) + m_\nu^{(\lambda)}(\omega) + m_\nu^{(\lambda)}(S), \tag{4.34}$$

which are defined, respectively, by the expressions (4.20), (4.25), (4.30), and (4.32).

Thomas Correction. The Hamiltonian (4.1) gives the hfs energy to an accuracy of $(v/c)^2$, if retardation effects are neglected, as well as effects associated with the noninertiality of the coordinate system in which the considered nucleus lies in the rotating molecule. Thus, in order to compute the hfs energy in the Pauli approximation, we need to consider effects caused by noninertiality to an accuracy of $(v/c)^2$. For this we should add to the Hamiltonian of the purely electromagnetic interaction between nucleus and molecular field (4.1) an energy

$$\mathcal{H}_T = -\hbar\,\mathbf{I}_0\cdot\boldsymbol{\omega}_T, \tag{4.35}$$

associated with the Thomas precession that a nucleus of angular momentum $\hbar\mathbf{I}_0$ performs in a noninertial coordinate system. The Thomas precession, as is well known, is an effect of relativistic kinematics, and its frequency is [29]

$$\omega_T = \frac{\mathbf{v}_0 \times \mathbf{a}}{2c^2}, \tag{4.36}$$

where $\mathbf{a}$ is the acceleration and $\mathbf{v}_0$ is the velocity of the subject nucleus in the rotating molecule. In order to estimate the order of magnitude of the energy $\mathcal{H}_T$, we note that the acceleration $\mathbf{a}$ is a result of the deviation of the nucleus from its equilibrium vibrational position. This displacement is caused by centrifugal deformations in the rotating molecule and is extremely small in comparison to the structural distances between nuclei in the molecule. Hence we can consider that the centripetal acceleration equals

$$\mathbf{a} = -\omega_0^2 \mathbf{r}_0, \tag{4.37}$$

and the linear velocity of the nucleus is

$$\mathbf{v}_0 = \boldsymbol{\omega} \times \mathbf{r}_0, \tag{4.38}$$

where $\boldsymbol{\omega}$ is the angular velocity of the molecule and $\mathbf{r}_0$ is the coordinate of the equilibrium position of the nucleus in the system of the center of inertia of the molecule. A typical frequency of a rotational transition in a molecule $\nu = \omega/2\pi$ is of order 10^{11} cps, and interatomic separations are 1-2 A. From (4.33) it follows that the frequency shift engendered by Thomas precession is of order 10^{-2} cps. Since we are considering effects capable of leading to frequency shifts exceeding tens of cycles per second, we shall henceforth neglect Thomas precession.

The Effective hfs Hamiltonian. The molecular multipole operators are defined in essence in the coordinate representation. To calculate these operators in concrete form it is necessary to know the exact expressions for the vectors of the electronic and vibrational states of the molecule in the coordinate representation, which we have denoted by ψ. We have very little information about these functions, but in every case two extremely general properties of these functions are firmly established. One is the separation of the electronic, vibrational, and rotational states (adiabatic approximation); the other is the separation of the spin and "orbital" states of the electrons (Russel-Saunders approximation). We can make use of these properties in the formal averaging of the multipole moment operators over the electronic and vibrational states.

An immediate consequence of the second property is that operators containing electron spins act only on electronic spin states. Since we are considering only molecules in singlet states, the action of the spin operator $\mathbf{S}$ will yield zero as a result. Operators of this type are given by (4.32), and their average value, consequently, equals zero:

$$\langle m_\nu^{(\lambda)}(S)\rangle = 0. \tag{4.39}$$

The consequencesof the approximately adiabatic character of the molecular motion when account was taken of the rotational-electronic perturbation were considered in detail in the preceding section. Here we shall formulate the results of this consideration insofar as they are applicable to all the remaining multipole moment operators.

In averaging the electric multipole operators the rotational-electronic interaction shows up only in the third order of perturbation theory, which can be neglected. For the calculation of the mean value of the electric multipole operators it is sufficient to take the diagonal matrix elements over the ground electronic and vibrational states,

$$\langle q_\nu^{(\lambda)}\rangle = \langle 0, v\,|\,q_\nu^{(\lambda)}\,|\,0, v\rangle \equiv q_\nu^{(\lambda)}. \tag{4.40}$$

The same applies also to the magnetic multipole operators $m_\nu^{(\lambda)}(I)$, which according to (4.25) do not contain the electronic orbital momentum operator $\mathbf{L}'$ at all:

$$\langle m_\nu^{(\lambda)}(I)\rangle = \langle v\,|\,m_\nu^{(\lambda)}(I)\,|\,v\rangle \equiv m_\nu^{(\lambda)}(I). \tag{4.41}$$

Averaging of the operators $m_\nu^{(\lambda)}(L)$ was considered in detail in Part 1 of Section 3, from which it follows that even-order perturbation terms enter into the expressions for $\langle m_\nu^{(\lambda)}(L)\rangle$. It is sufficient for us to consider only the second-order term, which can be represented in the form (3.17):

$$\langle m_\nu^{(\lambda)}(L)\rangle = [J^{(1)} \times G^{(\lambda,\,\lambda+1)}(L)]_\nu^{(\lambda)} + [J^{(1)} \times G^{(\lambda,\,\lambda-1)}(L)]_\nu^{(\lambda)}, \tag{4.42}$$

where the "g-factors" $G_\nu^{(\lambda,\lambda+1)}(L)$ are defined via the operator by means of (3.16).

The operator $m_\nu^{(\lambda)}(\omega)$, as follows from (4.30), contains expressions analogous to electric multipoles of even rank, and therefore in finding its average value it is sufficient to use the first-order approximation, which is nonzero:

$$\langle m_\nu^{(\lambda)}(\omega)\rangle = \langle 0v \mid m_\nu^{(\lambda)}(\omega) \mid 0v\rangle. \tag{4.43}$$

Since it is proportional to the vector $\boldsymbol{\omega}$, it also can be presented in the form (4.42), having separated out the operator $J_\mu^{(1)}$. To this end, we write the expression $\omega_k = J_i B_{ik}$ in the form

$$\omega_\nu^{(1)} = [J^{(1)} \times B^{(2)}]_\nu^{(1)} + B^{(0)} J_\nu^{(1)}, \tag{4.44}$$

where the components of the irreducible tensors of the rotational constants are expressed in the following way:

$$\begin{aligned} B^{(0)} &= \frac{1}{3}(B_{xx} + B_{yy} + B_{zz}); \\ B_0^{(2)} &= \left(\frac{5}{3}\right)^{1/2}(B^{(0)} - B_{zz}); \\ B_{\pm 1}^{(2)} &= \mp\left(\frac{5}{3}\right)^{1/2}(B_{xz} \pm iB_{yz}); \\ B_{\pm 2}^{(2)} &= \frac{1}{2}\left(\frac{5}{3}\right)^{1/2}(B_{yy} - B_{xx} \mp 2iB_{xy}). \end{aligned} \tag{4.45}$$

Defining further the quantities

$$\begin{aligned} G_\nu^{(\lambda,\,k)}(\omega) = \frac{e}{c}\int \langle \Sigma v \mid r\rangle r^{-\lambda} \{a_k B^{(0)} Y_\nu^{(k)}(\theta,\varphi) + \\ + b_{\lambda+1}^k [B^{(2)} \times Y^{(\lambda+1)}(\theta,\varphi)]_\nu^{(k)} + b_{\lambda-1}^k [B^{(2)} \times Y^{(\lambda-1)}(\theta,\varphi)]_\nu^{(k)}\} \times \\ \times \left[\sum_k Z_k \delta(\mathbf{r} - \mathbf{r}_k) - \sum_i \delta(\mathbf{r} - \mathbf{r}_i)\right] \langle r \mid \Sigma v\rangle\, dv, \end{aligned} \tag{4.46}$$

where

$$\left.\begin{aligned} a_{\lambda-1} &= -\frac{[(\lambda-1)(2\lambda-1)]^{1/2}}{2\lambda+1}; & a_{\lambda+1} &= \frac{[\lambda(2\lambda+3)]^{1/2}}{2\lambda+1}; \\ b_{\lambda+1}^{\lambda-1} &= -\frac{2\lambda-1}{2\lambda+1}\left[\frac{3}{5}\frac{(\lambda+1)}{(2\lambda+1)}\right]^{1/2}; & b_{\lambda+1}^{\lambda+1} &= \left[\frac{\lambda(\lambda+2)(2\lambda+3)(2\lambda+5)}{10(\lambda+1)(2\lambda+1)^3}\right]^{1/2}; \\ b_{\lambda-1}^{\lambda-1} &= -\left[\frac{(\lambda^2-1)(2\lambda-1)(2\lambda+3)}{10\lambda(2\lambda+1)^3}\right]^{1/2}; & b_{\lambda-1}^{\lambda+1} &= \frac{2\lambda+3}{2\lambda+1}\left[\frac{2\lambda}{5(2\lambda+1)}\right]^{1/2}, \end{aligned}\right\} \tag{4.47}$$

we can write

$$\langle m_\nu^{(\lambda)}(\omega)\rangle = [J^{(1)} \times G^{(\lambda,\,\lambda+1)}(\omega)]_\nu^{(\lambda)} + [J^{(1)} \times G^{(\lambda,\,\lambda-1)}(\omega)]_\nu^{(\lambda)}. \tag{4.48}$$

Summing this with (4.42), we introduce an expression for the effective magnetic multipole,

$$m_\nu^{(\lambda)}(J) \equiv \langle m_\nu^{(\lambda)}(L)\rangle + \langle m_\nu^{(\lambda)}(\omega)\rangle = [J^{(1)} \times G^{(\lambda,\,\lambda+1)}]_\nu^{(\lambda)} + [J^{(1)} \times G^{(\lambda,\,\lambda-1)}]_\nu^{(\lambda)}, \tag{4.49}$$

where

$$G_q^{(\lambda,\,\lambda\pm1)} \equiv G_q^{(\lambda,\,\lambda\pm1)}(L) + G_q^{(\lambda,\,\lambda\pm1)}(\omega). \tag{4.50}$$

The Hamiltonian (4.10), in which the nuclear multipole operators are averaged over the nuclear ground state and the molecular multipole operators are averaged over the vibrational and electronic state (taking into account electron-rotational interaction), will be called the effective hfs Hamiltonian. The effective Hamiltonian operator

$$\langle\mathcal{H}\rangle = (q^{(2)}\cdot Q^{(2)}) + (q^{(4)}\cdot Q^{(4)}) - (m^{(1)}(J)\cdot M^{(1)}) - \\ - (m^{(3)}(J)\cdot M^{(3)}) - (m^{(1)}(I)\cdot M^{(1)}) \qquad (4.51)$$

contains terms of three types. The first two represent electrical (quadrupole and hexadecapole) interactions. The second two represent the so-called $\mathbf{I}\cdot\mathbf{J}$-interactions of the dipole and octupole type. The last term represents the dipolar spin-spin interaction. A similar term, representing spin-spin interactions of the octupole type, has been omitted because of its small size (frequency shifts less than 10^{-1} cps).

The operator (4.51) is convenient because it acts only on rotational states of the molecule and spin states of the nuclei. Note that the method presented here of calculating the effect of electron-rotational interaction on hfs by means of an effective Hamiltonian is very similar to the method of the so-called spin Hamiltonian, which has received wide application in the theory of paramagnetic resonance [3].

2. Calculation of Hyperfine Interactions

Perturbation Theory and Interval Rule. All the hfs interactions represented by the different terms of the effective Hamiltonian (4.51), with the exception of the quadrupolar, are so small in comparison to the rotational energy that they can be satisfactorily handled by first-order perturbation theory. Higher orders must be used to calculate the quadrupole splitting. Each of the five types of hfs interaction involved in (4.51) is presented in the form of a scalar product of irreducible tensor operators of the type $(A^{(\lambda)}\cdot B^{(\lambda)})$ of rank λ.

In the representation $E_{J\tau}IFM_F$, where $\mathbf{F} = \mathbf{J} + \mathbf{I}$, and M_F is the projection of $\mathbf{F}$ on the laboratory z axis, the matrix element for the energy of a 2^λ-pole interaction is expressed by the formula

$$(E_{J\tau}IF|(A^{(\lambda)}\cdot B^{(\lambda)})|E_{J\tau}IF) = \\ = (-1)^{\lambda+J'+I+F}\begin{Bmatrix}\lambda J J'\\ F I I\end{Bmatrix}(E_{J\tau}\|A^{(\lambda)}\|E_{J'\tau'})(I\|B^{(\lambda)}\|I), \qquad (4.52)$$

where the dependence on the quantum number F is concentrated in the 6j-symbol. The calculation of the reduced matrix element $(E_{J\tau}\|A^{(\lambda)}\|E_{J'\tau'})$ was considered in Section 3, and the reduced matrix element $(I\|B^{(\lambda)}\|I)$ is associated with the "magnitude of nuclear multipole":

$$\langle B_0^{(\lambda)}\rangle = \langle\gamma I m_I = I|B_0^{(\lambda)}|\gamma I m_I = I\rangle \equiv B_\lambda \qquad (4.53)$$

by the relation

$$(I\|B^{(\lambda)}\|I) = \frac{[(2I-\lambda)!\,(2I+\lambda+1)!]^{1/2}}{(2I)!}B_\lambda, \qquad (4.54)$$

which comes directly from the Wigner-Eckart theorem (1.112).

More complicated coupling cases arising in the hfs calculation for several nuclei were considered in Section 2.

To first order, the splitting caused by the 2^λ-pole interaction is characterized by a definite interval rule for a given value of λ. According to (4.52), this interval rule is described by the function $\begin{Bmatrix}\lambda J J\\ F I I\end{Bmatrix}$, which also contains within itself the selection rule $2J, 2I \geq \lambda$. Since the 6j-symbols, by their very nature, represent a complete system of functions in the sense of interval rules (orthogonality property), from a given system of levels E_{JIF} observed in an experiment it is easy to separate in first order the constant for the 2^λ-pole interaction, namely 2,

$$(J\tau\|A^{(\lambda)}\|J\tau)(I\|B|I) = \sum_{F=|J-I|}^{F=J+I}(-1)^{\lambda+I+J+F}(2\lambda+1)(2F+1)\begin{Bmatrix}\lambda J J\\ F I I\end{Bmatrix}E(JIF). \qquad (4.55)$$

In higher orders the 2^λ-pole interaction contains shifts of hfs components characterized by λ', different in general from λ. For separating out the corrections of multipole-type λ, the product of the corresponding 6j-symbols can be written in the form

$$\begin{Bmatrix} \lambda J J' \\ F I I \end{Bmatrix} \begin{Bmatrix} \lambda J' J'' \\ F I I \end{Bmatrix} = \sum_{\lambda'=0}^{2\lambda} (-1)^{\lambda'+I+J''+F} \begin{Bmatrix} \lambda' J J'' \\ F I I \end{Bmatrix} a \begin{pmatrix} \lambda' J J'' \\ \lambda J' I \end{pmatrix}, \tag{4.56}$$

where the expansion coefficients, which are independent of F, are given by

$$a \begin{pmatrix} \lambda' J J'' \\ \lambda J' I \end{pmatrix} = (-1)^{2I+J+J'} (2\lambda+1) \begin{Bmatrix} \lambda' J J \\ J' \lambda \lambda \end{Bmatrix} \begin{Bmatrix} \lambda' I I \\ I \lambda \lambda \end{Bmatrix}. \tag{4.57}$$

We turn now directly to individual consideration of each type of interaction.

Electrical Interactions. The magnitude of the quadrupole interaction ($\lambda = 2$) in molecules is 10^6-10^8 cps [1]. The second-order correction to the rotational levels is of the order of 10^4-10^5 cps, and third-order corrections can be comparable to the hexadecapole interaction ($\lambda = 4$) and amount to hundreds of cycles per second.

We shall use the definitions of [1] for the quadrupole constants used in spectroscopic literature; these are related to (4.53) and (4.40) by

$$\left. \begin{aligned} eQ &= -2B_2; \\ q &= -2q_0^{(2)}. \end{aligned} \right\} \tag{4.58}$$

The diagonal matrix element of the quadrupole interaction, calculated from (4.52), (4.54), and (3.28) for molecules of the symmetric top type,

$$E_1^{(2)} = \frac{eQq}{4} (-1)^{J-K} \begin{pmatrix} J & J & 2 \\ K & -K & 0 \end{pmatrix} \left[\frac{(2I+3)(2I+2)(2I+1)}{2I(2I-1)} \right]^{1/2} (2J+1) \times$$
$$\times (-1)^{J+I+F} \begin{Bmatrix} 2 & J & J \\ F & I & I \end{Bmatrix} \tag{4.59}$$

agrees, of course, with the well-known result of Bardeen and Townes [1] if, in place of $\begin{pmatrix} J & J & 2 \\ K & -K & 0 \end{pmatrix}$ and $\begin{Bmatrix} 2 J J \\ F I I \end{Bmatrix}$, the corresponding analytical expressions are substituted [15].

The correction for multipole-type λ ($0 \leq \lambda \leq 4$) occasioned by the quadrupole interaction is, in second order (symmetric top molecule),

$$\Delta E_2^{(\lambda)} = (-1)^{\lambda+J+I+F} \begin{Bmatrix} \lambda J J \\ F I I \end{Bmatrix} A_{JK}^{\lambda}, \tag{4.60}$$

where

$$A_{JK}^{(\lambda)} = \sum_{J'K'}{}' (2J+1)(2J'+1)\, a \begin{pmatrix} \lambda J J \\ 2 J' I \end{pmatrix} \left[\begin{pmatrix} J' & J & 2 \\ K' & -K & \nu \end{pmatrix} \right]^2 \times \frac{(eQ)^2 \,|\langle q_\nu^{(2)} \rangle|^2}{4(E_{JK} - E_{J'K'})}, \tag{4.61}$$

and the quantity $a \begin{pmatrix} \lambda J J \\ 2 J' I \end{pmatrix}$ is given by (4.57).

In third order it makes sense to write out the correction of hexadecapole character, which is found to be

$$\Delta E_3^{(4)} = (-1)^{J+I+F} \begin{Bmatrix} 4 J J \\ F I I \end{Bmatrix} B_{JK}^{(4)}, \tag{4.62}$$

where

$$B^4_{JK} = \left[\frac{(I+1)(2I+3)(2I+1)}{I(2I-1)}\right]^{1/2} \Bigg\{ \sum_{J'K'}{}' \sum_{J''K''}{}' (-1)^{K+K'+K''}(2J+1)(2J'+1)(2J''+1)\times$$
$$\times b(IJ;J'J'') \begin{pmatrix} J' & J & 2 \\ K' & -K & \nu_1 \end{pmatrix} \begin{pmatrix} J'' & J' & 2 \\ K'' & -K' & \nu_2 \end{pmatrix} \begin{pmatrix} J & J'' & 2 \\ K & -K'' & \nu_3 \end{pmatrix} \times$$
$$\times \frac{(eQ)^3 \langle q_{\nu}^{(2)} \rangle \langle q_{\nu_2}^{(2)} \rangle \langle q_{\lambda_3}^{(2)} \rangle}{8(E_{J'K'}-E_{JK})(E_{J''K''}-E_{JK})} - (-1)^k (2J+1)^2 \begin{pmatrix} J & J & 2 \\ K & -K & 0 \end{pmatrix} \sum_{J'K'}{}' b(IJ;JJ') \times$$
$$\times \begin{pmatrix} J' & J & 2 \\ K' & -K & \nu \end{pmatrix} \frac{2(eQ)^3 q \,|\langle q_{\nu}^{(2)} \rangle|^2}{16(E_{J'K'}-E_{JK})^2} \Bigg\}, \tag{4.63}$$

in which

$$b(IJ;J'J'') = \sum_{\varkappa=0}^{4} (-1)^{J+J''+2J}\, 9 \begin{Bmatrix} J & J & 4 \\ 2 & \varkappa & J'' \end{Bmatrix} \begin{Bmatrix} I & I & 4 \\ 2 & \varkappa & I \end{Bmatrix} a \begin{pmatrix} \varkappa J J'' \\ 2 J' I \end{pmatrix}. \tag{4.64}$$

The energy of the hexadecapole interaction in first order is

$$E_1^{(4)} = \langle q_0^{(4)} \rangle \langle Q_0^{(4)} \rangle (-1)^{J-K} \begin{pmatrix} J & J & 4 \\ K & -K & 0 \end{pmatrix} \left[\frac{(2I+5)(2I+3)(2I+1)(I+2)(I+1)}{I(I-1)(2I-1)(2I-3)}\right]^{1/2} \times$$
$$\times (-1)^{J+I+F} \begin{Bmatrix} 4 & J & J \\ F & I & I \end{Bmatrix} (2J+1). \tag{4.65}$$

For asymmetric top molecules we write out only the energy of the quadrupole interaction to first order. It differs from (4.59) only in the reduced matrix element $(J\tau \| q^{(2)} \| J\tau)$, which is calculated from (3.37):

$$E_1^{(2)}(J\tau) = -\frac{eQ}{2}\left[q_0^{(2)} b_0^{(2)}(J\tau;J\tau) + {}^1/_2 (q_2^{(2)} + q_{-2}^{(2)})\, b_2^2(J\tau;J\tau)\right] \times$$
$$\times (-1)^{J+I+F} \begin{Bmatrix} 2 & J & J \\ F & I & I \end{Bmatrix} \left[\frac{(2I+3)(2I+1)(I+1)}{I(2I-1)}\right]^{1/2}. \tag{4.66}$$

<u>I·J-Interactions.</u> The third term of the effective Hamiltonian (4.51) describes I·J-interaction of dipole character ($\lambda = 1$), which, as a rule, corresponds to an energy of 10^5-10^6 cps. The fourth term gives the octupole interaction ($\lambda = 3$), which only in exceptional cases can amount to tens of cycles per second as a result of a contribution to $m_{\nu}^{(3)}(J)$ of electrons close to the nucleus. The energy of the I·J-interaction is therefore satisfactorily calculated to first order. Each of the terms of the Hamiltonian of type $-(m^{(\lambda)}(J) \cdot M^{(\lambda)})$, as follows from (4.52), (4.53), and (3.35), gives a contribution (for symmetric top molecules)

$$(JKIF| - (m^{(\lambda)}(J) \cdot M^{\lambda})| JK'IF) =$$
$$= (-1)^{I+F+K} \begin{Bmatrix} \lambda & J & J \\ F & I & I \end{Bmatrix} \left[\frac{(2I-\lambda)!\,(2I+1+\lambda)!\,(2\lambda+1)\,\lambda}{(2I)!\,(2I)!\,(\lambda+1)}\right]^{\frac{1}{2}} \times$$
$$\times (2J+1)[J(J+1)(2J+1)]^{\frac{1}{2}} \left[\begin{Bmatrix} J & J & \lambda+2 \\ J & \lambda & J \end{Bmatrix} \begin{pmatrix} J & J & \lambda+1 \\ K' & -K & \nu \end{pmatrix} G_{\nu}^{(\lambda,\,\lambda+1)}(J) + \right.$$
$$\left. + \begin{Bmatrix} J & J & \lambda-1 \\ 1 & \lambda & J \end{Bmatrix} \begin{pmatrix} J & J & \lambda-1 \\ K' & -K & \nu \end{pmatrix} G_{\nu}^{(\lambda,\,\lambda-1)} \right] M_{\lambda}, \tag{4.67}$$

where M_{λ} denotes the magnitude of the multipole moment in correspondence with [18], which is related to (4.53) by

$$M_{\lambda} = -i \left(\frac{\lambda+1}{\lambda}\right)^{1/2} B_{\lambda}. \tag{4.68}$$

For the dipole interaction, (4.67) gives essentially the result of Henderson and Van Vleck [30],

$$(JKIF\,|-[m^{(1)}(J)\cdot M^{(1)}]\,|\,JKIF)=\mu_n g\left[G_0^{(1,0)}-\frac{1}{\sqrt{10}}\times\right.$$

$$\left.\times\left(\frac{3K^2}{J(J+1)}-1\right)G_0^{(1,2)}\right]\cdot{}^1/_2\,[F(F+1)-J(J+1)-I(I+1)], \tag{4.69}$$

where we set $M_1 = g\mu_n I$.

By means of (3.29) we can calculate the dipole interaction for molecules of the asymmetric top type. It depends on three molecular constants:

$$(J\tau IF|-[m^{(1)}(J)\cdot M^1]|J\tau IF)=\mu_n gG\cdot{}^1/_2\,[F(F+1)-J(J+1)-I(I+1)], \tag{4.70}$$

where

$$G=G_0^{(1,0)}-\left[\frac{(2J+3)(2J-1)}{10J(J+1)}\right]^{1/2}(2J+1)^{-1}\,[b_0^2(J\tau;\ J\tau)\cdot G_0^{(1,2)}+$$

$$+{}^1/_2\,(G_2^{(1,2)}+G_{-2}^{(1,2)})\,b_2^2(J\tau;\ J\tau)], \tag{4.71}$$

and the coefficients in $b_{0,2}^2(J\tau;\,J\tau)$ are given by (3.36).

Spin-Spin Interactions of the Nuclei. The last term of the Hamiltonian (4.51) gives the energy of the spin-spin interaction, which in order of magnitude does not exceed a few cycles per second. This is the only hfs interaction which can be calculated accurately, since the interaction constants depend only on the magnitudes of the nuclear dipole moments and the structural constants of the molecule.

The energy of the spin-spin interaction of a nucleus with spin I_1 with a nucleus having spin I_0 according to (4.25), for $\lambda = 1$, can be written in one of two forms:

$$\mathcal{H}(I_0 I_1)=\sqrt{10}\,\mu_n^2 g_0 g_1 r_1^{-3} I_0^{(1)}\,[Y^{(2)}(\theta\varphi)\times I_1^{(1)}]^{(1)} \tag{4.72}$$

or

$$\mathcal{H}(I_0 I_1)=\sqrt{6}\mu_n^2 g_0 g_1 r_1^{-3}\,Y^{-(2)}(\theta,\varphi)\cdot[I_1^{(1)}\times I_0']^{(2)}, \tag{4.73}$$

where r_1, θ_1, and φ_1 are the spherical coordinates of the nucleus with spin I_1 and g-factor g_1 in the system associated with the nucleus having spin I_0 and g-factor g_0.

The first form (4.72) is convenient for calculation in the representation $E_{J\tau}I_1F_1I_0FM_F$, where $F_1 = J + I_1$, $F = F_1 + I_0$. The matrix elements of the Hamiltonian (4.72) are calculated from (4.52) and (1.118), giving the result

$$(E_{J\tau}I_1F_1I_0F\,|\,\mathcal{H}(I_0I_1)\,|\,E_{J\tau}I_1'I_0F)=\sqrt{30}\,\mu_n^2 g_0 g_1\,(E_{J\tau}\,\|\,Y^{(3)}(\theta_1,\ \varphi_1)\,\|\,E_{J\tau})\times$$

$$\times\,[I_0(I_0+1)(2I_0+1)I_1(I_1+1)(2I_1+1)(2F_1+1)(2F_1'+1)]^{1/2}\times$$

$$\times(-1)^{F_1'+F+I_0}\begin{Bmatrix}1 & F_1 & F_1'\\ F & I_0 & I_0\end{Bmatrix}\begin{Bmatrix}I_0 & F_1 & J\\ I_0 & F_1' & J\\ 1 & 1 & 2\end{Bmatrix}. \tag{4.74}$$

For calculation in the representation $E_{J\tau}I_1I_0IFM_F$, where $F = J + I$; $I = I_0 + I_1$, it is convenient to use the second formula (4.73). Then

$$(E_{J\tau}I_1I_0IF\,|\,\mathcal{H}(I_0I_1)\,|\,E_{J\tau}I_1I_0I'F)=\sqrt{30}\,\mu_n^2 g_0 g_1 r_1^{-3}(E_{J\tau}\,\|\,Y^{(2)}(\theta,\varphi)\,\|\,F_{J\tau})\times$$

$$\times\,I_0(I_0+1)(2I_0+1)I_1(I_1+1)(2I_1+1)^{I+F+J}\begin{Bmatrix}2 & J & J\\ F & I' & I\end{Bmatrix}\begin{Bmatrix}I_1 & I_0 & I\\ I_1 & I^0 & I'\\ 1 & 1 & 2\end{Bmatrix}. \tag{4.75}$$

The reduced matrix element $(E_{J\tau}\|Y^{(2)}(\theta,\varphi)\|E_{J\tau})$, appearing in (4.74) and (4.75), is calculated from (3.38):

$$(E_{J\tau}\|Y^{(2)}(\theta,\varphi)\|E_{J\tau}) = Y_0^{2}(\theta,\varphi)\, b_0^{2}(J\tau;\ J\tau| + {}^1/_2\,[Y_2^{(2)}(\theta_1,\varphi_1) + \\ + Y_{\bar{2}}^{(2)}(\theta_1,\varphi_1)]\, b_{\bar{2}}^{2}\ (J\tau;\ J\tau). \tag{4.76}$$

For the case of molecules of the symmetric top type, it should be replaced by

$$(JK\|Y^{(2)}(\theta_1,\varphi_1)\|JK') = (-1)^{J-K}(2J+1)\begin{pmatrix} J & J & 2\\ K' & -K & \nu\end{pmatrix} Y_\nu^{(2)}(\theta_1,\varphi_1). \tag{4.77}$$

For matrix elements of (4.74) and (4.75) diagonal in F_1 or I, the calculation of the corresponding 9j-symbols can be simplified with the aid of the formula

$$\begin{Bmatrix} j_1 & j_2 & j\\ j_1 & j_2 & j\\ 1 & 1 & 2\end{Bmatrix} = \frac{(-1)^{2j}}{5\begin{Bmatrix} 1 & 1 & 2\\ j & j & j\end{Bmatrix}}\left[\begin{Bmatrix} j_1 & j_2 & j\\ 1 & j & j_2\end{Bmatrix}\begin{Bmatrix} j_2 & j_1 & j\\ 1 & j & j_1\end{Bmatrix} + \frac{(-1)^{j_1+j_2+j+1}}{3(2j+1)}\begin{Bmatrix} j & j_1 & j_2\\ 1 & j_2 & j_1\end{Bmatrix}\right]. \tag{4.78}$$

If the spins of the nuclei are equal, $I_1 = I_0 = j$, the 9j-symbols appearing in (4.74) can be further simplified:

$$\begin{Bmatrix} j & j & I\\ j & j & I'\\ 1 & 1 & 2\end{Bmatrix} = \frac{(-1)^{2\lambda}}{5\begin{Bmatrix} 1 & 1 & 2\\ I' & I & \lambda\end{Bmatrix}}\left[\begin{Bmatrix} j & j & I'\\ 1 & \lambda & j\end{Bmatrix}\begin{Bmatrix} j & j & I\\ 1 & \lambda & j\end{Bmatrix} + \delta_{II'}\frac{(-1)^{2j+I+1}}{3(2I+1)}\begin{Bmatrix} I & j & j\\ 1 & j & j\end{Bmatrix}\right], \tag{4.79}$$

where $\lambda = \frac{1}{2}(I + I')$. We also note that this 9j-symbol contains the selection rule $\Delta I = I - I' = 0, \pm 2$. The last two formulas have been derived from the relations presented in Appendix A.

Hyperfine Structure in the Case of Several Interacting Nuclei. In the case when the hfs is the result of the interaction of several nuclei, the total hfs Hamiltonian must be written as the sum of effective Hamiltonians of type (4.51) for each nucleus. In doing this, of course, pair (i.e., spin-spin) interactions must be counted only once in the general Hamiltonian. The case of several nuclei differs from the case of one nucleus by the more complicated representation corresponding to the coupling scheme of the larger number of moments. The procedure for calculating the matrix elements for this case was considered specially in Section 2. Sometimes it turns out that the hfs energies of two or more nuclei are of the same order of magnitude. In this case, the problem amounts to diagonalization of the matrix of the hfs Hamiltonian according to all possible values of the intermediate moments. If the nuclei are identical, then, as was indicated in Section 2, the permutation symmetry of the Hamiltonian of equivalent nuclei can be utilized to obtain the maximum possible simplification of the matrix of the Hamiltonian by means of the proper choice of the initial representation.

Hyperfine Interaction of Three Identical Nuclei. For the particular, but frequently occurring case of three identical nuclei, we calculated in Section 2 the wave functions and corresponding reduction coefficients that bring the matrix of the Hamiltonian into the simplest quasi-diagonal form. This method allows us to treat the three identical nuclei as a single particle, the spin I of which can be changed under the influence of the hfs interaction. The number of possible values of the spin is limited by the given type of permutation symmetry of the spin state $|j^3\alpha_s Im)$, and the hfs interaction matrix is diagonal in the index α_s.

Since the equivalent nuclei are not situated on the symmetry axis of the molecule, the hfs Hamiltonian of these nuclei has lower symmetry than the symmetry of the pure rotation Hamiltonian D_∞. Hence, the rotational levels with the same K, but with different symmetry, will experience, generally speaking, different hfs shifts. In the case of the inversion spectrum of ammonia, this effect is observed in the form of a splitting of the line belonging to $|K| = 1$. For a valid calculation of this effect it is necessary to use the state vectors of the molecule

(3.44)-(3.46), symmetrized according to the requirements of nuclear statistics.

In case K is a multiple of three, the state is described by the vector (3.43) or (3.44). The matrix element of the 2^{λ}-pole interaction of the form $H_{\lambda} = (A^{(\lambda)} B^{(\lambda)})$ is calculated with the aid of the coefficients $\Delta(\lambda j A_{1,2}; II')$, which are given by (2.42),

$$(E_{Jk} j^3 A_{1,2} IF \mid H_{\lambda} \mid E_{Jk} j^3 A_{1,2} I'F) = \Delta(\lambda j A_{1,2};\ II') \times \\ \times (-1)^{\lambda+I+J+F} \begin{Bmatrix} \lambda J J \\ F I' I \end{Bmatrix} (E_{Jk} \| A^{(\lambda)} \| E_{Jk}) (j \| B^{(\lambda)} \| j). \quad (4.80)$$

"Splitting" of levels with $|K| = 3$ can arise only through an interaction of the sixth order and is not allowed for in (4.80).

A somewhat different result is obtained for K not a multiple of 3. In this case, the nuclear statistics permit only states of the type (3.45). We obtain for the matrix elements between these states

$$(E_{Jk} j^3 EIF \mid H_{\lambda} \mid E_{Jk} j^3 EIF) = (-1)^{\lambda+I+J+F} \begin{Bmatrix} \lambda J J \\ F I' I \end{Bmatrix} (j \| B^{(\lambda)} \| j) \times \\ \times [\alpha_{\lambda}(II')(JK \| A^{(\lambda)} \| JK) + (-1)^{J+v+2j} \beta_{\lambda}(JJ') \left(JK \left\| \frac{1}{2}(A^{(\lambda)} + A^{(\lambda)+}) \right\| J - K \right)], \quad (4.81)$$

where

$$\left.\begin{aligned} \alpha_{\lambda}(II') &= \frac{1}{2}[\Delta(\lambda j E_1;\ II') + \Delta(\lambda j E_2;\ II')]; \\ \beta_{\lambda}(II') &= 1/2\,[\Delta(\lambda j E_1;\ II') - \Delta(\lambda j E_2;\ II')]. \end{aligned}\right\} \quad (4.82)$$

The second term in square brackets in (4.81) describes an additional displacement of the rotational levels. The case $\lambda = 2$ is of practical interest; then this term differs from zero for quadrupole and magnetic interactions, since, in this case, the matrix element between states with $K = \pm 1$ can be different from zero and proportional to the constant $\frac{1}{2}(A_2^{(2)} + A_{-2}^{(2)})$.

A result similar to (4.81) is obtained for the spin-spin interaction of a group of identical nuclei with some other nucleus. The small difference is connected with the presence of a 9j-symbol that enters in the same way as in (4.74).

A more complicated case is that of the spin-spin interactions of a group of identical nuclei with each other. Using the coefficients $\sigma(jd_s;\ II')$, given by (2.46) in Section 2, and (4.75) and (4.77), it is easy to obtain for K=3n

$$(JKj^3 A_{1,2} IF \mid \mathcal{H}(j \cdot j) \mid JKj^3 A_{1,2} I'F) = \\ = \frac{1}{2}\sqrt{30}(\mu_n g)^2 r^{-3} (-1)^{F+K-I} \sigma(jA_{1,2};\ II')\, j(j+1)(2j+1) \begin{pmatrix} J & J\,2 \\ K & -K0 \end{pmatrix} \begin{Bmatrix} 2JJ \\ FI'I \end{Bmatrix}. \quad (4.83)$$

For $K = 3n \pm 1$,

$$(E_{JK} j^3 EIF \mid \mathcal{H}(j \cdot j) \mid E_{JK} j^3 I'F) = \\ = \sqrt{30}(\mu_n g)^2 r^{-3} j(j+1)(2j+1)(-1)^{F+K-I} \begin{Bmatrix} 2JJ \\ FI'I \end{Bmatrix} (2J+1) \left[\frac{1}{2}\sigma(II') \begin{Bmatrix} J & J2 \\ K & -K0 \end{Bmatrix} - \right. \\ \left. - \delta_{K,1}(-1)^{J+v+2j} \left(\frac{3}{2}\right)^{\frac{1}{2}} \rho(II') \begin{pmatrix} JJ & 2 \\ 1\,1 & -2 \end{pmatrix} \right], \quad (4.84)$$

where

$$\left.\begin{aligned} \sigma(II') \\ \rho(II') \end{aligned}\right\} = \frac{1}{2}[\sigma(jE_1;\ II') \pm \sigma(j_2;\ II')]. \quad (4.85)$$

In (4.83) and (4.84) the quantity r is the equilibrium distance between identical nuclei and, in addition, we used the fact that

$$Y_0^{(2)}\left(\frac{\pi}{2}, \varphi\right) = \frac{1}{2} \quad \text{and} \quad \frac{1}{2}\left[Y_2^{(2)}\left(\frac{\pi}{2}, 0\right) + Y_{\bar{2}}^{(2)}\left(\frac{\pi}{2}, 0\right)\right] = -\sqrt{\frac{3}{2}}.$$

3. Intensity of the Hyperfine Structure Lines

Relative Intensities of hfs Transitions. In distinction from the usual conditions of observation of optical spectra, the transitions observed in radiospectroscopy are induced by waves with a specific (linear or circular) polarization. The probability of such a transition is proportional to the square of the absolute magnitude of the matrix element of the electrical dipole moment operator of the molecule $d_\mu^{(1)}$, where the index μ takes one of the values $\mu = 0$ for linear polarization (z axis chosen along the direction of the electric field) and $\mu = \pm 1$ respectively for right- or left-circularly polarized light. We use the indices $E_{J_\tau}\gamma FM$ to denote the hfs level, where F is the resultant angular momentum of the molecule, M is its projection on the axis, and γ is the set of all intermediate moments.

The relative intensity of the transition $i \to f$ is equal to

$$\sum_{\text{all } M^i, M^f} |(E^i_{J\tau}\gamma^i F^i M^i \,|\, d_\mu^{(1)} \,|\, E^f_{J\tau}\gamma^f F^f M^f)|^2 =$$

$$= |(E^i_{J\tau}\gamma^i F^i \,\|\, d^{(1)} \,\|\, E^f_{J\tau}\gamma^f F^f)|^2 \cdot \sum_{M^i, M^f} \left|\begin{pmatrix} F^f & F^i & 1 \\ M^f & M^i & \mu \end{pmatrix}\right|^2 =$$

$$= \frac{1}{3} |(E^i_{J\tau}\gamma^i \,\|\, d^{(1)} \,\|\, E^f_{J\tau}\gamma^f F^f)|^2. \tag{4.86}$$

The equality of the latter sum to $^1/_3$ comes from the orthogonality property of the 3j-symbols (Appendix A). The square of the reduced matrix element, which is the so-called strength of the hyperfine transition, is easily calculated in a completely defined coupling scheme by means of applying (1.120) the required number of times. For example, if γ stands for $(I_1F_1I_2)$, where $\mathbf{F}_1 = \mathbf{J} + \mathbf{I}_1$ and $\mathbf{F}_1 + \mathbf{I}_2 = \mathbf{F}$, then

$$S(F_1^i F^i \,|\, F_1^f F^f) = |(E_{J\tau} I_1 F_1^i I_2 F^i \,\|\, d^{(1)} \,\|\, E^f_{J\tau} I_1 F_1^f I_2 F^f)|^2 =$$

$$= (2F^i + 1)(2F^f + 1)\begin{Bmatrix} 1 & F_1^i & F_1^f \\ I_2 & F^f & F^i \end{Bmatrix}^2 (2F_1^i + 1)(2F_1^f + 1)\begin{Bmatrix} 1 & J^i & J^f \\ I & F_1^f & F_1^i \end{Bmatrix} |(E^i_{J\tau} \,\|\, d^{(1)} \,\|\, E^f_{J\tau})|^2. \tag{4.87}$$

The last square of a reduced matrix element represents the strength of a purely rotational transition. These have been tabulated in [1], and we do not consider them here.

If the hfs levels $E_{J_\tau}\gamma FM$ are obtained by diagonalization of the hfs Hamiltonian with respect to the intermediate moment F_1, then in place of (4.87) we have

$$S(\gamma^i F^i \,|\, \gamma^f F^f) = |(E^i_{J\tau} \,\|\, d^{(1)} \,\|\, E^f_{J\tau})|^2 (2F^i + 1)(2F^f + 1) \times$$

$$\times \left|\sum_{F_1 F_1'} (\gamma^i | F_1)(F_1'\gamma^f)(-1)^{F_1 + F_1'}[2F_1 + 1)(2F_1' + 1)]^{1/2} \begin{Bmatrix} 1 & F_1 & F_1' \\ I_2 & F^f & F^i \end{Bmatrix} \begin{Bmatrix} 1 & J^i & J^f \\ I_1 & F_1' & F_1 \end{Bmatrix}\right|^2, \tag{4.88}$$

where $(F_1 | \gamma)$ is the matrix of the diagonalizing transformation.

Sum Rule. The total intensity of the lines of a hyperfine multiplet is given by

$$\frac{1}{3} S(E^i_{J\tau} \,|\, E^f_{J\tau}) = \frac{1}{3} \sum_{\gamma^i F_1^i \gamma^f F^f} S(\gamma^i F^i \,|\, \gamma^f F^f). \tag{4.89}$$

The last sum is the trace of the matrix of the operator $d_\mu^{(1)} d_\mu^{(1)\dagger}$, i.e.,

$$S(E^i_{J\tau}\,|\,E^f_{J\tau}) = \sum_{\gamma^i F^i} (E^i_{J\tau}\gamma^i F^i \,\|\, d^{(1)}d^{(1)\dagger} \,\|\, E^i_{J\tau}\gamma^i F^i) = \sum_{\gamma^f F^f} (E^f_{J\tau}\gamma^f F^f \,\|\, d^{(1)}d^{(1)\dagger} \,\|\, E^1_{J\tau}\gamma^f F^f). \tag{4.90}$$

The property of invariance of the trace with respect to choice of representation is based on the sum rule. In the final analysis, the physical reason for fulfillment of the sum rule is the conservation of the total number of particles participating in the given transition, i.e., the same reason for the unitarity of the S-matrix in scattering theory. For this reason the trace of the matrix (4.90) can be calculated in any representation, regardless of whether it is the energy representation.

It is convenient, of course, to choose for this representation the one in which definite values are given for all the angular momenta, including the intermediate moments.

We can, for example, write

$$S(E^i_{J\tau}\,|\,E^f_{J\tau}) = \sum_{F_1,F'_1}\sum_{F^i,F^f} S(F_1F^i\,|\,F'_1F^f) =$$

$$= \sum_{F_1,F'_1}\sum_{F^i,F^f} |(E^i_{J\tau}I_1F_1I_2F^i \,\|\, d^{(1)} \,\|\, E^f_{J\tau}I_1F'_1I_2F^f)|^2. \tag{4.91}$$

The summations can be performed using (4.87) and the orthogonality property of the 6j-symbols (Appendix A). Performing the sums over all indices of the final states F^f (or initial states F^i), we obtain the well-known sum rule,

$$\sum_{\gamma^f F^f} S(\gamma^i F^i\,|\,\gamma^f F^f) = (2F^i+1)\sum_{F_1,F'_1} \frac{|(\gamma^i\,|\,F_1)|^2}{2F_1+1}\,|(E^i_{J\tau}I_1F_1 \,\|\, d^{(1)} \,\|\, E^f_{J\tau}I_1F'_1)|^2, \tag{4.92}$$

according to which the total intensity of the lines having a given initial (or final) level is proportional to the statistical weight of this initial (or final) level $2F^i + 1$ (or $2F^f + 1$).

Summing further over F'_1, we obtain the expression

$$\sum_{\gamma^f F^f} S(\gamma^i F^i\,|\,\gamma^f F^f) = \frac{(2F^i+1)}{(2J^i+1)}\,|(E^i_{J\tau} \,\|\, d^{(1)} \,\|\, E^f_{J\tau})|^2. \tag{4.93}$$

The total strength of the hyperfine multiplet is easily obtained by summing (4.93) over all possible F^i, keeping in mind that the statistical weight of the entire term equals $(2J^i+1)(2I_1+1)(2I_2+1)$,

$$S(E^i_{J\tau}\,|\,E^f_{J\tau}) = (2I_1+1)(2I_2+1)\,|(E^i_{J\tau} \,\|\, d^{(1)} \,\|\, E^f_{J\tau})|^2. \tag{4.94}$$

Finally, we take note of still another sum rule, which is convenient for calculating the intensities of hfs transitions of three identical nuclei. Let the diagonalizing transformation from the representation $j^3\alpha_s IF$ to the energy representation $I_{\alpha_s}F$ be $(j^3\alpha_s I\,|\,I_{\alpha_s})$. Transitions can occur between levels with the same symmetry α_s. We calculate the intensity of all possible transitions for given J^iF^i and J^fF^f:

$$S(E^i_{J\tau}\alpha_s F^i\,|\,E^f_{J\tau}\alpha_s F^f) = \sum_{I^i_{\alpha_s} I^f_{\alpha_s}} |(E^i_{J\tau}I^i_{\alpha_s}F^i \,\|\, d^{(1)} \,\|\, E^f_{J\tau}I^f_{\alpha_s}F^f)|^2 =$$

$$= \sum_I^{\alpha_s} |(E^i_{J\tau}\alpha_s IF^i \,\|\, d^{(1)} \,\|\, E^f_{J\tau}\alpha_s IF^f)|^2 =$$

$$= (2F^i+1)(2F^f+1)\,|(E^i_{J\tau} \,\|\, d^{(1)} \,\|\, E^f_{J\tau})|^2 \sum_I^{\alpha_s} \begin{Bmatrix} 1 & F^i & F^f \\ I & J^f & J^i \end{Bmatrix}^2. \tag{4.95}$$

In the last two sums, the sign $\sum_{I}^{\alpha_s}$ indicates summation over all values I permitted for a given representation α_s. The total strength of a multiplet characterized by a given α_s equals

$$S(E^i_{J\tau}\alpha_s \mid E^f_{J\tau}\alpha_s) = S(E^i_{J\tau} \mid E^f_{J\tau}) \sum_{I}^{\alpha_s} (2I+1). \tag{4.96}$$

Section 5

HYPERFINE STRUCTURE OF THE ND_3 MOLECULE

As an application of the theory set forth in the preceding sections, we shall in this section carry out a calculation of the hfs of the inversion transition of the line J = K = 6 of the heavy ammonia molecule ND_3. The rotation-inversion spectrum of ND_3 is like the well-studied spectrum of ordinary ammonia NH_3; however, the hfs of the inversion transition in ND_3 is considerably more complex than in NH_3. This complication is caused by the fact that each deuterium nucleus has a spin equal to 1 and a quadrupole moment (whereas protons have spin $\frac{1}{2}$ and zero quadrupole moment).

At present the most thorough experimental investigation of the inversion spectrum of ND_3 is that of [31], where, with the aid of a radiospectrometer with molecular modulation, lines ~25 kc wide have been obtained successfully. However, through the use of a maser constructed at FIAN, a preliminary investigation of the hfs of the inversion transition J = K = 6 has recently been carried out with an absolute resolving power of the order of a few hundred cycles per second [10].

The theoretical consideration of the hfs of the inversion spectrum of NH_3 [32] was generalized to the case of ND_3 in [12]. In this paper neither the rotational symmetry of the hfs interactions nor the permutation symmetry of the identical deuterium nuclei was used to a sufficient extent. Hence, the formulas and secular equations obtained in [12] are extremely cumbersome and inconvenient for practical calculations, although in principle they are satisfactory for calculating spectra with an accuracy of a few kilocycles per second. The theory set forth in the preceding chapters utilizes this symmetry in an essential way, and this permits a marked simplification of the calculation in this case. Besides this, additional simplification is attained on account of the use of the tabulated magnitudes of the 6j-symbols, as well as the reduction coefficients presented in Appendix B.

1. Calculation of Hyperfine Structure

General Scheme of the Calculation. The program of the calculation is divided into four steps.

1. Establishment of an effective Hamiltonian of the type (4.51) for the calculation of the interactions of all nuclei in the ND_3 molecule.

2. Selection of representation (momentum coupling scheme) taking into account molecular symmetry and calculation of the matrix elements of the effective Hamiltonian in this representation from the formulas in Part 2 of Section 4.

3. Diagonalization of the matrix of the effective Hamiltonian with respect to the total spin of the deuterium nuclei.

4. Calculation of the intensities of the hfs transitions on the basis of the formulas presented in Part 3 of Section 4.

Each step will be considered briefly below.

Effective Hamiltonian. The spin of each nucleus in the molecule equals 1 and, therefore, the D and N^{14} nuclei do not have octupole or hexadecapole moments. Consequently, the hfs Hamiltonian contains six kinds of terms:

$$\mathcal{H} = \mathcal{H}_N^{(2)} + \mathcal{H}_D^{(2)} + \mathcal{H}(I_N \cdot J) + \mathcal{H}(I \cdot J) + \mathcal{H}(I_N \cdot I) + \mathcal{H}(I \cdot I). \tag{5.1}$$

The first two of these describe the quadrupole interactions of the nitrogen nucleus and the deuterium nuclei, respectively, and are expressed in the form

$$\mathcal{H}_{\mathrm{N}}^{(2)} = (q_{\mathrm{N}}^{(2)} \cdot Q_{\mathrm{N}}^{(2)}) \quad (5.2)$$

and

$$\mathcal{H}_{\mathrm{D}}^{(2)} = \sum_{k=1,2,3} (q_{\mathrm{D}k}^{(2)} \cdot Q_{\mathrm{D}k}^{(2)}), \quad (5.3)$$

where $Q_N^{(2)}$ is the quadrupole moment of the nitrogen nucleus, $q_N^{(2)}$ is the quadrupole gradient of the molecular field at the site of the nitrogen nucleus, $Q_{Dk}^{(2)}$ and $q_{Dk}^{(2)}$ are the corresponding quantities for the k-th deuteron.

$\mathcal{H}(I_N \cdot J)$ and $\mathcal{H}(I \cdot J)$ represent the dipole I · J-interactions of the nitrogen nucleus and the deuterium nuclei, respectively:

$$\mathcal{H}(I_{\mathrm{N}} \cdot J) = -(m_{\mathrm{N}}^{(1)}(J) \cdot M_{\mathrm{N}}^{(1)}); \quad (5.4)$$

$$\mathcal{H}(I \cdot J) = -\sum_{k=1}^{3} (m_{\mathrm{D}k}^{(1)}(J) \cdot M_{\mathrm{D}k}^{(1)}). \quad (5.5)$$

The last two terms in (5.1) represent the interactions of the dipole moments of the nitrogen nucleus with the deuterium nuclei,

$$\mathcal{H}(I_{\mathrm{N}} \cdot I) = r_{\mathrm{ND}}^{-3} \sqrt{10}\, g_{\mathrm{N}} g_0 \mu_n^2 \sum_{k=1}^{3} ([I_{\mathrm{N}}^{(1)} \times Y^{(2)}(\theta_k, \varphi_k)]\, I_{\mathrm{D}}^{(1)}), \quad (5.6)$$

and the dipole moments of the deuterons with each other,

$$\mathcal{H}(I \cdot I) = \sqrt{6}\,(g_0\mu_n)^2 r_{\mathrm{DD}}^{-3} \sum_{k=2,3} [Y^2(\theta_{1k}, \varphi_{1k})]\, [I_{\mathrm{D}_1}^{(1)} \times I_{\mathrm{D}k}^{(1)}]^2. \quad (5.7)$$

In the last two formulas we have introduced the symbols: g_N, g_D, the g-factors of the nitrogen and deuterium nuclei; I_N and I_D, the spins of these nuclei; r_{ND}, the mean distance between the nitrogen and deuterium nuclei; r_{DD}, the mean distance between deuterons; θ_k and φ_k, the spherical coordinates of the k-th deuteron in a system centered on the nitrogen; θ_{1k} and φ_{1k}, the spherical coordinates of the second and third deuterons in a system with its origin at the center of mass of the first deuteron, whereby the axes of both systems coincide with the symmetry axis of the molecule, and the deuteron with label 1 lies in the zy plane.

Order of Magnitude of the Different Interactions. The NH_3 molecules and its isotopic modification ND_3 have completely analogous electronic structures. This gives a basis for estimating the magnitudes of the quadrupole interaction of nitrogen (eQq = −4084 kc), as well as the I · J-interactions of nitrogen ($\mu_n g_N G$ = 3.4 kc) and the deuteriums ($\mu_n g_D G_D \approx -4.5$ kc) in the ND_3 molecule from the experimental data obtained for NH_3 [33, 12]. The quadrupole coupling constant of the deuterium (eqQ = 50 kc) can be estimated on the basis of the data in [31]; however, this does not appear to us to be sufficiently reliable, since, in that work, the magnitude $\frac{1}{2}(q_2^{(2)} + q_{-2}^{(2)})$ was measured from the splitting of the levels $|K| = 1$, and then converted to the direction of the N − D chemical bond, relative to which axial symmetry was assumed. The calculations of the spin interaction constants $g_N g_D \mu_n^2 \langle r_{ND}^{-3} \rangle$ and $(g_0 \mu_n)^2 \langle r_{DD}^{-3} \rangle$ can be carried out starting from the known values of the internuclear spacings. These constants equal 1.24 and 0.62 kc, respectively.

Selection of the Initial Representation. The nitrogen nucleus interacts with the molecular field most strongly, leading to a firm coupling of the nitrogen spin $\mathbf{I}_N$ with the angular momentum $\mathbf{J}$.

Since the energies of the strongest interactions – the quadrupole interactions of nitrogen and the deuterons – differ by approximately 80 times, the quantity $\mathbf{F}_1 = \mathbf{J} + \mathbf{I}_N$ is a sufficiently accurate integral of the motion.

hen coupled (with the same strength) to the momentum $\mathbf{F}_1$. Hence, it
e deuterons in the representation proposed in Section 2, wherein the

total spin of the deuterons I and the type of permutation symmetry α_s characterized by one of the irreducible representations A_1, A_2 and the rows of the two-dimensional representation $E(E_1$ and $E_2)$ are given. Despite the fact that the total spin of the deuterons, which can take on the values 0, 1, 2, 3, is not in general a "good quantum number," the total momentum $F = F_1 + I$ should be an accurate integral of the motion.

The initial representation, therefore, can be symbolized as $vJKI_NF_1j^3\alpha_sIFM_F$, where v is the quantum number of the inversion state. The corresponding symmetrized state vectors in this representation are given by (3.44)-(3.46).

The Energy Representation. For a complete characterization of the energy representation it is necessary for us to turn to a more detailed consideration of the possible spin states of the deuterons. We are interested in the inversion state $J = K = 6$, i.e., the case where K is a multiple of 3. In this case only spin states of the type (3.44) and (3.45) are allowed. In case $j = 1$ for states of type A_2, the one value $I = 0$ is possible (see Appendix B), and for states of type A_1 two values, $I = 1$ and 3. Thus, for states of type A_2 we have $F = F_1$, and, since $I = 0$, all deuteron hfs interactions are mutually compensated.

For states of type A_1 for a given $F_1 = |J + I_N| = 5, 6, 7$ there are seven possible values of $F = F_1 - 3, F_1 - 2, \ldots, F_1 + 3$, since $|F_1 - 1| \leq F \leq F_1 + 1$. Since the hfs energy is diagonal in F_1 and F, then in the four cases $|F - F_1| > 1$ the single value $I = 3$ is possible, which is, consequently, a "good" quantum number. The corresponding representation is given by the quantum numbers $v; J = K = 6; I_N = 1; F_1; A_1I = 3FM_F$. The three cases $|F - F_1| \leq 1$, when two values $I = 1, 3$ are possible, are more complicated. Here we need to diagonalize a matrix of the second order. In place of the quantum number I, we can use the label $I_i^{A_1}$, which takes on the two values $I_1^{A_1}$ and $I_2^{A_1}$.

If we make the condition that in case $|F - F_1| > 1$ the quantum number I^{A_1} takes on the single value $I^{A_1} = 3$ and behaves like an ordinary angular momentum, then both cases can be united by characterizing the energy representation by the set of quantum numbers $v; J = K = 6; I_N = 1; F; I^{A_1}F; M_F$.

Calculation of Energy. The hfs Hamiltonian does not depend on the projection M_F and, as pointed out in Part 1 of Section 3, is diagonal in the inversion state index v. It can be conveniently divided into two terms $\mathcal{H} = \mathcal{H}_1 + \mathcal{H}_2$:

$$\mathcal{H}_1 = \mathcal{H}_N^{(2)} + \mathcal{H}(I_N \cdot J) \tag{5.8}$$

and

$$\mathcal{H}_2 = \mathcal{H}_D^{(2)} + \mathcal{H}(I \cdot J) + \mathcal{H}(I_N \cdot I) + \mathcal{H}(I \cdot I). \tag{5.9}$$

The first part $\mathcal{H}_1$ does not contain interactions of the deuterium nuclei and depends only on F_1 (for given $J = K = 6$, $I_N = 1$). The matrix elements of $\mathcal{H}_1$ are obtained by means of (4.59) and (4.69), those of $\mathcal{H}_2$ from (4.80), (4.81), and (4.82). Reduced matrix elements of the type $(vJKI_NF_1 \| A^{(\lambda)} \| vJKI_NF_1)$ are calculated as indicated in (2.3).

The dependence of the hfs constant on inversion state is extremely weak and need be taken into account only for the nitrogen quadrupole interaction, which we discuss below. For the representation A_2 all matrix elements $\mathcal{H}_2$ equal zero, since $I = 0$. In this case, each inversion line is split into three, according to the possible values $F_1 = 5, 6, 7$. For the representation A_1 each of these levels is split into ten sublevels according to the possible values of F and I^{A_1}.

Selection Rules and Intensities. The electrical dipole moment of the molecule is odd and, therefore, a transition can occur only with a change in inversion state index $v^i = 0 \rightarrow v^f = 1$. The magnitude of $|\langle v^i | d | v^f \rangle|^2$ is 1.49 D [10]. We are interested in the pure inversion transition $\Delta v = 1$ and $\Delta J = \Delta K = 0$ $(J = K = 6)$.

Further selection rules are:

$$\Delta F_1,\ \Delta F = 0, \pm 1;\ \alpha_s^i \rightarrow \alpha_s^f = \alpha_s^i. \tag{5.10}$$

Because of the fact that the type of irreducible representation cannot change during a transition, it is as if there were two kinds of molecule – type A_1 and type A_2 – and in the latter the deuterium nuclei do not give hfs. The statistical weights of the levels A_1 and A_2 are in the ratio 10 : 1. The same applies to the total intensities of all hfs transitions of type A_1 and A_2.

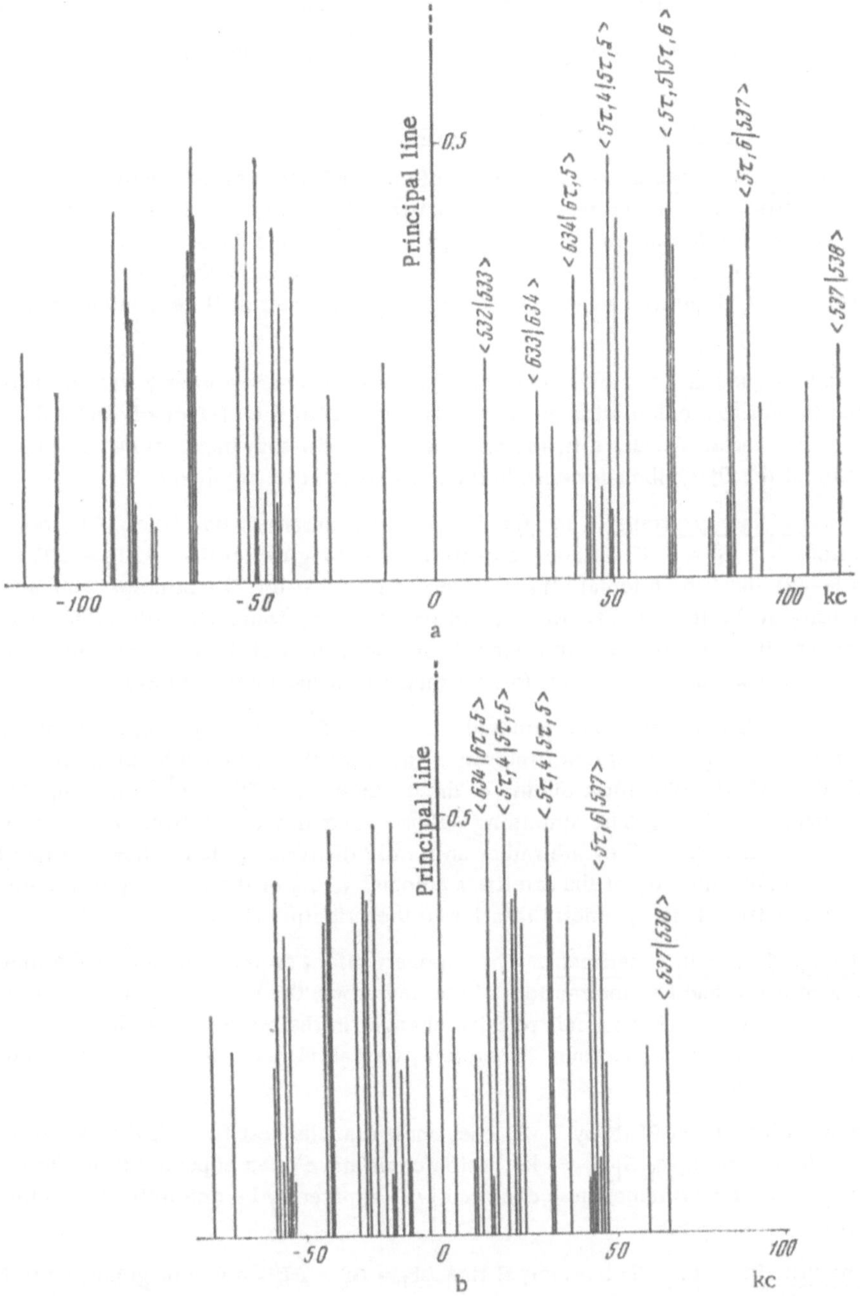

Fig. 2. Calculated spectrum of the satellites to the principal line $\Delta F_1 = \Delta F = 0$ of the hfs of the inversion transition J = K = 6 in the ND_3 molecule. (The scales of eqQ and $R_{DJ} = \mu_n g_D G_D$ are doubled.) a) eqQ = −100 kc, R_{DJ} = −10 kc; b) eqQ = −100 kc, R_{DJ} = −5 kc.

In the majority of cases it is sufficient to use Eqs. (4.86) and (4.87) for the intensity calculations. Only for those transitions which include levels with $|F - F_1| \leq 1$ is it required to know the diagonalizing transformation $(F_1A_1IF|E_1I_i^AF)$; the intensity is calculated from (4.88), where J must be set equal to F_1, and then from (1.120).

The intensities of all hfs transitions of type A_1 and A_2 can be normalized as 10 : 1 by means of the sum rules given in Part 3 of Section 4.

2. Discussion of the Results. Some Conclusions.

Comparison with Handley's Work. The superiority of our methods over the traditional ones used by Handley [12] is quite clearly demonstrated. In our case, the calculation of the hfs energy amounted to the calculation of 30 matrix elements and the solution of ten secular equations of the second degree. To calculate the same energy levels according to [12], it is necessary to solve cubic equations instead of quadratics. This outstanding simplification arises because, in [12], the initial representation involved the intermediate momentum $\mathbf{I}_{23}$ of two deuterium spins.

In the calculation of quadrupole interactions in [12] their rotational symmetry was not utilized, so that their calculation amounted in fact to a calculation of the standard values of the 6j-symbols and reduction coefficients $\Delta(21; II')$ for some special cases. Besides this, the calculation of spin-spin interactions according to [12] amounts in fact to a calculation of the 9j-symbols for certain special values of its argument.

General Character of the Spectrum. The type A_2 spectrum consists of five lines. The principal line ($\Delta F_1 = 0$) corresponds to transitions $F_1^i = F_1^f = 5, 6, 7$. Four satellite lines belong to transitions $\Delta F_1 = \pm 1$, two "external" ($F_1 = 6 \rightarrow 5$ and $F_1 = 6 \rightarrow 7$) and two "internal" ($F_1 = 7 \rightarrow 6$ and $F_1 = 5 \rightarrow 6$). The principal line accounts for almost 97% of the total intensity of the inversion transition, and its frequency coincides with the inversion frequency. Each of the satellites, which are distributed symmetrically on both sides of the principal line at separations of ±2600 kc ($6 \rightleftharpoons 5$) and ±2200 kc ($6 \rightleftharpoons 7$) accounts for less than 1% of the total intensity.

The type A_1 spectrum is much more complicated. Each line $F_1^i \rightarrow F_1^f$ is split into 20-30 fine lines. The splitting of the principal line $\Delta F_1 = 0$ is of chief interest. The transitions $\Delta F_1 = 0$ break up into several tens of lines $\Delta F = 0 \pm 1, (I^{A_1})^i \rightarrow (I^{A_1})^f$. The most intense of these, $\Delta F = 0$ and $I_i^{A_1} \rightarrow I_f^{A_1}$, make up the principal line. The total intensity amounts to ≈86%. The remaining ≈13% occurs in the transitions $\Delta F = \pm 1$ and $I_i^{A_1} \neq I_f^{A_1}$. There are about 40 lines of this kind. They are rather uniformly distributed around the principal line in an interval of ±50 kc. The maximum intensity of the satellites amounts to $^1/_{40}$ of the intensity of the principal line. An essential feature of the spectrum is its symmetry relative to the principal line.

Parameters. Comparison with Experiment. The strongest effect on the structure of the main transition $\Delta F_1 = 0$ comes from the quadrupole and $\mathbf{I}\cdot\mathbf{J}$-interactions of the deuterons, the magnitudes of which, unfortunately, are not known with sufficient accuracy. Even small relative changes in these magnitudes have a pronounced effect on the general picture of the complex spectrum. Therefore, we calculated a few of the parameters of this spectrum, and the results are shown in Figs. 2 and 3.

Preliminary results obtained at FIAN by V. S. Zuev show that the best fit with the experimental data is obtained with $eQ_Dq_D = -50$ kc and $\mu_N g_D G_D = -5$ kc, which could have been expected from the estimates given above. However, in the final experiment, these constants will apparently be determined with much greater accuracy (~300-500 cps).

Structure of the Principal Line. The principal line $\Delta F_1 = \Delta F = \Delta I^{A_1} = 0$ is of great practical interest, since it is used in masers. In first approximation all transitions participating in the principal line have the same frequency, equal to the inversion frequency. However, under closer examination, the principal line is found to consist of a number of closely spaced lines, i.e., it has its own structure. The main reason for the splitting is the small difference between the quadrupole coupling constant of nitrogen for the upper and lower inversion levels [6]. As a result, the frequencies of the transitions $F_1^i = F_1^f = 5, 6, 7$ differ somewhat and yield three components. Since the intensities of the individual components depend on the parameters of the generator, then, being incompletely resolved, they affect the shape and position of the peak of the line, which in its turn limits the absolute stability of the maser frequency [26]. An estimate of this effect, as well as a more detailed consideration of the question, is contained in [10]. In particular, on the basis of the hfs calculation carried out in this paper, the absolute

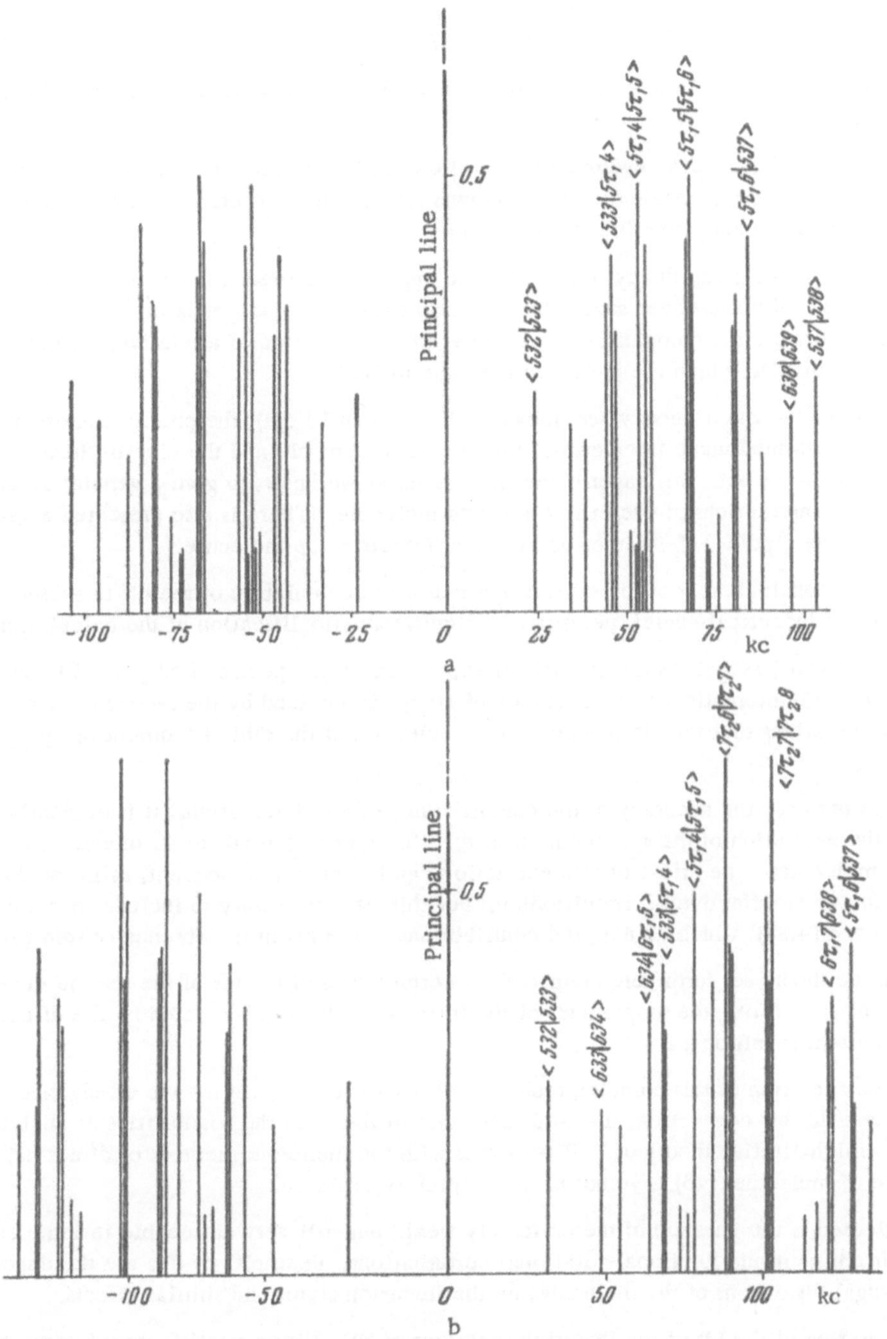

Fig. 3. Calculated spectrum of the satellites to the principal line $\Delta F_1 = \Delta F = 0$ of the hfs of the inversion transition J = K = 6 in the ND_3 molecule. a) eqQ = −50 kc, R_{DJ} = −10 kc; b) eqQ = −100 kc, R_{DJ} = 15 kc.

frequency stability of an ND_3 maser is given there as $2.5 \cdot 10^{-10}$ (i.e., lower than in the usual NH_3 maser), and ways to increase this stability are indicated.

CONCLUSION

In summing up, certain conclusions can be drawn about the current state of the theory of the hfs of the rotational spectra of molecules.

1. A theory of the hfs has been developed which allows the calculation of hyperfine effects in rotational spectra of molecules (in a nondegenerate electronic ground state) with an accuracy not lower than 10 cps, which is the level of precision of modern experimental technique.

By applying methods of group theory, especially the apparatus of irreducible tensor operators and the 3nj-symbols, the calculation of hfs has been successfully greatly simplified and standardized. The effectiveness of these methods can be seen from a comparison of examples of a calculation of a relatively simple case, the hfs of the inversion spectrum of ND_3, which are presented here and in [12].

2. In order to provide the necessary accuracy (not lower than 10 cps), the present treatment includes, besides the usual dipole and quadrupole interactions, the magnetic octupole and the electric hexadecapole interactions. Besides the last two effects, this paper is the first, as far as we know, to give a general calculation of the energy of the spin-spin interactions of nuclei in a rotating molecule. There is also presented a general expression for the energy of the dipole $\mathbf{I} \cdot \mathbf{J}$-interaction in the asymmetric top molecule.

The method of genealogical coefficients was applied to the calculation of the hfs in molecular rotational spectra caused by three indentical nuclei, permitting a significant simplification of the calculation.

3. As estimates show, as well as the analysis of experimental hfs spectra of NH_3 [6, 32] and ND_3, a treatment of the magnetic $\mathbf{I} \cdot \mathbf{J}$-interactions to an accuracy of 10 cps is provided by the second (nonvanishing) approximation of those perturbations that take into account the excitation of the orbital moment of electrons in a rotating molecule.

4. In order to increase the accuracy of the calculations of these interactions, it is evidently necessary to take into account the excitation of the spin moment of the electrons as a result of its interaction with the angular momentum of the molecule. The effect of spin excitation can be taken into account, as in the case of orbital excitation, by setting up an effective hfs Hamiltonian. For this, it is necessary to include in the initial Hamiltonian the terms (4.32) and (4.33), which do not give contributions to the hfs in the absence of spin excitations.

5. Although the theory set forth here permits the interpretation of the hfs of the spectra of singlet molecules, the problem of calculating the magnitudes of the intramolecular fields created by the distribution of electronic charge and current is left open.

The satisfactory solution of this complex problem will evidently require the use of advanced methods using electronic computers. In this connection, it would seem that in this case the possibilities of analytical methods that combine the semiqualitative theory of Daily-Townes with the numerous methods of direct calculation of the electronic structure of molecules [25] have not been completely explored.

6. Also neglected is the question of the extremely weak, but still very noticeable in some experiments, dependence of the hfs constant on vibrational-rotational perturbations. Examples of this are the dependence of the constant on centrifugal distortions of the molecule, on the inversion state, and similar effects.

7. The calculation of the hfs of the inversion spectrum of ND_3 allows a satisfactory interpretation to be made of the experimentally observed spectral transition $J = K = 6$. Based on this calculation, a theoretical estimate of the absolute frequency stability of an ND_3-beam maser is given ($2.5 \cdot 10^{-10}$), and a way to improve this stability is also indicated.

In conclusion, the author takes pleasure in expressing his gratitude to N. G. Basov, A. M. Prokhorov, and A. M. Baldin for a number of valuable discussions that were of great benefit to the work, as well as to A. N. Oraevskii for helpful advice and comments and A. A. Pimenov for assistance in the calculations.

A. Some Properties of the 3nj-Symbols

In Section 4 some relations between the 3j-, 6j-, and 9j-symbols were used that were not given in Section 1.

In calculating intensities (4.86) the orthogonality relation for the 3j-symbols was used:

$$\sum_{m_1 m_2} \begin{pmatrix} j_1 & j_2 & j_3 \\ m_1 & m_2 & m_3 \end{pmatrix} \begin{pmatrix} j_1 & j_2 & j_3' \\ m_1 & m_2 & m_3' \end{pmatrix} = (2j_3 + 1)^{-1} \delta_{j_3 j_3'} \cdot \delta_{m_3 m_3'} \cdot \delta(j_1 j_2 j_3), \tag{A.1}$$

where $\delta(j_1 j_2 j_3) = 1$ if j_1, j_2, and j_3 satisfy the triangle rule and is equal to zero otherwise. This property comes from the unitarity of the Clebsch-Gordan transformation (1.30) and (1.67).

In establishing the sum rule (4.91) and in following relations the orthogonality of the 6j-symbols was used:

$$\sum_{j} (2j + 1)(2j'' + 1) \begin{Bmatrix} j_1 & j_2 & j' \\ j_3 & j_4 & j \end{Bmatrix} \begin{Bmatrix} j_1 & j_2 & j'' \\ j_3 & j_4 & j \end{Bmatrix} = \delta_{j', j''}, \tag{A.2}$$

which comes from the unitarity of the matrix for the recoupling of momenta and (1.71). This relation was used for the interval rule (4.55).

Various more complex relations between the symbols can be established by means of the diagrammatic technique developed in [16]. Some of these, e.g., Eq. (4.56), we used directly in the text. Equations (4.78) and (4.79) were obtained starting from

$$\sum_{\mu} (2\mu + 1) \begin{Bmatrix} j_{11} & j_{12} & \mu \\ j_{21} & j_{22} & j_{23} \\ j_{31} & j_{32} & j_{33} \end{Bmatrix} \cdot \begin{Bmatrix} j_{11} & j_{12} & \mu \\ j_{23} & j_{33} & \mu \end{Bmatrix} = (-1)^{2\lambda} \begin{Bmatrix} j_{21} & j_{22} & j_{23} \\ j_{12} & \lambda & j_{32} \end{Bmatrix} \cdot \begin{Bmatrix} j_{31} & j_{32} & j_{33} \\ \lambda & j_{11} & j_{21} \end{Bmatrix}, \tag{A.3}$$

and setting $j_{11} = j_{12} = 1$ and $\lambda = \frac{1}{2}(j_{23} + j_{33})$.

B. Genealogical Coefficients; Reduction Coefficients

In Section 2 we introduced the genealogical coefficients (2.28)-(2.31) and the reduction coefficients (2.42) and (2.46). Here we present their numerical values for the cases of identical nuclei with spins $j = 1$ and $^3/_2$. The case of spin $^1/_2$ is trivial, since the specification of total spin I at once determines the permutation symmetry, namely: $I = {}^3/_2$, type A_1; $I = {}^1/_2$, type E. In the cases $j = 1$ and $^3/_2$, each of the irreducible representations is contained in the product $D^{(j)} \times D^{(j)} \times D^{(j)}$ not more than once. Hence, there is no necessity for introducing an additional index (β). The phases of the matrix elements are specified such that the least possible value is chosen as the initial intermediate moment (L).

The reduction coefficients have the symmetry properties

$$\Delta(\lambda j \alpha_s;\ II') = \Delta(\lambda j \alpha_s;\ I'I), \tag{B.1}$$

$$\sigma(j \alpha_s;\ II') = \sigma(j \alpha_s;\ I'I), \tag{B.2}$$

whereby only the values for $I \leq I'$ are given in the tables.

Reduction Coefficients $\Delta(\lambda j\ \alpha_s;\ \text{II}')$

$\lambda = 1$ $\lambda = 2$

$j = 1$

$\Delta(11\ A_1;\ 11) = 1$
$\Delta(11\ A_1;\ 13) = 0$
$\Delta(11\ A_1;\ 33) = 3.741\ 66$
$\Delta(11\ A_2;\ 00) = 0$
$\Delta(11\ E_1;\ 11) = 0.5$
$\Delta(11\ E_1;\ 12) = 1.936\ 52$
$\Delta(11\ E_1;\ 22) = -1.118\ 03$
$\Delta(11\ E_2;\ 11) = 1.5$
$\Delta(11\ E_2;\ 12) = 1.936\ 52$
$\Delta(11\ E_2;\ 22) = -3.354\ 10$

$\Delta(21\ A_1;\ 11) = 1.8$
$\Delta(21\ A_1;\ 12) = 1.833\ 28$
$\Delta(21\ A_1;\ 22) = 2.245\ 00$
$\Delta(21\ A_2;\ 00) = 0$
$\Delta(21\ E_1;\ 11) = 1.5$
$\Delta(21\ E_1;\ 12) = -0.866\ 03$
$\Delta(21\ E_1;\ 22) = -2.291\ 29$
$\Delta(21\ E_2;\ 11) = -1.5$
$\Delta(21\ E_2;\ 12) = -2.598\ 07$
$\Delta(21\ E_2;\ 22) = -2.291\ 29$

$j = 3/2$

$\Delta(1\ 3/2\ A_1;\ 3/2\ 3/2) = 1$
$\Delta(1\ 3/2\ A_1;\ 3/2\ 5/2) = 0$
$\Delta(1\ 3/2\ A_1;\ 5/2\ 5/2) = -1.870\ 83$
$\Delta(1\ 3/2\ A_2;\ 3/2\ 3/2) = 1$
$\Delta(1\ 3/2\ E_1;\ 1/2\ 1/2) = -1.581\ 14$
$\Delta(1\ 3/2\ E_1;\ 1/2\ 3/2) = 0.77460$
$\Delta(1\ 3/2\ E_1;\ 3/2\ 3/2) = 0.6$
$\Delta(1\ 3/2\ E_1;\ 3/2\ 5/2) = 0.916\ 52$
$\Delta(1\ 3/2\ E_1;\ 5/2\ 5/2) = -1.657\ 02$
$\Delta(1\ 3/2\ E_1;\ 5/2\ 7/2) = -2.028\ 37$
$\Delta(1\ 3/2\ E_1;\ 7/2\ 7/2) = -2.070\ 20$
$\Delta(1\ 3/2\ E_2;\ 1/2\ 1/2) = 0.948\ 68$
$\Delta(1\ 3/2\ E_2;\ 1/2\ 3/2) = 0.774\ 60$
$\Delta(1\ 3/2\ E_2;\ 3/2\ 3/2) = 2.6$
$\Delta(1\ 3/2\ E_2;\ 3/2\ 5/2) = -0.916\ 52$
$\Delta(1\ 3/2\ E_2;\ 5/2\ 5/2) = -2.084\ 64$
$\Delta(1\ 3/2\ E_2;\ 5/2\ 7/2) = 2.022\ 14$
$\Delta(1\ 3/2\ E_2;\ 7/2\ 7/2) = 3.726\ 35$

$\Delta(2\ 3/2\ A_1;\ 3/2\ 3/2) = 0.6$
$\Delta(2\ 3/2\ A_1;\ 3/2\ 5/2) = -1.959\ 59$
$\Delta(2\ 3/2\ A_1;\ 5/2\ 5/2) = -0.458\ 28$
$\Delta(2\ 3/2\ A_1;\ 5/2\ 9,2) = -1.5$
$\Delta(2\ 3/2\ A_2;\ 3/2\ 3/2) = -1$
$\Delta(2\ 3/2\ E_1;\ 1/2\ 1/2) = 0$
$\Delta(2\ 3/2\ E_1;\ 1/2\ 3/2) = -1.039\ 23$
$\Delta(2\ 3/2\ E_1;\ 3/2\ 3/2) = 0.6$
$\Delta(2\ 3/2\ E_1;\ 1/2\ 5/2) = -0.648\ 07$
$\Delta(2\ 3/2\ E_1;\ 3/2\ 5/2) = -1.8$
$\Delta(2\ 3/2\ E_1;\ 3/2\ 7/2) = 2.091\ 85$
$\Delta(2\ 3/2\ E_1;\ 5/2\ 5/2) = -0.130\ 93$
$\Delta(2\ 3/2\ E_1;\ 5/2\ 7/2) = -0.907\ 12$
$\Delta(2\ 3/2\ E_1;\ 7/2\ 7/2) = -0.925\ 82$
$\Delta(2\ 3/2\ E_2;\ 1/2\ 1/2) = 0$
$\Delta(2\ 3/2\ E_2;\ 1/2\ 3/2) = -0.346\ 41$
$\Delta(2\ 3/2\ E_2;\ 1/2\ 5/2) = -1.944\ 22$
$\Delta(2\ 3/2\ E_2;\ 3/2\ 3/2) = 2.2$
$\Delta(2\ 3/2\ E_2;\ 3/2\ 5/2) = -0.6$
$\Delta(2\ 3/2\ E_2;\ 3/2\ 7/2) = 0.692\ 82$
$\Delta(2\ 3/2\ E_2;\ 5/2\ 5/2) = 0.392\ 79$
$\Delta(2\ 3/2\ E_2;\ 5/2\ 7/2) = 2.721\ 34$
$\Delta(2\ 3/2\ E_2;\ 7/2\ 7/2) = 2.777\ 46$

Reduction Coefficients $\sigma(j\ \alpha_s;\ \text{II}')$

$j = 1$

$\sigma(1\ A_1;\ 11) = 0.133\ 33$
$\sigma(1\ A_1;\ 13) = 0.101\ 84$
$\sigma(1\ A_1;\ 33) = 0.748\ 33$
$\sigma(1\ A_2;\ 11) = 0$
$\sigma(1\ E_1;\ 11) = 0.216\ 66$

$\sigma(1\ E_1;\ 12) = -1.338\ 93$
$\sigma(1\ E_1;\ 22) = -0.381\ 88$
$\sigma(1\ E_1;\ 11) = -0.75$
$\sigma(1\ E_1;\ 12) = 1.299\ 04$
$\sigma(1\ E_1;\ 22) = -1.145\ 64$

$j = {}^3/_2$

$\sigma(3/2\ A_1;\ 3/2\ 3/2) = 0.24310$
$\sigma(3/2\ A_1;\ 3/2\ 5/2) = -0.05243$
$\sigma(3/2\ A_1;\ 5/2\ 5/2) = 0.24112$
$\sigma(3/2\ A_1;\ 5/2\ 9/2) = 0.12197$
$\sigma(3/2\ A_1;\ 9/2\ 9/2) = 0.70356$
$\sigma(3/2\ A_2;\ 3/2\ 3/2) = -0.013\,19$
$\sigma(3/2\ E_1;\ 1/2\ 1/2) = 0$
$\sigma(3/2\ E_1;\ 1/2\ 3/2) = -0.07748$
$\sigma(3/2\ E_1;\ 1/2\ 5/2) = 0.15896$
$\sigma(3/2\ E_1;\ 3/2\ 3/2) = -0.13131$
$\sigma(3/2\ E_1;\ 3/2\ 5/2) = 0.13027$
$\sigma(3/2\ E_1;\ 5/2\ 5/2) = 0.33081$

$\sigma(3/2\ E_1;\ 7/2\ 7/2) = -0.45356$
$\sigma\ 3/2\ E_1;\ 3/2\ 7/2) = -0.12702$
$\sigma(3/2\ E_1;\ 5/2\ 7/2) = -0.31451$
$\sigma(3/2\ E_2;\ 1/2\ 1/2) = 0$
$\sigma(3/2\ E_2;\ 1/2\ 3/2) = -0.03243$
$\sigma(3/2\ E_2;\ 1/2\ 5/2) = 0.15866$
$\sigma(3/2\ E_2;\ 3/2\ 3/2) = 0.01319$
$\sigma(3/2\ E_2;\ 3/2\ 5/2) = 0.06559$
$\sigma(3/2\ E_2;\ 3/2\ 7/2) = -0.27373$
$\sigma(3/2\ E_2;\ 5/2\ 5/2) = 0.16036$
$\sigma(3/2\ E_2;\ 5/2\ 7/2) = 0.33333$
$\sigma(3/2\ E_2;\ 7/2\ 7/2) = 0.45356$

Tables of Genealogical Coefficients

$j = 1$

α_s	A_1		A_2	E_1		E_2	
L \ I	1	3	0	1	2	1	2
0	$\sqrt{5/3}$	0	0	2/3	0	0	0
1	0	0	1	0	0	1	1
2	2/3	1	0	$-\sqrt{5/3}$	1	0	0

$j = {}^3/_2$

α_s	A_1			A_2		$E_2 - E_1$				E_2			
L \ I	1/2	3/2	5/2	9/2	3/2	1/2	3/2	5/2	7/2	1/2	3/2	5/2	7/2
0	0	0	0	0	0.4082	0	0	0	0	0	0.9128	0	0
1	1	0.8367	0.7303	0	0	1	0.5478	0.6831	0	0	0	0	0
2	2	0	0	0	0.9130	0	0	0	0	1	0.4082	1	1
3	3	0.5478	0.6832	1	0	0	0.8367	0.7303	1	0	0	0	0

LITERATURE CITED

1. C. H. Townes and A. Shawlow, Radiospectroscopy [Russian translation] IL (1959).
2. N. F. Ramsey, Molecular Beams [Russian translation] IL (1960).
3. A. Abragam and M. H. L. Pryce, Proc. Roy. Soc. (London), A205, 135 (1951).
4. P. Kursch and H. N. Folly, Phys. Rev., 72, 1256 (1947).
5. P. Thaddeus and J. Loubser, Nuovo cimento, 10, 1060 (1959).
6. J. P. Gordon, Phys. Rev., 99, 1253 (1955).
7. P. Thaddeus, J. Loubser, L. Kriser, and H. Lecar, J. Chem. Phys., 31, 1677 (1959).
8. Quantum Electronics, Columbia University Press, New York (1960), p. 47.
9. N. G. Basov and A. N. Oraevskii, Radiotekhnika i élektronika, 4, 1185 (1959).
10. N. G. Basov, V. S. Zuev, and K. K. Svidzinskii, Molecular Beam Generators, Repts. FIAN (1959).
11. E. Condon and G. Shortley, Theory of Atomic Spectra [Russian translation] IL (1948).
12. C. F. Handley, J. Chem. Phys., 26, 1482 (1957).
13. E. P. Wigner, Gruppentheorie, Vieweg (1931).
14. U. Fano and G. Racah, Irreducible Tensorial Sets, Academic Press, New York (1959).
15. A. R. Edmonds, Angular Momentum in Quantum Mechanics, Princeton University Press (1957).
16. A. P. Yutsis, I. B. Levinson, and V. V. Vanagas, Mathematical Apparatus of the Theory of Angular Momentum, Gos. izdat. polit. i nauchn. lit., Lithuanian SSR, Vilna (1960).
17. J. H. Van Vleck, Revs. Modern Phys., 23, 213 (1951).
18. C. Schwartz, Phys. Rev., 97, 380 (1955).
19. N. G. Basov and K. K. Svidzinskii, ND_3 Maser, Izv. vyssh. ucheb. zav., radiofizika, 1, No. 2 (1958).
20. K. K. Svidzinskii, Calculations of the hfs of Molecules, Optika i spektroskopiya, 6, 254 (1959).
21. P. A. M. Dirac, Principles of Quantum Mechanics [Russian translation] Fizmatgiz (1960).
22. L. Landau and E. Lifshits, Quantum Mechanics, Gostekhizdat (1949).
23. P. Met'yus, Relativistic Quantum Theory of the Interactions of Elementary Particles [Russian translation] IL (1959).
24. P. J. Redmond, Proc. Roy. Soc. (London), A222, 84 (1954).
25. C. A. Coulson, Revs. Modern Phys., 32, 170 (1960).
26. C. H. Townes and B. P. Daily, Journ. Phys., 17, 782 (1949).
27. J. Blatt and V. Weisskopf, Theoretical Nuclear Physics [Russian translation] IL (1954).
28. E. P. Wigner, Nachr. Ges. Wiss. Gottingen (1932), p. 546.
29. L. H. Thomas, Nature, 117, 514 (1926).
30. R. S. Henderson, Phys. Rev., 74, 107 (1948).
31. G. R. Herrman, J. Chem. Phys., 29 , 875 (1958).
32. G. R. Gunter-Mohr, C. H. White, and J. H. Van Vleck, Phys. Rev., 94, 1191 (1954).
33. G. R. Gunter-Mohr, R. L. White, A. L. Shawlow, W. E. Good, and D. R. Coles, Phys. Rev., 94, 1184 (1954).
34. W. E. Lamb, Repts. Progr. Phys., 14, 19 (1951).
35. N. G. Basov and A. M. Prokhorov, Uspekhi Fiz. Nauk, 12, 485 (1955).

THE ND_3 MASER

N. G. Basov, V. S. Zuev, and K. K. Svidzinskii

INTRODUCTION

The present work was aimed at investigating the possibility of making a maser using inversion transitions of heavy ammonia (ND_3).

As a result, we have succeeded in making a working model of a maser using a beam of ND_3 molecules. The power output of the maser at 1656.18 Mc (J = 6, K = 6 line in the inversion spectrum of ND_3) amounts to 10^{-11} watt. The absolute stability of the line, according to preliminary data, is of the order of 10^{-9}.

The possibility of making such a maser was indicated by calculations [1] given in the first and second sections of this paper. In the first section it is shown that a ND_3 maser whose principle of operation and constructional details are analogous to the NH_3 maser [2-5] can be made to self-oscillate for reasonable values of the various parameters.

The second section contains the results of calculations on the hyperfine inversion spectrum of ND_3. Analysis of the hyperfine structure leads to an estimate for the absolute stability of the J = 6, K = 6 line ($\sim 10^{-9}$).

Further, we discuss the possibility of increasing the absolute stability by means of choosing other lines (J = 3, K = 2, and J = 5, K = 5), or by means of substituting the isotope N^{15} for N^{14}.

The third section contains a description of the apparatus and gives the results obtained by studying the working model. There is satisfactory agreement between the measured results and the calculated data.

The power radiated by the ND_3 maser is about 100 times smaller than that of the NH_3 maser. Since the maximum sensitivity of microwave detectors in the region of 1600 Mc is substantially greater than that of detectors in the 2400-Mc range, it was no harder to detect a signal from the ND_3 maser than from the NH_3 maser. In our experiments, the signal-to-noise ratio exceeded 100 when the power from the maser resonator was 10^{-12} watt. Such a maser is particularly interesting from the point of view of frequency division, since in the region of 1600 Mc one can probably amplify directly before dividing the maser frequency.*

In a ND_3 maser, one can use a beam of large cross section with diameter about 100 mm. This eases constructional problems when obtaining a beam of slow molecules for making a maser of ultrastable frequency [6]. Also, as in the case of the ordinary ammonia maser [7,8], this maser may be used as a spectroscope of very high resolving power (of the order of several hundred cycles) [26]. The high spectroscopic resolving power enables one to study more completely the hyperfine interaction in the ammonia molecule. Detailed investigation of the hyperfine structure in the ND_3 spectrum is necessary also for evaluating the stability of the NH_3 maser.

* At present, in connection with making a system for phase stabilization of a klystron frequency with respect to a NH_3 maser, a possibility has arisen for direct division of the ammonia maser frequency by means of stabilizing the klystron frequency on a harmonic.

Section 1

MASER CALCULATIONS

1. Choice of Inversion Transition

Frequencies of the inversion transitions of ND_3 are given by the empirical formula [9]

$$\nu = 1595.69 - 7.155\,(J+1)\,J + 10.03K^2 \quad \text{Mc.}$$

Measurements and identification for lines of the principal series (K = J) are given in Table I [9].

The electrical dipole moment of the inversion transition is 1.49(6) Debye units according to measurements taken on the polarization of ND_3 in the gaseous state as a function of temperature [10].

TABLE I. Frequencies of Inversion Transitions of ND_3 [9]

J,K	Frequency, Mc	J,K	Frequency, Mc
1	1598.10	10	1815.37
2	1591.72	11	1872.43
3	1599.53	12	1937.31
4	1612.99	13	2010.57
5	1631.82	14	2092.32
6	1656.18	15	2183
7	1686.45	16	2285
8	1722.85	17	2403
9	1765.80	18	2540

The intensity of the inversion line is proportional to the number N_{JKM_J} of molecules on the particular rotational level characterized by the numbers J, K, M_J, to the nuclear statistical weight q_I, and to the square of the matrix element of the dipole moment,

$$|d_{JKM_J}|^2 = d_0^2\left[\frac{KM_J}{J(J+1)}\right]^2.$$

The nuclear statistical weight q_I for ND_3 has the value [11]

$$q_I = \begin{cases} \frac{11}{19} & \text{for K a multiple of 3;} \\ \frac{8}{19} & \text{for K not a multiple of 3.} \end{cases}$$

Since the dipole moment is proportional to K^2, the line to be chosen is the one with the largest value of K for given J, i.e., the line K = J.

The frequency of an inversion transition is independent of M_J and, as a result, molecules with all possible projections M_J add to the intensity of the line. The intensity provided by a molecule is proportional to M_J^2. In the beam of sorted molecules, however, molecules with high M_J predominate, since the efficiency of separating (as will be shown in a further calculation) is proportional to M_J. Hence, the intensity provided by molecules with various values of M_J is proportional to M_J^3. This leads to the intensity of the line (for the choice K = J), depending only on J and being proportional to

$$\exp\left[-\frac{hc}{kT}(BJ + AJ^2)\right]\cdot J\cdot q_I,$$

where $\exp\left[-\frac{hc}{kT}(BJ + AJ^2)\right]$ characterizes the population of the symmetric top rotational level K = J, and the factor J appears due to molecules with various M_J's. The rotational constants of ND_3 have the following values [11]: $A = 3.15\ \text{cm}^{-1} = 94.5 \cdot 10^3$ Mc, $B = 5.138\ \text{cm}^{-1} = 154 \cdot 10^3$ Mc, $hc/kT = 4.83 \cdot 10^{-3}$.

This expression has its maximum for J equal to about 5 or 6. Therefore, we made the choice of the J = 6, K = 6 line, and all further calculations will be for this line. The frequency of the J = 6 , K = 6 inversion transition is 1656.18 Mc.

2. Resonator

As in the case of the NH_3 maser [2], the resonator was chosen in cylindrical form, in which a wave of type TM_{010} was excited. In such a resonator the molecular beam lies along the axis, and there is minimum Doppler broadening of the radiated lines.

The stability of a maser is improved, and the threshold is reduced, on increasing the resonator length up to

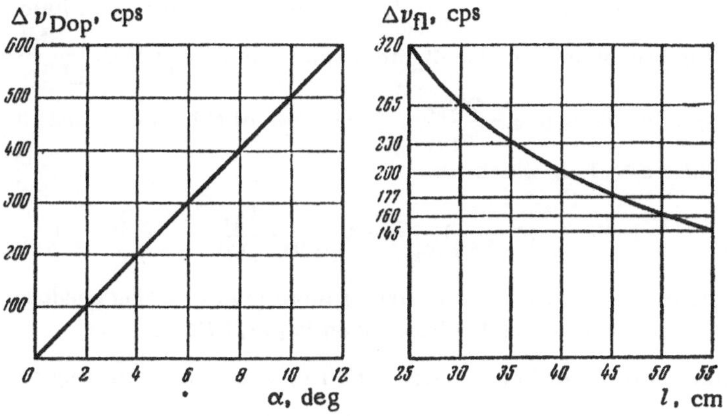

Fig. 1.

the point when the "flight" (or natural) width of the line $\Delta\nu_{fl} = 1/2\pi\tau$ becomes narrower than the Doppler width (here $\tau = l/\bar{v}$ is the mean time of flight of the molecules across the resonator, l is the resonator length, and $\bar{v}$ is the mean speed). Figure 1 is a graph of $\Delta\nu_{fl}$ versus l and of the Doppler width $\Delta\nu_{Dop}$ versus the angle α between the direction of motion of a molecule and the axis of the beam.

The calculation is carried out for $\bar{v} = 5 \cdot 10^4$ cm/sec. In our apparatus the resonator length is l = 40 cm. This length is apparently maximal in the above sense, since for reasonable beam apertures (~4°), the Doppler line width is equal to the "flight" width.

The resonator radius R_{res} is determined by the frequency of the line, and equals about 7 cm.

3. Beam Source

The maximum molecular flux which may be obtained from an oven with a grid of radius R_n is [2]

$$N_n = \frac{\pi\bar{v}R_n}{4\sqrt{2}\sigma} = 0.46 \cdot 10^{19} R_n \text{ molecules/sec,}$$

where σ is the self-collision cross section for ND_3 molecules, and equals $5.9 \cdot 10^{-15}$ cm^2 [12].

Of these molecules in the J = 6, K = 6 state, the number on the upper inversion level is given by

$$N_{JKM_J} = 0.48 \cdot 10^{16} R_n \text{ molecules/sec,}$$

where M_J is the projection of J on the electric field direction in the separator. The following numerical values for the molecular parameters of ND_3 [11] were used in evaluating N_{JKM_J}: rotational constants A = 3.15 cm^{-1}, B = 5.138 cm^{-1}; rotational statistical sum taking account of inversion doubling Q = $1.184 \cdot 10^3$; nuclear statistical weight $q_I = {}^{11}/_{19}$ (K = 6); rotational statistical weight q_{JK} = 2 (J = 6, K = 6).

4. Separating Molecules According to Inversion States

If separating the molecules according to inversion states is carried out by means of an inhomogeneous electrical field, the number $(N_{JKM_J})_{sep}$ is given by [1]

$$(N_{JKM_J})_{sep} \quad N_{JKM_J}\frac{W_m}{kT}, \quad (W_m \ll kT). \tag{1}$$

W_m is the maximum displacement of the inversion level of the molecules in the field,

$$W_m = \frac{W_{inv}}{2}\left(\sqrt{1 + 4\left(\frac{W_{st}}{W_{inv}}\right)^2} - 1\right). \tag{2}$$

N_{JKM_J} is the number of molecules in the level characterized by the numbers J, K, M_J ; $W_{inv} = h\nu_{inv}$ is the energy of the inversion interaction; $W_{st} = E_m |d_{JKM_J}|$, where E_m is the maximum field intensity of the separator and $|d_{JKM_J}|$ is the matrix element of the dipole moment, which equals $d_0 \left[\frac{KM_J}{J(J+1)}\right]$.

Expression (2) holds for any value of $a = \frac{2W_{st}}{W_{inv}}$. In the case of NH_3, $a \ll 1$ and $(N_{JKM_J})_{sep} = N_{JKM_J} \cdot \frac{W_{st}^2}{W_{inv} kT}$. For ND_3, usually $a \gg 1$, in which case

$$(N_{JKM_J})_{sep} \approx N_{JKM_J} \cdot \frac{W_{st}}{kT}. \tag{3}$$

The separator may be either a condenser with four or more rods, or a ring condenser. In our apparatus a ring condenser was used, the theory for which has been given in [13, 27].

Reference [14] contains a numerical calculation for a ring condenser consisting of eight rings.

According to [13], the expression for the potential of a condenser consisting of a large number of rings has the form

$$\varphi(\rho z) = \begin{cases} \sum_{n=0}^{\infty} A_n I_0 \left[\frac{(2n+1)\pi\rho}{l}\right] \sin\left[\frac{(2n+1)}{l}\pi z\right] & \text{for } \rho \leqslant R, \\ \sum_{n=0}^{\infty} B_n K_0 \left[\frac{(2n+1)\pi\rho}{l}\right] \sin\left[\frac{2n+1}{l}\pi z\right] & \text{for } \rho > R, \end{cases} \tag{4}$$

where $I_n(x)$ and $K_n(x)$ are the modified Bessel function and the MacDonald function, respectively, for $n = 0$; R is the radius of the ring; l is the distance between rings; and ρ, z are cylindrical coordinates with the axial coordinate z lying along the axis of the separator. The coefficients A_n and B_n are determined by the boundary conditions and are rapidly decreasing functions of the index n. Hence, to an accuracy of a few percent, the expression (4) for the potential may be written in the form

$$\varphi(\rho, z) = \begin{cases} \frac{0.6U}{I_0(\pi R/l)} I_0(\pi\rho/l) \sin\frac{\pi z}{l} + \frac{0.1U}{I_0(3\pi R/l)} I_0(3\pi R/l) \sin\frac{3\pi z}{l} & \text{for } \rho \leqslant R, \\ \frac{0.6U}{K_0(\pi R/l)} K_0(\pi\rho/l) \sin\frac{\pi z}{l} + \frac{0.1U}{K_0(3\pi R/l)} K_0(3\pi R/l) \sin\frac{3\pi z}{l} & \text{for } \rho > R, \end{cases} \tag{5}$$

where U is the potential difference between neighboring rings.

From (5) the electric field intensity of a ring condenser may easily be evaluated. To calculate the number of active molecules, it is necessary to know only the maximum field intensity E_m, which for the ring condenser used by us ($R = 2$, $l = 2$ cm) is given by

$$E_m = 1.41U. \tag{6}$$

The general expression for the field intensity is too cumbersome to be given here.

With the aid of (6), (1), and (2), we obtain the following expression for the number of separated molecules:

$$(N_{JKM_J})_{sep} = \frac{N_{JKM_J} W_{inv}}{2\, kT}\left(\sqrt{1 + 4\left(\frac{1.41U\,|d_{JKM_J}|}{W_{inv}}\right)^2} - 1\right) =$$

$$= \chi 0.48 \cdot 10^{16} R_n \frac{h\nu_{inv}}{2kT}\left(\sqrt{1 + 4\left(\frac{1.41Ud_0}{h\nu_{inv}} \cdot \frac{KM_J}{J(J+1)}\right)^2} - 1\right), \tag{7}$$

where for N_{JKM_J} we have used the expression obtained in Section 1, Part 3,

$$N_{JKM_J} = 0.48 \cdot 10^{6} R_n,$$

R_n being the radius of the beam source. The factor χ is to allow for the fact that molecules leaving peripheral points of the source are focused weakly [2]. In our case, χ = 10%, since the distance from the beam source to the separator was several times the beam source diameter.

The minimum length for the separator may be calculated by solving the equation of motion for a molecule in the separator field. For a quadrupole condenser the calculation has been carried out in [1,2]. The potential barrier of a ring condenser is a parabola of order higher than 2 [13, 14]. Hence, this barrier may be taken as different from zero only close to the rings and the minimum length of the ring condenser must be calculated from the mean angle of capture.

Capture angle is the name we give to the maximum angle α_{max} between the direction of travel of a molecule and the condenser axis through which the molecule will still be reflected from the potential barrier. α_{max} is given by the relation

$$W_m = \frac{m\bar{v}^2 \sin^2 \alpha_{max}}{2}, \tag{8}$$

where W_m is the maximum interaction energy of a molecule in the condenser field, and $\bar{v}$ is the average speed of molecules in the beam.

The minimum length L_{min} for the condenser is derived from the condition that a molecule entering the condenser at an angle α_{max} experiences at least two reflections. Accordingly,

$$L_{min} = 3R \cot \alpha_{max}, \tag{9}$$

where R is the radius of the ring.

5. Minimum Quality Factor and Power

The direction of the field in the condenser is the same as that of the electric field in the resonator where a wave of type TM_{010} is excited. Hence, the number of active molecules (with a fixed value of $|M_J|$) is given by

$$N_{JK|M_J|} = 2\,(N_{JKM_J})_{sep}. \tag{10}$$

The minimum value of the resonator Q_0 which will allow self-oscillation follows from the relation [2]

$$\frac{1}{Q_0} = \frac{4l_{res}}{R^2_{res}\, v^2 \hbar} \sum_{M_J=1}^{6} N_{JK|M_J|} \cdot |d_{JKM_J}|^2. \tag{11}$$

Taking into account that the field E in the resonator is homogeneous, we may write the power radiated by molecules in the resonator in the form

$$P = \frac{V_{res}\,\omega}{8\pi Q} |\mathbf{E}|^2, \tag{12}$$

where ω is the generated frequency, V_{res} is the volume of the resonator, and Q is the measured quality factor of the resonator including the losses caused by power leaking out. The amplitude of the electric field $|\mathbf{E}|$ may be expressed in terms of the saturation parameter

$$F = \frac{\tau^2 d_0^2}{\hbar^2} |\mathbf{E}|^2, \tag{13}$$

which may be found from the following equation [2]:

$$\frac{1}{Q} = \frac{4l_{res}}{R^2_{res}\, v^2 \hbar} \sum_{M_J=1}^{6} \frac{N_{JK|M_J|}\,|d_{JKM_J}|^2}{1 + \frac{|d_{JKM_J}|^2}{d_0^2} F}.$$

Calculation shows that F may be written in the form

$$F \approx 1.35\left(\frac{Q}{Q_0} - 1\right). \tag{14}$$

By means of (12)-(14) and (11) we find that

$$P = \left(1 - \frac{Q_0}{Q}\right) P_0,$$

where

$$P_0 = 0.67 h\nu_{\text{inv}} \sum_{M_J=1}^{6} N_{JK\,|M_J|}\, |d_{JKM_J}|^2/d_0^2.$$

for the line J = 6, K = 5, ν_{inv} = 1656.18 Mc, and d_0 = 1.5 · 10^{-18} e.s.u. For beam source radius R_n equal to 1.75 cm and U = 100 kV, the minimum quality factor Q_0 = 9750. For Q = 20,000, P = 2.6 · 10^{-11} watt.

Section 2

HYPERFINE STRUCTURE OF THE INVERSION SPECTRUM AND ESTIMATE OF THE ABSOLUTE STABILITY OF THE MASER

1. Hyperfine Structure of the Inversion Transition J = 6, K = 6

The hyperfine structure of the inversion spectrum of NH_3 has been the subject of a number of investigations [15, 16], and it was studied in particular detail in connection with the creation of a maser [17, 8]. In the case of ND_3, the quadrupole interaction of the deuterium nucleus appreciably complicates the hyperfine structure, and calculation of the spectrum is unwieldy. In [2] the calculation of the spectrum for NH_3 [16] was generalized for the case of ND_3, and the secular equations determining the energy levels were derived. However, the formulas obtained are rather cumbersome and inconvenient for numerical evaluation. The general theory of the hyperfine structure developed in [18] makes significantly simpler calculations possible. Calculation of the hyperfine structure of the inversion transition J = K = 6 of the ND_3 molecule is described in Section 5 of [18]. There also are to be found various calculations of satellites about the main line, which are compared with preliminary experimental data.

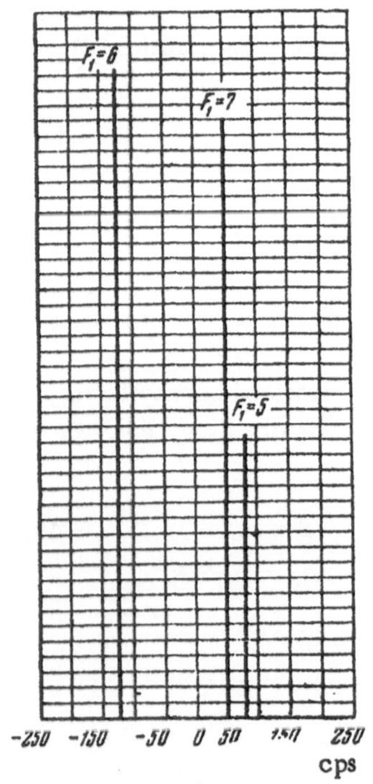

Fig. 2.

2. Structure of the Main Line

In a maser, oscillation takes place at the most intense main line in the inversion spectrum formed by eight transitions of the type v = 1 → 0, $\Delta F_1 = \Delta F = 0$, $I_i^{A1} \rightarrow I_i^{A1}$ [18]. To a first approximation all these lines have the same frequency as the inversion transition. However, on closer inspection, it is found that the main line consists of three closely spaced lines, i.e., has its own structure [17]. In the case of NH_3 this structure is not resolved, as a result of which the position of the peak of the main line depends on the relative intensities of the individual unresolved lines. The intensities of the unresolved lines in turn depend on the separator voltage, which leads to displacement in the peak of the main line and, consequently, in maser frequency on changing the separator voltage. This circumstance (at present) sets the most serious limit on the absolute frequency stability of the maser, and therefore will be analyzed in more detail.

In the ND_3 molecule the splitting of the main line may be brought about by two effects: 1) the dependence of the hyperfine structure constants on the inversion state; 2) removal of degeneracy according to the sign of K. Removal of degeneracy according to the sign of K may result from a) the

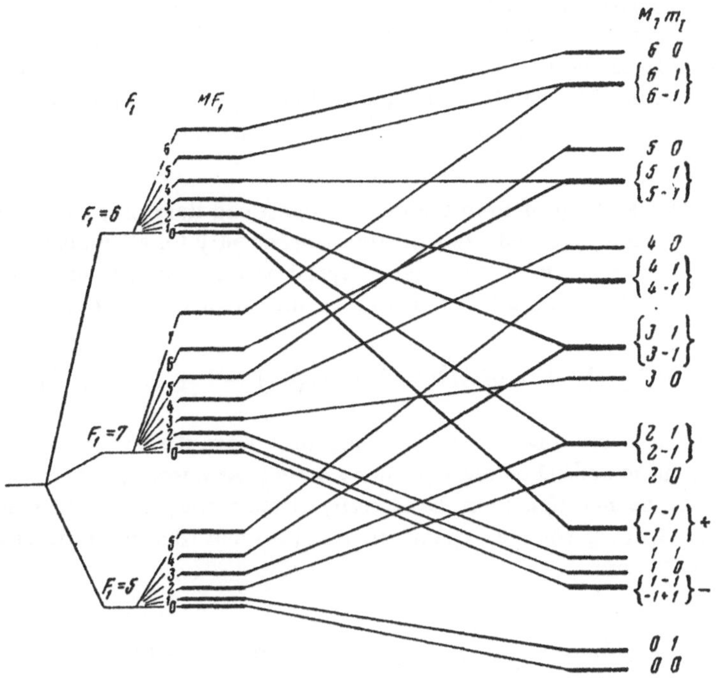

Fig. 3.

hyperfine interaction of nuclei situated not on the molecule's axis of symmetry; b) rotational-vibrational interaction.

For ND_3 the splitting of the $|K| = 1$ line is $\Delta\nu_1 \approx 50$ kc. Estimates show that this splitting, caused by interactions of nonaxial nuclei, must amount to 2-10 cps for $|K| = 2$, and is negligibly small ($\sim 10^{-15}$ cps) for $|K| = 6$.

Rotational-vibrational interaction may remove the degeneracy between the levels A_1 and A_2, i.e., for $|K| = 3n$. The corresponding splitting of the inversion line for $J = 5$, $|K| = 3$ amounts to the fairly large value $\Delta\nu_3 = 1.728$ Mc [23].

However, for $K = 6$, the magnitude of $\Delta\nu_6$ is appreciably smaller, and according to our estimates amounts to about 0.05 cps. We note that $\Delta\nu_6$ is proportional to J^{12}, and for $J = 12$ is already approximately 200 cps.

The above estimates show that removal of degeneracy according to the sign of K may be neglected. We turn now to effects of the second type which lead to the appearance of structure in the main line, namely, dependence of the hyperfine interaction on the inversion state.

The transition from the upper inversion state to the lower involves a change in the average value of the angle β between the axis of the molecule and a ND bond, and this leads to a rather weak dependence of the hyperfine structure constants on the inversion state. The structure of the main line is perceptibly affected only by a change in the constant of the quadrupole nitrogen bond, which is at least one order of magnitude greater than the other constants. The theory of [18] leads to an expression for the change in the second derivative of the electric field,

$$(\Delta q)_{ND_3} \simeq \left(q\frac{\nu_{inv}}{\nu_{rot}}\right)_{ND_3}\left(\frac{\nu_{rot}}{\nu_{inv}}\frac{\Delta q}{q}\right)_{NH_3},$$

which gives $\Delta|eqQ|_{ND_3} \simeq 500$ cps, i.e., the magnitude of the quadrupole splitting in the lower inversion state is 500 cps greater than in the upper state. This leads to the structure for the main line shown in Fig. 2. For a line width of approximately 200-300 cps this must result in noticeable distortion of the form of the line.

3. Influence of the Separator on the Intensity of the Hyperfine Structure Lines

For calculating the effect of the separator on transitions of the type $F_1 \rightarrow F_1'$ we may neglect the weak interaction of the deuterons and reckon that, in the absence of a field or in very weak fields ($W_{st} \ll eqQ$), the energy of a molecule is determined by given quantum numbers F_1, M_{F_1}. On entering the separator, the levels are equally populated. Within the separator the energy levels undergo strong Stark splitting ($W_{st} \gg eqQ$). As a result, the relation $\mathbf{F}_1 = \mathbf{J} + \mathbf{I}_N$ is violated and to each level $E(F_1, M_{F_1})$ in the weak field there will correspond a definite strong field level $E(M_J, M_I)$, where M_J and M_I are the projections of $\mathbf{J}$ and $\mathbf{I}_N$ on the direction of the electric field in the separator. Correlations between energy levels in weak and strong fields are shown schematically in Fig. 3.

In view of the fact that the transition from the weak to the strong field may be considered adiabatic, and also that separation of the molecules proceeds statistically independently for all molecular energy states, the populations of the F_1, M_{F_1} states on leaving the separator depend only on the appropriate value of M_J in the strong field. According to what was said above (Part 4 of Section 1), the effectiveness of separating the level M_J is given by the factor

$$\varphi(M_J) = \sqrt{1 + a^2 M_J^2} - 1 \simeq a|M_J| - 1 + \ldots (a > 1),$$

where a is a factor proportional to the voltage U on the separator. Thus, the populations of the F_1, M_{F_1} levels on exit from the separator turn out to depend on the separator voltage. As a result, the intensity of the $F_1 \to F_1'$ transition in the case of a separated beam differs from the intensity of the corresponding transition in a gas under equilibrium, and depends on the separator voltage. Calculations in [18] show that the ratio of intensities I_{sep}/I_{gas} is proportional to the following factors:

$$I_5(a) = \frac{1}{2F_1 + 1}(a - 1) \quad \text{for} \quad F_1 = 5;$$

$$I_6(a) = \frac{1}{2F_1 + 1}(2.37a - 1) \quad \text{for} \quad F_1 = 6;$$

$$I_7(a) = \frac{1}{2F_1 + 1}(2.1a - 1) \quad \text{for} \quad F_1 = 7.$$

At sufficiently large field ($a \gg 1$) the ratios of intensities I_{F_1} of the lines are independent of the magnitude of the field and are equal to

$$I_5 : I_6 : I_7 = 0.183 : 0.425 : 0.392 \quad (I_5 + I_6 + I_7 = 1).$$

This distribution of intensities is shown in Fig. 2, Now it is not difficult to calculate the shift of the line peak as a function of separator voltage. If we ignore saturation effects and reckon that the structure of the main line is unresolved, the position of the maximum is determined by the center of gravity of the unresolved components:

$$\nu_m = \frac{\nu_5 I_5(a) + \nu_6 I_6(a) + \nu_7 I_7(a)}{I_5(a) + I_6(a) + I_7(a)}.$$

Taking $\nu_5 = 80$ cps, $\nu_6 = 125$ cps, and $\nu_7 = 50$ cps, and assuming that the intensities are determined by the formulas given above, one easily obtains

$$\delta\nu_m = \frac{13.6}{a} \cdot \frac{\delta a}{a} \text{ cps} = \frac{13.6}{a}\left(\frac{\delta U}{U}\right) \text{cps}.$$

On changing the voltage U from 100 to 60 kV, the parameter a changes from 18.0 to 10.2, which corresponds to a displacement of the peak by $\delta\nu_m = 0.4$ cps. Accordingly, the absolute stability, which is determined by the shift of the spectral line for a given change of separator voltage, is given by

$$\frac{\delta\nu_m}{\nu} = 2.5 \cdot 10^{-10},$$

i.e., is not lower than for a usual ammonia maser (where $\delta\nu_m/\nu = 10^{-9}$ for the J = 3, K = 3 line).

4. Methods for Increasing Stability

To increase the absolute stability the same methods may be used as were pointed out for the NH_3 maser [19]. The principal possibilities for increasing the absolute stability are: 1) choice of the J = 3, K = 2 inversion line for maser action, since this line has no nitrogen quadrupole interaction; * 2) replacement of N^{14} by N^{15}, which has no quadrupole moment.

In spite of the fact that these steps lead to no simplification of the hyperfine structure spectrum, the structure will be determined by hyperfine interaction constants which are at least an order of magnitude smaller than eqQ for nitrogen. The resulting shift of the main line peak, caused by the difference in these constants for the upper and lower inversion levels, must also be an order of magnitude smaller.

We note that the intensity of the J = 3, K = 2 ND_3 line is only half that of the J = 6, K = 6 line. If N^{15} is substituted for N^{14}, it is expedient to use the J = 5, K = 5 line. This line is as intense as the J = 6, K = 6 line, but has a rather simpler hyperfine structure, since its spectrum is of type E (see [18], Section 5).

Section 3

CONSTRUCTION OF THE ND_3 MASER

Constructional details of the maser are shown in Fig. 4. The source of the ND_3 molecular beam is indicated by 1, the separator by 2, and the resonator by 3. The whole apparatus is enclosed in a hermetically sealed chamber. The beam length in our apparatus was about 1.5 m; hence, the apparatus was maintained at a vacuum of (6-7) · 10^{-6} mm Hg, at which the mean free path is about 1.7 m. As a result of the constant flow of ammonia into the apparatus, it was necessary to use either high-speed pumps or to cool large areas of the surface by means of liquid nitrogen in order to freeze out the ammonia (which was done in our apparatus). The number 4 indicates the cold jacket, 5 is the trap for freezing out the beam when it has passed through the resonator, and 6 is the diaphragm for freezing out molecules which are traveling at large angles to the axis.

A constant voltage of 100 kV was applied to the separator. A highly sensitive superheterodyne receiver was used to indicate maser radiation.

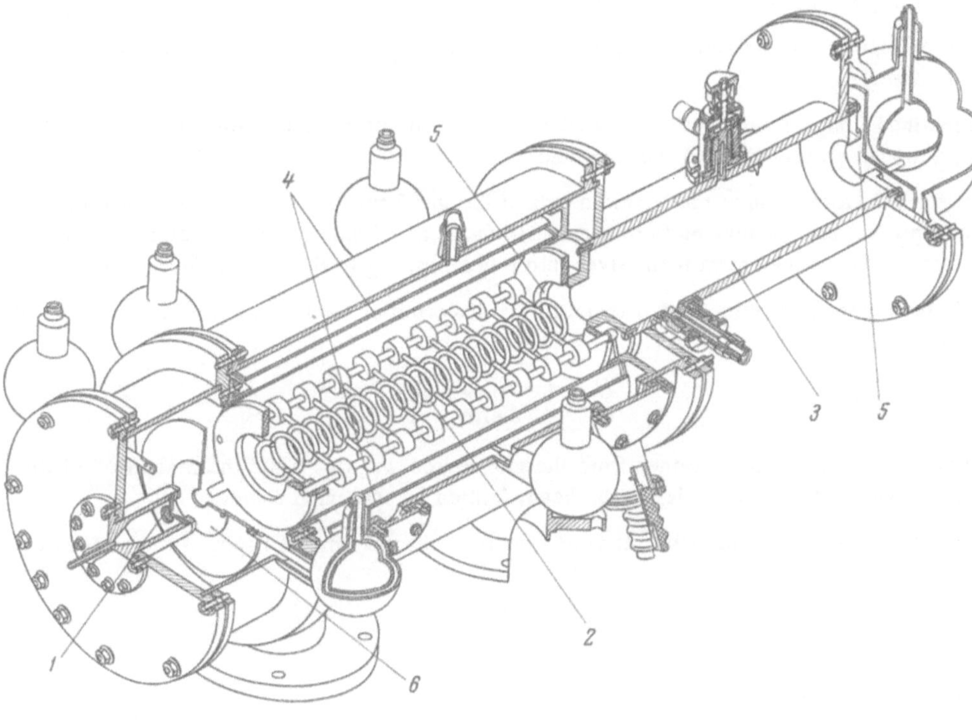

Fig. 4.

* In first-order perturbation theory.

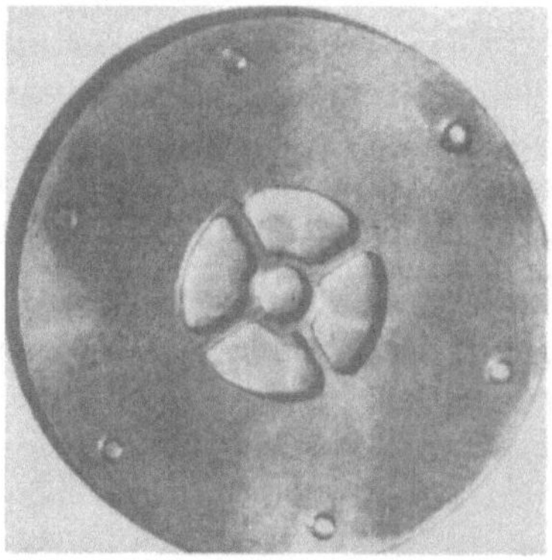

Fig. 5.

Fig. 6.

In what follows we shall consider in separate parts the construction of the maser and the superheterodyne receiver. The final section of the paper contains experimental results and discussion.

1. Beam Source and Structure for Launching the ND_3

For creating a beam of ND_3 molecules we used a grid source in the form of a copper foil with square openings measuring 0.1 × 0.1 mm. The distance between neighboring holes was 0.1 mm. The porosity factor of such a grid (i.e., the ratio of the hole area to the total area of the grid) is equal to 0.25. The thickness of the foil was 0.05 mm and the radius R of the grid was 17.5 mm. A grid of R = 10 mm was also used.

Experiments carried out with an NH_3 beam maser have shown [17] that a canal may be used as beam source, the diameter of the canal being appreciably greater than the mean free path of molecules in the oven. Accordingly, we used a source having 19 canals arranged uniformly in a circle of radius R = 17.5 mm. The diameter of the canals was 1.5 mm, and their length was 10 mm.

The beams obtained from the above-mentioned sources were sufficiently intense for maintaining self-oscillation of the maser.

A beam obtained from a source with one canal of 1.5-mm diameter and 10-mm length did not achieve self-oscillation, although a radiation line was observed.

The mechanism of formation of the beam on exit of the gas from the canals is not known in detail. We may probably assume that the gas flowing from a canal forms a ray of molecules accelerated in the direction of the canal, and that these rays then form themselves into a beam owing to their slight divergence.

Increase in the component of molecular speed in the direction of the beam axis leads to a rise in the number of separated molecules:

$$N \simeq \left(\frac{\bar{v}}{v_{\text{therm}}}\right)^2 N_0,$$

where N_0 is the number of molecules obtained from the grid source, v_{therm} is the thermal speed of the molecules, and $\bar{v}$ is the mean speed of molecules in the beam formed by the canal source.

Increase in the speed of the molecules leads to a decrease in the time of flight of the molecules through the resonator:

$$\tau = \frac{l}{v},$$

where l is the length of the resonator.

However, the threshold condition for the maser remains unchanged, since the decrease in the time τ is compensated by an increase in the number of active molecules,

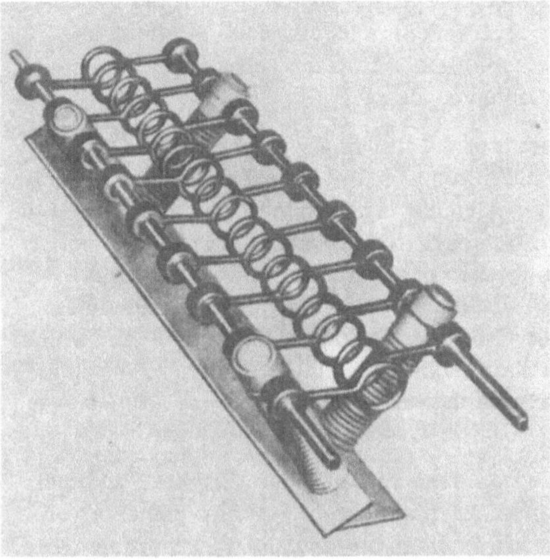

Fig. 7.

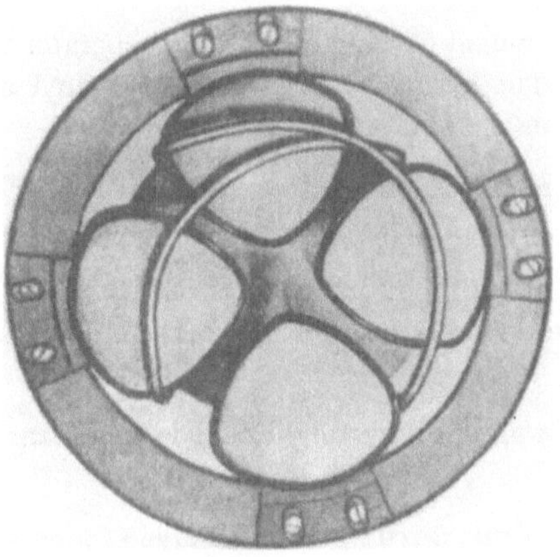

Fig. 8.

$$\beta \sim N\tau^2,$$

where β is the parameter of self-oscillation [2].

The power of the maser rises somewhat:

$$P = P_0 \frac{N}{N_0}.$$

Here P_0 is the power of the maser with the grid as beam source.

The increase in molecular speed, and the resulting decrease in interaction time between the molecules and the resonator field, leads to a broadening of the radiated line.

$$\Delta \nu = \frac{1}{2\pi\tau}.$$

Canals probably give a more directional beam than a grid. This could lead to a reduction in the quantity of ND_3 expended.

Our measurements show that the gas pressure in the beam source chamber for which an optimal beam is formed is identical for both sources* (grid radius R = 17.5 mm and 19 canals). The self-oscillation threshold in terms of voltage for both cases was 80 kV. We did not measure the line widths.

The ammonia was let into the beam chamber from a balloon in which it was stored in the liquid state (the saturated vapor pressure of ND_3 is 11 atm at 0°C). The balloon was connected to the expansion volume by means of a needle valve with a teflon seating. From the expansion volume the gas passed through the accumulator into the beam source chamber.

The accumulator was a constrained piece of vacuum rubber tubing, inside of which was placed a length of 0.5-mm diameter wire with the object of forming a narrow canal.

The whole system, right from the balloon up to where the ammonia was launched, was evacuated to a pressure of $5 \cdot 10^{-3}$ mm Hg.

We did not use standard reducers, since these cannot be well evacuated unless they are radically altered.

The spent ammonia was removed from the apparatus by freezing it into a glass ampule which could be taken off.

2. The Separator

For separating the molecules according to inversion states we used a focuser in the form of rings. The metal rings were arranged in a single row as if they had been threaded on a rod. The separation of the rings was of the order of one ring radius. All even rings were maintained at plus the source voltage, and all odd rings at minus the same voltage.

*Pressure measured by means of an LT-2 manometer tube in the tubing feeding ammonia to the beam source chamber.

Fig. 9.

As the beam travels along the axis of such a system, the molecules are separated according to their inversion levels, since the field is inhomogeneous and increases from the axis to the periphery in the same way as the field in a quadrupole condenser.

At a ring radius in front of the first ring a diaphragm was installed which was cooled with liquid nitrogen and which passed a narrow pencil from the wide beam of molecules leaving the oven. The diameter of the aperture in the diaphragm was equal to a ring diameter. Behind the last ring of the separator, at a distance of about a half-ring radius, was a second cold diaphragm. This diaphragm evidently froze out molecules which were on the lower level of the transition being studied and which were focused into the region of maximum field, i.e., close to the rings. The diameter of the aperture in the second diaphragm was also equal to a ring diameter.

The rings were made from brass rods of 6-cm diameter, and their inside diameter was 38.5 mm. Each of them had an arm branching off in the plane of the ring, 6 mm in diameter and about 50 mm long. The arms ended in bushings, by means of which the rings were secured to the crossbeams. All even rings were secured to one crossbeam, and all odd rings to the other. Altogether, 20 rings were installed. The distance between them was 20 mm, and the length of the whole system was 50 cm.

The crossbeams were fitted to two ring insulators with lugs by means of which the whole separator was mounted in the vacuum jacket between the oven and the resonator. The insulators were made of a vinyl resin. Figure 7 shows the separator assembled (with a somewhat different construction of the insulators).

The voltage was brought to the crossbeams via vacuum-soldered lead-ins. All the metal parts were free of sharp points. The rings were polished.

The insulators were washed in alcohol and rinsed with acetone. During assembly their surfaces were not touched by hand. Only by observing such cleanliness were we able to avoid discharges between the crossbeams along the surfaces of the insulators.

The high-voltage leads were insulated by means of plexiglas insulators with large breakdown distances. The same precautions were taken with regard to this insulation.

The system, once made, withstood 110 kV. On exceeding this limit, discharge was observed along a ring insulator.

We also tried a separator in the form of a quadrupole condenser (Fig. 8). The minimum clearance between the electrodes was about 5 mm. The electrodes were secured to two glass ring insulators. The distance along the ring between the electrodes was about 100 mm. Owing to the awkwardness of the whole system, we did not succeed in bringing the surfaces of the electrodes and insulators to the required state of cleanliness. Discharge between the electrodes and along the insulators took place at voltages of the order of 30-40 kV. For this reason the maser was operated with the ring-type separator.

3. The Resonator

The resonator of the maser (Fig. 9) was a cylindrical volume, in which a wave of type TM_{010} was excited. The electric field vector in such a resonator is directed along the geometrical axis of the cylinder, perpendicular to the ends. The electric field is maximum at the resonator axis and is zero at the walls. The magnetic field lines form closed circles in cross-sectional planes of the resonator.

Coupling between the resonator and the high-frequency line was accomplished by means of a loop projecting through its side and in a plane passing through its axis. Since we were using a transmission-type resonator, extraction of power was accomplished by means of a similar loop on the opposite side of the resonator. The loops

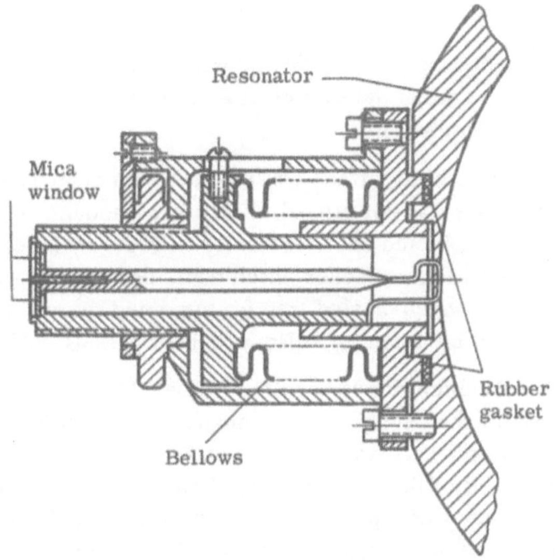

Fig. 10.

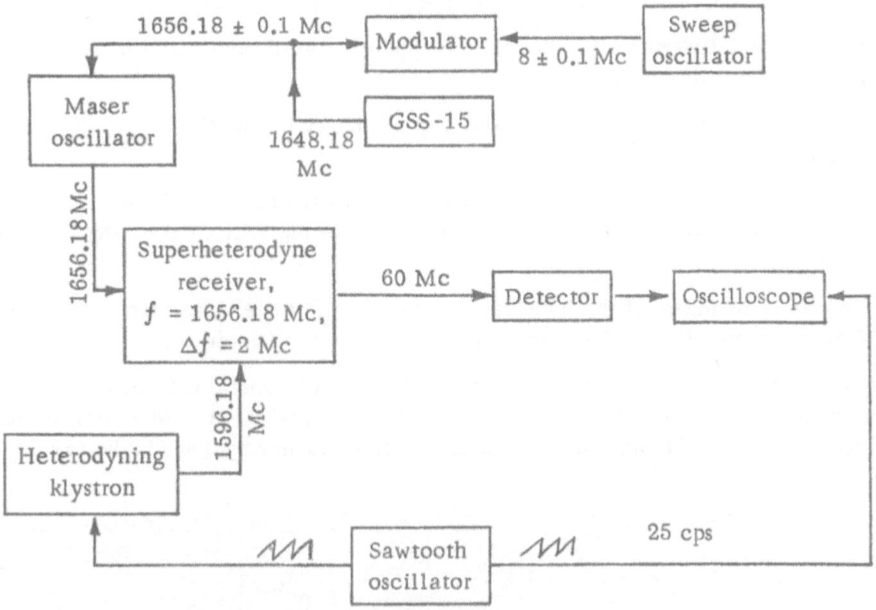

Fig. 11.

were connected to coaxial cables ending in standard couplings. Coupling between the resonator and the cables was adjusted by varying the depth of insertion of the loops.

The resonator consisted of three parts – a cylinder and two end covers, the latter being bolted to the former when the resonator was assembled. Since currents flow along generators of the cylinder and the current paths are closed by means of the end covers, it is necessary to have good contact at the junction in order to achieve high Q. In order to improve the contact, a ring of fine copper wire (diameter 0.3 to 0.5 mm) was placed under the ends. On tightening the bolts the ring was crushed, and good contact was obtained. The end covers had channels by which the beam could enter and leave, these channels coinciding with the axis of the resonator. The diameter of the entry channel was 35 mm, and its length was 40 mm; the corresponding figures for the exit channel were 60 and 80 mm, respectively. These channels act as cutoff attenuators and prevent the escape of radiation from the resonator.

Owing to the divergence of the beam and the possible noncoincidence of the beam and resonator axes, the "spent" molecules will be reflected from the rear wall of the resonator and will accumulate in it. Hence, the diameter of the exit channel must be made large in comparison with that of the entrance channel. The effect of accumulation was observed in our apparatus when the exit channel was of 20 mm diameter and 25 mm length and the entrance channel had the same measurements.

The resonator was made of brass. Its inside surface was covered with copper and was polished to a mirror-like finish. In our construction the resonator was not placed wholly in a vacuum. Only its inner volume was evacuated. The resonator had two flanges soldered to it, coaxially with the cylinder. By means of one of the flanges the resonator was connected to the vacuum jacket which contained the oven and the separator. The other flange served to connect with the liquid-nitrogen-cooled trap for removing the spent molecules. The loops were also vacuum sealed. Sealing was achieved by means of bellows, rubber gaskets, and mica windows (Fig. 10).

The resonance wavelength of a cylindrical cavity excited in the TM_{010} mode is given by the relation [20] $\lambda_{res} = 2.61a$, where $2a$ is the diameter of the cavity. The resonance condition is independent of the length. However, in a cavity of this diameter, TM_{011} and TE_{111} waves may also be excited, though their resonance conditions do depend on the distance between the ends of the cylinder. By choice of this distance we may avoid these unwanted resonances. The diameter of a cavity of type TM_{010} without ports for a frequency of 1656.18 Mc is 138.57 mm. For a length of 400 mm the nearest resonance of type TM_{011} is close to a frequency of 1700 Mc, and the nearest resonance of type TE_{111} is close to 1300 Mc. A TM_{010} resonance is always observed at a lower frequency than a TM_{011} resonance. Therefore, these two types of oscillation are easily distinguishable.

A TE_{111} wave is not axially symmetric; hence, this type of wave is recognized by means of a dipole rotated in a plane of diametral section of the cavity.

Rough adjustment of the resonator frequency was carried out by means of a length of quartz tube introduced through the exit channel.

The external diameter of the tube was ~60 mm, and its wall thickness was 2 mm. The length of the tube was equal to the length of the exit channel. For a 30-mm length of the quartz projecting into the cavity, the resonance was observed at 1656 Mc.

Tuning of the cavity by means of the tube was carried out with the end trap removed. A tuned vacuum resonance of 1656.18 Mc changed to 1655.66 Mc on filling the cavity with air.

Fine tuning of the evacuated cavity was performed by means of a metal rod introduced halfway along its side. The rod was vacuum sealed by means of bellows and a rubber gasket. The diameter of the rod was 10 mm, and the depth of its insertion was 15 mm. Such a rod shifts the frequency by 300-350 kc toward higher values.

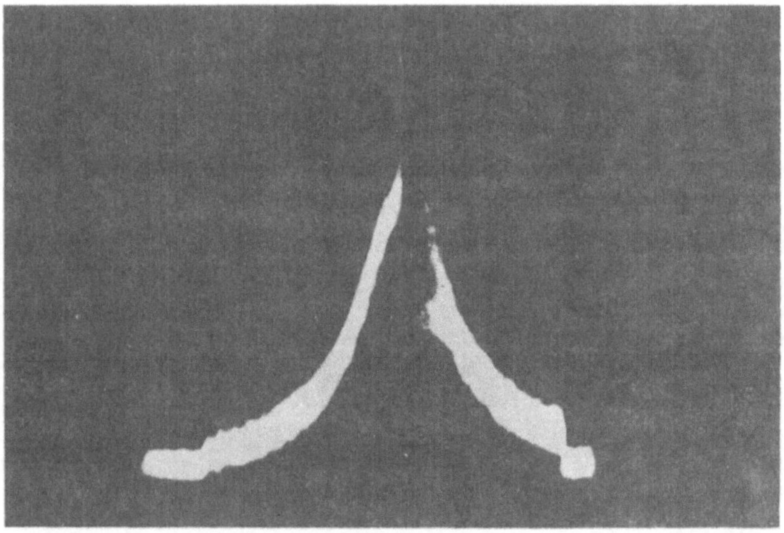

Fig. 12.

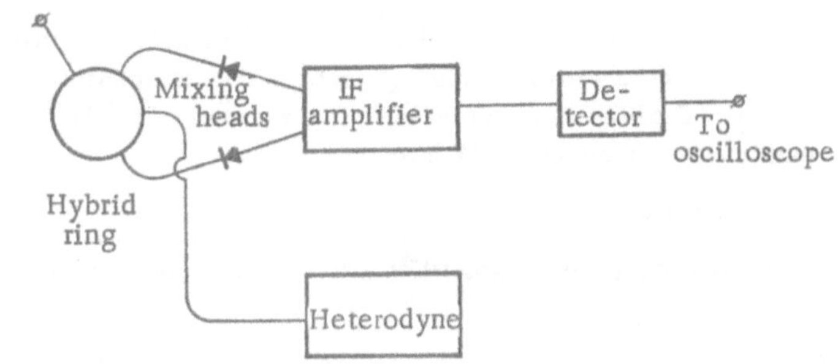

Fig. 13.

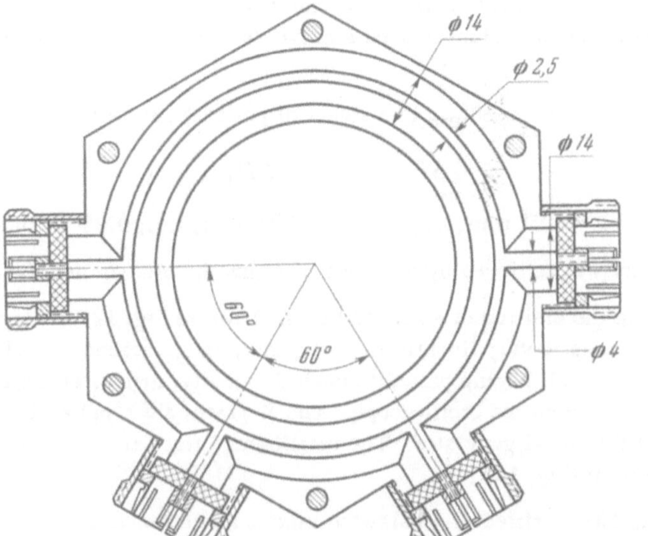

Length of central conductor $6\frac{\lambda}{4}$

Fig. 14.

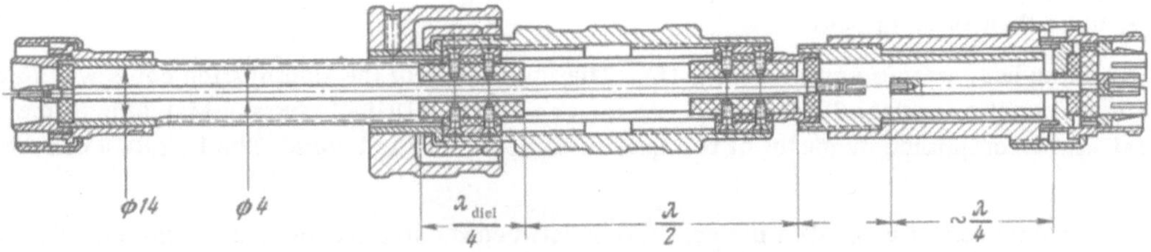

Fig. 15.

The calculated Q of the resonator was ~34,000. For an area of the feed loop of less than 5 mm^2 and of the pickup loop of less than 15 mm^2, its Q was ~20,000. The Q was determined by measuring the frequency width of the cavity for half-power transmission. The maximum value of Q (~25,000) was determined apparently by parasitic radiation from the cavity. For a Q of 20,000, the resonance width measured between the half-power points is 80 kc.

Tuning of the cavity and measuring its width was carried out with the aid of a ShGV-S heterodyne wavemeter. The frequency of the wavemeter was calibrated against an internal crystal to a relative accuracy of 10^{-5}.

4. The Vacuum System

The oven and the separator were housed in a chamber evacuated by means of two TsVL-100 pumps. To freeze out the ammonia there was a liquid-nitrogen-cooled jacket within the chamber. The resonator was connected to the chamber by a flange. Oscillation occurred only if a vacuum of $6 \cdot 10^{-6}$ mm Hg was attained. At this pressure the mean free path of the molecules is ~1.5 m.

5. Detection of the Radiation

The power radiated by the ND_3 beam maser was 10^{-12} watt. Hence, the sensitivity of the receiver had to be better than that value.

In the process of setting up the maser it was also necessary to observe the absorption line. Since the absorption coefficient of the line corresponding to the inversion transition of the ND_3 J = 6, K = 6 rotational transition is $\sim 10^{-6}$ cm^{-1}, 10^{-8} watt is the power at which the line becomes saturated in a cavity of quality factor 10,000. The length of waveguide equivalent to such a cavity is ~100 m [21]. For this length the absorption is 1% of the total power.

The minimum observable change ΔP_{min} in a signal of power P due to receiver noise P_n is given by [21]

$$\Delta P_{min} = \sqrt{4PP_n}\,.$$

For $P_n = 10^{-14}$ watt and $P = 10^{-8}$ watt, $\Delta P_{min} = 2 \cdot 10^{-11}$ watt, i.e., 0.2% of the power $P = 10^{-8}$ watt.

The required sensitivity was achieved by us by use of a superheterodyne receiver.

The setup for observing the emitted or absorbed line is shown in Fig. 11. As is seen from the figure, the signal from the generator is made to vary linearly in a frequency range around 1656.18 Mc, and is applied to the feed loop in the cavity. The signal, having passed through the cavity, arrives at the receiver, where it is amplified, detected, and then displayed on the oscilloscope. The X-plates are supplied by the same sawtooth voltage as controls the frequency of the signal generator. The oscilloscope trace shows the response curve of the cavity, with the line superimposed on it (Fig. 12).

The receiver (see Fig. 13) consisted of a balanced mixer, a heterodyning klystron, and an i.f. amplifier.

The balanced mixer consisted of a hybrid ring, two mixing heads, and a constant attenuator.

The hybrid ring was of coaxial type (Fig. 14). The output ports were coaxial, of 75-ohm impedance, and were terminated in standard 75-ohm couplings. The impedance of the ring coaxial was 106 ohms.

The standing-wave ratio of such a ring with matched loads is 1.2-1.3. The signal and heterodyne were decoupled by more than 10 db in power.

The mixing head was also coaxial (Fig. 15). The impedance of the transmission cable was 75 ohms, and the couplings were also standard 75-ohm ones. As mixer we used a DGS-2 crystal. Matching of the crystal and the coaxial was accomplished by means of two quarter-wave dielectric layers. The best SWR of such a head is 1.05.

The attenuator served to weaken the connection between the heterodyne and the mixer. We used a fixed 15-db attenuator from the complete set belonging to the GSS-15 signal generator.

As a heterodyne we used a K-12 klystron with an external toroidal-type cavity. The klystron supplies were stabilized.

The i.f. amplifier was tuned to 60 Mc, and its bandwidth was 2 Mc. The input to the amplifier was balanced, the first two stages being connected in cascade fashion (a triode with a grounded cathode + a triode with a grounded grid). The noise figure of the i.f. amplifier was 1.5, and its amplification was about 3000. The RC time constant at the output of the detector was $5 \cdot 10^{-5}$ sec. The sensitivity of the overall receiver was about $5 \cdot 10^{-14}$ watt with a bandwidth of 2 Mc and a detector RC output of $5 \cdot 10^{-5}$ sec.

The sensitivity of the receiver was estimated by means of a signal generator with an output calibrated by the method of adjacent signals.

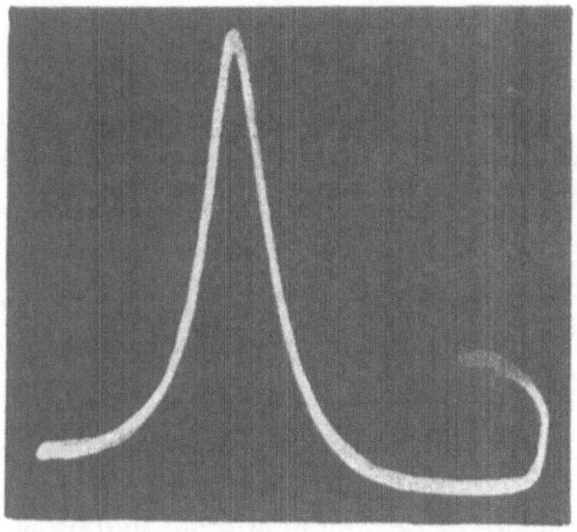

Fig. 16.

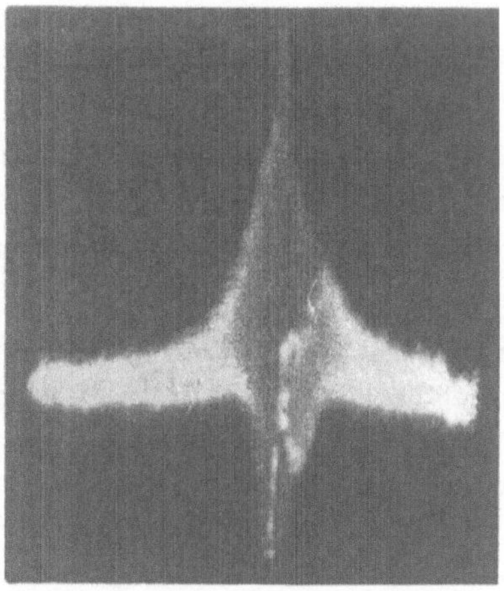

Fig. 17.

We obtained a signal for observing the radiated or absorption line in the following fashion (see Fig. 11). The signal from the GSS-15 signal generator at a frequency of 1659 Mc was modulated by a DGS-3 crystal detector at 7.18 Mc from a PNT-3M sweep oscillator.

The cavity responded to a sideband of the GSS-15 modulated signal. The frequency of the PNT-3M oscillator was swept, and this allowed the signal transmitted through the cavity to be displayed on the oscilloscope.

By working on a sideband we avoided the receiver picking up the carrier of the signal generator. An oscillogram of the emitted line is shown in Fig. 12.

To observe maser action the signal generator was switched off and the heterodyning klystron was switched over to working at a scanning frequency. The oscilloscope, as before, was synchronized with the sawtooth oscillator. In this case, the oscilloscope showed the transmission curve of the i.f. amplifier (Fig. 16).

The changeover to self-oscillation could be observed also with the signal generator switched on. Figures 12 and 17 illustrate the changeover to self-oscillation.

The power of the maser was measured by means of calibrating the receiver with a signal from the GSS-15 oscillator.

The maser power was $\sim 10^{-11}$ watt for a voltage of about 100 kV on the separator, a resonator Q of $\sim$20,000, and for an optimal beam obtained from the grid of radius R = 17.5 mm or from the source with 19 canals.

When the beam originated from the oven with a grid of radius 10 mm, the voltage being 100 kV, the power was several times 10^{-13} watt.

6. Estimating the Absolute Frequency Stability of the Maser

The frequency stability of the maser was estimated by comparing its frequency with that of a harmonic of a quartz resonator. A harmonic of the quartz sufficiently close to the maser radiation frequency was subtracted from it.

If the frequency stability of the quartz exceeded the stability of the maser, then the difference frequency would contain only the frequency instability of the maser (in absolute units). However, the drift of the quartz resonator used by us amounted to about 1 cps in a time of 5-10 seconds at a frequency of 1656 Mc. As a result, we were able to detect only fast variations of the maser frequency, which were caused by sharp changes of separator voltage or of the pressure in the beam source chamber.

The 8-Mc quartz frequency was multiplied up to 72 Mc by standard tubes, and then 23 times by a DGS-3 crystal detector. Thus, we obtained a signal of sufficient power close to the radiation frequency of the maser.

The quartz resonator could be pulled by 0.02% by means of a variable capacity connected in parallel with the quartz. By means of this capacity the frequency of the maser and the quartz harmonic were adjusted to differ by 1 kc. Since the signals being compared were of such low power, they were detected by a superheterodyne receiver.

Having been amplified at the intermediate frequency, the signals were mixed at the detector of the receiver. The extracted difference frequency was measured by means of an audio-oscillator and Lissajous figures. We were interested not in the absolute magnitude of the difference, but only in changes in speed of rotation of a figure.

We note that instability of the heterodyning frequency was not included in the difference frequency, since both signal frequencies – that of the maser and that of the harmonic generator – were heterodyned simultaneously with one frequency.

TABLE II. Intensities of Inversion Transitions of ND_3 [19]

J	K	Frequency, Mc	Intensity, $10^{-7} cm^{-1}$
5	3	1509.22	6.6
4	3	1558.18	9.2
3	2	1560.78	5.6
5	4	1561.15	9.2
6	5	1569.05	8.7
7	6	1582.22	10.6
2	2	1591.69	7.2
3	3	1599.70	12.0
8	7	1600.58	7.5
4	4	1613.00	11.5
6	6	1656.18	18.6
5	5	1631.82	13.3

On changing the separator voltage within the limits 80 to 100 kV, and on varying the pressure in the beam source by a factor of two from optimum, variations of the maser frequency were no greater than 10-20 cps.

In relative units this amounts to between $6.3 \cdot 10^{-9}$ and $1.2 \cdot 10^{-8}$, which is the same to an order of magnitude as the frequency changes of a maser using the 3,3 line of NH_3.

We did not succeed in studying the maser structure for minimum dependence of frequency on beam intensity owing to the above-mentioned stability of the quartz resonator.

Final conclusions about stability can be drawn only after studying the results of comparing two independent masers.

CONCLUSION

As a result of the present work, we have shown that a maser can be made using the J = 6, K = 6 inversion transition of the ND_3 molecule, for which the intensity of absorption is $\sim 10^{-6}$ cm^{-1}. The radiated power is 10^{-11} watt, and the absolute frequency stability is of the order of 10^{-9}. The measured values of the power and stability agree as to order of magnitude with the calculated data. On the basis of the present work, it can be asserted that a number of other inversion transitions of sufficient intensity of the ND_3 molecule could be made to oscillate. A list of these transitions, together with their intensities, is given in Table II.

The sensitivity of microwave detectors in the 18-cm range exceeds that of detectors in the 1.25-cm range. Thus, the small power output of the ND_3 maser need present no obstacle to its use as a frequency standard. In our experiments the signal-to-noise ratio exceeds 100 at an output from the maser resonator of 10^{-12} watt. Hence, the problem of transferring the frequency stability of the maser to a klystron or quartz resonator may be solved by using the methods developed for the NH_3 maser [22]. Furthermore, in the region of 1600 Mc it is probably possible to carry out direct division of the maser frequency.

The maximal frequency stability of a maser may be finally determined as a result of studying the hyperfine structure of the inversion spectrum of ND_3. The calculations in [15] on the J = 6, K = 6 line allow one to estimate the stability of a maser using this line.

For the J = 3, K = 2 line in the spectrum of ND_3, as also in the NH_3 spectrum, the quadrupole splitting appears only in the second-order perturbation theory owing to symmetry properties. Use of this transition leads to increased frequency stability of the ND_3 maser. The intensity of the J = 3, K = 2 inversion transition of ND_3 is only half that of the J = 6, K = 6 transition, while for NH_3 these transitions differ in intensity by a factor of 4.

As in the case of the NH_3 maser, the frequency stability of the ND_3 maser may be improved also by replacing $N^{14}D_3$ by $N^{15}D_3$, whose inversion spectrum has no quadrupole hyperfine structure.

Studies of the spectral hyperfine structure of a maser may be carried out by using it as a spectroscope. The resolving power of such a spectroscope is very high, amounting to several hundreds of cycles. Since to record spectra it is not necessary that the conditions for self-oscillation be fulfilled, we may study by means of a maser the hyperfine structure not only of the lines given in Table II, but also that of a number of appreciably weaker ones.

It is interesting no doubt to compare the dimensions and weights of masers using light and heavy ammonia. This is done most easily by comparing the diameters of the resonators and the lengths of the beams.

The diameter of the resonator for the 3,3 NH_3 line is about 1 cm, while the length of the beam is roughly 20 to 30 cm. The resonator diameter for the 6, 6 ND_3 line is 14 cm, and the beam length is 1.2 to 1.5 m.

The dimensions of our apparatus were 1.5 × 0.5 × 0.5 m (excluding diffusion pumps), and the weight was about 150 kg. However, since this was only an experimental prototype, both the dimensions and weight could certainly be reduced.

Finally, the authors count it their pleasant duty to express their indebtedness to A. M. Prokhorov for useful discussions.

LITERATURE CITED

1. N. G. Basov and K. K. Svidzinskii, Izv. vyssh. ucheb. zav., radiofizika, I, No. 2 (1958).
2. N. G. Basov, Masers (Molekulyarnyi generator), Doctoral thesis, FIAN (1956).
3. G. P. Gordon, H. J. Zeiger, and C. H. Townes, Phys. Rev., 95, 282 (1954); 99, 1264 (1955).
4. N. G. Basov and A. M. Prokhorov, Zhur. Éksp. Teor. Fiz., 27, 431 (1954); Uspekhi Fiz. Nauk, 7, 485 (1955).
5. N. G. Basov, Pribory i Tekh. Éksp., 1, 71 (1957); 1, 77 (1957).
6. N. G. Basov and A. N. Oraevskii, Zhur. Éksp. Teor. Fiz., 32, 1068 (1959).
7. K. Shimoda and T. C. Wang, Rev. Sci. Instr., 26, 1148 (1955).
8. J. P. Gordon, Phys. Rev., 99, 1253 (1955).
9. R. G. Nuckolls, L. J. Rueger, and H. Lyons, Phys. Rev., 89, 1101 (1953).
10. H. Landolt and R. Bornstein, Zahlenwerte und Funktionen. Berlin (1950).
11. G. Herzberg, Infrared and Raman Spectra of Polyatomic Molecules, Van Nostrand, New York (1945).
12. C. H. Townes and A. H. Schawlow, Microwave Spectroscopy, New York (1955).
13. V. A. Shcheglov, Izv. vyssh. ucheb. zav., radiofizika, IV, 648 (1961).
14. Ch. M. Briskina, Theory of a Ring Separator (Raschet sortirovki v vide kolets) Report, IRÉAN (1959).
15. G. R. Gunter-Mohr, R. L. White, A. H. Schawlow, W. E. Good, and D. K. Coles, Phys. Rev., 94, 1184 (1954).
16. G. R. Gunter-Mohr, C. H. Townes, and J. H. Van Vleck, Phys. Rev., 94, 1191 (1954).
17. K. Shimoda, J. Phys. Soc. Japan, 13, 939 (1958).
18. K. K. Svidzinskii (in this volume).
19. N. G. Basov and A. N. Oraevskii, Radiotekhnika i Élektronika, 4, 7, 1185 (1959).
20. S. Ramo and D. Whinnery, Fields and Waves in Modern Radio Technology, Wiley, New York, 1953.
21. W. Gordy, W. V. Smith, and R. F. Trambarulo, Microwave Spectroscopy, Wiley, New York (1953).
22. I. D. Murin, Molecular Clocks (Molekulyarnye chasy) Dissertation FIAN (1959).
23. A. F. Krupnov, Izv. vyssh. ucheb. zav., radiofizika, II, 658 (1959).
24. G. F. Hadley, J. Chem. Phys., 26, 1482 (1957).
25. G. Herrmann, J. Chem. Phys., 29, 875 (1958).
26. V. S. Zyev, Optika i Spektroskopiya, 12, 641 (1962).
27. A. A. Sokova, Trudy Institutov Komiteta SMIP, Vych. 59 (110), 101 (1962).

BIBLIOGRAPHY OF PAPERS PUBLISHED BY MEMBERS OF THE OSCILLATION LABORATORY OF THE PHYSICS INSTITUTE, ACADEMY OF SCIENCES (FIAN), USSR, 1935-1961*

ELECTRON PARAMAGNETIC RESONANCE

V. V. Antonov-Romanovskii, V. G. Dubinin, A. M. Prokhorov, Z. A. Trapeznikova, and M. V. Fok. Detection of Ionization of Eu^{++} in the SrS–Eu, Sm Phosphor by the Paramagnetic Absorption Method. Zhur. Eksp. i Teoret. Fiz., Vol. 37, No. 5, pp. 1466-1467, 1959. Bibliography: 1 title. [Soviet Physics – JETP, Vol. 37(10), p. 1039.]

A. I. Gorbanev, S. D. Kaitmazov, A. M. Prokhorov, and A. B. Tsentsiper. Paramagnetic Resonance of the Products Formed at Low Temperature During the Dissociation of H_2O, H_2O_2, and D_4O Vapors in a Gas Discharge. Zhur. Fiz. Khim., Vol. 31, No. 2, p. 515, 1957. Bibliography: 5 titles.

S. D. Kaitmazov and A. M. Prokhorov. Paramagnetic Resonance of the Free Radicals Obtained by Freezing a Plasma of H_2S. Zhur. Eksp. i Teoret. Fiz., Vol. 35, No. 2, p. 551, 1958. Bibliography: 1 title. [Soviet Physics – JETP, Vol. 35(8), p. 381.]

S. D. Kaitmazov and A. M. Prokhorov. Cavities for Observing Electronic Paramagnetic Resonance at Low Temperatures. Pribory i Tekh. Eksp., No. 5, pp. 107-110, 1959. Bibliography: 4 titles. [Instr. and Exptl. Tech., No. 5, p. 808, 1959.]

S. D. Kaitmazov and A. M. Prokhorov. Electronic Paramagnetic Resonance Spectra of Frozen OH Radicals. Zhur. Eksp. i Teoret. Fiz., Vol. 36, No. 4, pp. 1331-1332, 1959. Bibliography: 3 titles. [Soviet Physics – JETP, Vol. 36(9), p. 944.]

S. D. Kaitmazov and A. M. Prokhorov. Electronic Paramagnetic Resonance Spectrum of the Free Radical Obtained During the Ultraviolet Irradiation of H_2O_2. Zhur. Fiz. Khim., Vol. 34, No. 1, pp. 227-228, 1960. Bibliography: 3 titles.

N. V. Karlov, Yu. P. Pimenov, and A. M. Prokhorov. Saturation of Paramagnetic Amplifiers and Restoration Time. Radiotekh. i Elektron., Vol. 6, No. 3, pp. 410-415, 1961. Bibliography: 6 titles.

N. V. Karlov, Yu. P. Pimenov, and A. M. Prokhorov. The Sensitivity of Paramagnetic Amplifier Radio Receivers. Radiotekh. i Elektron., Vol. 6, No. 3, pp. 416-421, 1961. Bibliography: 4 titles.

N. V. Karlov, Yu. P. Pimenov, and A. M. Prokhorov. 10-cm Paramagnetic Amplifier Utilizing Fe^{3+} Corundum Ions. Radiotekh. i Elektron., Vol. 6, No. 5, p. 846, 1961. Bibliography: 2 titles.

L. S. Kornienko and A. M. Prokhorov. Fine Structure of Electron Paramagnetic Resonance of the Fe^{+++} Ions in the Al_2O_3 Lattice. Zhur. Eksp. i Teoret. Fiz., Vol. 33, No. 3, pp. 805-807, 1957. [Soviet Physics – JETP, Vol. 33(6), p. 620.]

L. S. Kornienko and A. M. Prokhorov. Paramagnetic Amplifier and Generator, Using Fe^{+++} Ions in Corundum. Zhur. Eksp. i Teoret. Fiz., Vol. 36, No. 3, pp. 919-920, 1959. Bibliography: 9 titles. [Soviet Physics – JETP, Vol. 36(9), p. 649.]

L. S. Kornienko and A. M. Prokhorov. Electronic Paramagnetic Resonance of the Ti^{+++} Ion in Corundum. Zhur. Eksp. i Teoret. Fiz., Vol. 38, No. 5, pp. 1651-1652, 1960. Bibliography: 4 titles. [Soviet Physics – JETP, Vol. 38(11), p. 1189.]

*Compiled by senior bibliographer of the Library of the Physics Institute, Academy of Sciences, USSR, A. A. Sakova. Editor: Cand. Phys.-Math. Sciences K. K. Svidzinskii.

L. S. Kornienko and A. M. Prokhorov. Electronic Paramagnetic Resonance of the Fe^{+++} Ion in Corundum. Zhur. Eksp. i Teoret. Fiz., Vol. 40, No. 6, pp. 1594-1601, 1961. Bibliography: 10 titles. [Soviet Physics – JETP, Vol. 40(13), p. 1120.]

V. I. Lushchikov, A. A. Manenkov, and Yu. V. Taran. Dynamic Polarization of Protons in Neutron Irradiated Polyethylene. Fiz. Tverd. Tela, Vol. 3, No. 11, pp. 3503-3508, 1961. Bibliography: 5 titles. [Soviet Physics–Solid State, Vol. 3, p. 2541.]

A. A. Manenkov. Paramagnetic Resonance in Some Iron and Rare Earth Compounds. Candidate's Thesis, Mathematical-Physical Sciences, Moscow, 1955, 121 pp. and ill. (P. N. Lebedev Institute of Physics, Academy of Sciences, USSR). Typewritten.

A. A. Manenkov and A. M. Prokhorov. The Fine Structure of the Spectrum of the Paramagnetic Resonance of the Ion Cr^{+++} in Chromium Corundum. Zhur. Eksp. i Teoret. Fiz., Vol. 28, No. 6, p. 762, 1955. [Soviet Physics – JETP, Vol. 28(1), p. 611.]

A. A. Manenkov, A. M. Prokhorov, Z. A. Trapeznikova, and M. V. Fok. The Application of Paramagnetic Resonance in Studying the States of Phosphor Activators, in Papers Presented at the Fifth Conference on Luminescence (Crystal Phosphors), June 25-30, 1956. Tartu, Academy of Sciences, Estonian SSR, 1956, pp. 51-52.

A. A. Manenkov and A. M. Prokhorov. Radiotelescope for Observing Electronic Paramagnetic Resonance in the Centimeter Range. Radiotekh. i Elektron., Vol. 1, No. 4, pp. 469-477, 1956. Bibliography: 9 titles.

A. A. Manenkov and A. M. Prokhorov. Hyperfine Structure of the Paramagnetic Resonance Spectrum of $^{53}Cr^{+++}$ in Al_2O_3. Zhur. Eksp. i Teoret. Fiz., Vol. 31, No. 2, pp. 346-347, 1956. [Soviet Physics – JETP, Vol. 31 (4), p. 288.]

A. A. Manenkov and A. M. Prokhorov. Fine and Hyperfine Structure of Paramagnetic Resonance in Bivalent Europium. Doklady Akad. Nauk SSSR, Vol. 107, No. 3, pp. 402-404, 1956. Bibliography: 7 titles [Soviet Physics – Doklady, Vol. 1, p. 196.]

A. A. Manenkov and A. M. Prokhorov. Determination of Nuclear Moments of Gd^{155} and Gd^{157} from the Hyperfine Structure of Magnetic Resonance. Zhur. Eksp. i Teoret. Fiz., Vol. 33, No. 5, pp. 1116-1118, 1957. Bibliography: 11 titles. [Soviet Physics – JETP, Vol. 33(6), p. 860.]

A. A. Manenkov, A. M. Prokhorov, Z.A. Trapeznikova, and M. V. Fok. The Application of Paramagnetic Resonance in Studying the States of Phosphor Activators, in Papers Presented at the Fifth Conference on Luminescence (Crystal Phosphors), June 25-30, 1956, Tartu Izv. Akad. Nauk SSSR, Ser. Fiz., Vol. 21, No. 5, pp. 779-780, 1957.

A. A. Manenkov, A. M. Prokhorov, Z. A. Trapeznikova, and M. V. Fok. The Application of Paramagnetic Resonance in Studying the States of Phosphor Activators. Optika i Spektroskopiya, Vol. 2, No. 4, pp. 470-474, 1957. Bibliography: 4 titles.

A. A. Manenkov, A. M. Prokhorov, P. S. Trukhlyaev, and G. N. Yakovlev. Paramagnetic Resonance Hyperfine Structure. Nuclear Spin and the Magnetic Moment of the 5.3-Year Radioactive Isotope Eu^{152}. Doklady Akad. Nauk SSSR, Vol. 112, No. 4, pp. 623-625, 1957. Bibliography: 6 titles. [Soviet Physics – Doklady, Vol. 2, p. 64.]

A. A. Manenkov and V. B. Fedorov. Investigation of the Linewidth and Shape in the Paramagnetic Resonance Spectrum of the Cr^{+++} Ion in Corundum Single Crystals. Zhur. Eksp. i Teoret. Fiz., Vol. 38, No. 4, pp. 1042 to 1046, 1960. Bibliography: 3 titles. [Soviet Physics – JETP, Vol. 38(11), p. 751.]

A. A. Manenkov and A. M. Prokhorov. Spin-Lattice Relaxation in Chromium Corundum. Zhur. Eksp. i Teoret. Fiz., Vol. 38, No. 3, pp. 729-733, 1960. Bibliography: 7 titles. [Soviet Physics – JETP, Vol. 38(11), p. 527.]

A. A. Manenkov and A. M. Prokhorov. Paramagnetic Resonance of Mn^{++} in SrS. Zhur. Eksp. i Teoret. Fiz., Vol. 40, No. 6, pp. 1606-1609, 1961. Bibliography: 5 titles. [Soviet Physics – JETP, Vol. 40(13), p. 1129.]

A. A. Manenkov and V. A. Milyaev. Relaxation Phenomena in the Paramagnetic Resonance of Mn^{2+} Ions in the Cubic Crystal Field of SrS. Zhur. Eksp. i Teoret. Fiz., Vol. 41, No. 1, pp. 100-105, 1961. Bibliography: 9 titles. [Soviet Physics – JETP, Vol. 41(14), p. 75.]

P. P. Pashinin and A. M. Prokhorov. Measurements of the Spin–Lattice Relaxation Times of Cr^{+++} in Corundum. Zhur. Eksp. i Teoret. Fiz., Vol. 34, No. 3, p. 777, 1958. Bibliography: 2 titles. [Soviet Physics – JETP, Vol. 34(7), p. 535.]

P. P. Pashinin and A. M. Prokhorov. Measurement of the Spin–Lattice Relaxation Time in Compounds with Strong Covalent Bonding. Zhur. Eksp. i Teoret. Fiz., Vol. 40, No. 1, pp. 49-51, 1961. Bibliography: 8 titles. [Soviet Physics – JETP, Vol. 40(13), p. 33.]

A. K. Piskunov, A. A. Manenkov, and Z. A. Bagdasaryan. Paramagnetic Resonance in Potassium Ozonide. Zhur. Eksp. i Teoret. Fiz., Vol. 37, No. 1, pp. 302-304, 1959. Bibliography: 4 titles. [Soviet Physics – JETP, Vol. 37(10), p. 213.]

A. M. Prokhorov and A. A. Manenkov. Fine and Hyperfine Structure of Paramagnetic Resonance in Bivalent Europium. Doklady Akad. Nauk SSSR, Vol. 107, No. 3, pp. 402-404, 1957. Bibliography: 7 titles. [Soviet Physics – Doklady, Vol. 1, p. 196.]

D. K. Réi. Analysis of the Electron Paramagnetic Resonance Spectrum of V^{+++} Ions in Corundum (Al_2O_3). Fiz. Tverd. Tela., Vol. 3, No. 8, pp. 2214-2222, 1961. Bibliography: 12 titles. [Soviet Physics – Solid State, Vol. 3, p. 1606.]

D. K. Réi. Concerning the Paramagnetic Resonance Spectrum of V^{++++} Ions in Rutile (TiO_2). Fiz. Tverd. Tela., Vol. 3, No. 9, pp. 2535-2539, 1961. Bibliography: 7 titles. [Soviet Physics – Solid State, Vol. 3, p. 1845.]

D. K. Réi. Theoretical Analysis of the g-Factors of Ti^{+++} Ions in Certain Compounds. Fiz. Tverd. Tela., Vol. 3, No. 9, pp. 2525-2534, 1961. Bibliography: 10 titles. [Soviet Physics – Solid State, Vol. 3, p. 1838.]

D. K. Réi. Theory of Covalent Binding for Cobalt Salts and Analysis of the Paramagnetic Resonance Spectrum of Co^{++} in Corundum. Fiz. Tverd. Tela, Vol. 3, No. 8, pp. 2223-2239. Bibliography: 5 titles. [Soviet Physics – Solid State, Vol. 3, p. 1613.]

T. V. Rode and A. A. Manenkov. The Valence of Chromium Ions in Compounds Formed During the Thermal Decomposition of Chrome Anhydride. Zhur. Fiz. Khim., Vol. 33, No. 2, p. 503, 1959.

G. M. Zverev, L. S. Kornienko, A. A. Manenkov, and A. M. Prokhorov. A Chromium Corundum Paramagnetic Amplifier and Generator. Zhur. Eksp. i Teoret. Fiz., Vol. 34, No. 6, pp. 1660-1661, 1958. Bibliography: 10 titles. [Soviet Physics – JETP, Vol. 34(7), p. 1141.]

G. M. Zverev and A. M. Prokhorov. Fine Structure and Hyperfine Structure of Paramagnetic Resonance of Cr^{+++} in Synthetic Ruby. Zhur. Eksp. i Teoret. Fiz., Vol. 34, No. 2, pp. 513-514, 1958. [Soviet Physics – JETP, Vol. 34(7), p. 354.]

G. M. Zverev and A. M. Prokhorov. Electron Paramagnetic Resonance of the V^{+++} Ion in Sapphire. Zhur. Eksp. i Teoret. Fiz., Vol. 34, No. 4, pp. 1023-1024, 1958. Bibliography: 4 titles. [Soviet Physics – JETP, Vol. 34(7), p. 707.]

G. M. Zverev and A. M. Prokhorov. Electron Paramagnetic Resonance of Co^{++} in Corundum. Zhur. Eksp. i Teoret. Fiz., Vol. 36, No. 2, pp. 647-648, 1959. [Soviet Physics – JETP, Vol. 36(9), p. 451.]

G. M. Zverev and A. M. Prokhorov. Electron Paramagnetic Resonance Spectrum of V^{+++} in Corundum. Zhur. Eksp. i Teoret. Fiz., Vol. 38, No. 2, pp. 449-454, 1960. Bibliography: 11 titles. [Soviet Physics – JETP, Vol. 38(11), p. 330.]

G. M. Zverev and A. M. Prokhorov. Cross Spin Relaxation in the Hyperfine Structure of Electron Paramagnetic Resonance of Co^{++} in Corundum. Zhur. Eksp. i Teoret. Fiz., Vol. 39, No. 3, pp. 545-547, 1960. Bibliography: 4 titles. [Soviet Physics – JETP, Vol. 39 (11), p. 330.]

G. M. Zverev and A. M. Prokhorov. Electron Paramagnetic Resonance of Vanadium in Rutile. Zhur. Eksp. i Teoret. Fiz., Vol. 39, No. 1, pp. 222-223, 1960. [Soviet Physics – JETP, Vol. 39(12), p. 160.]

G. M. Zverev and A. M. Prokhorov. Electron Paramagnetic Resonance and Spin – Lattice Relaxation of Co^{++} in Corundum. Zhur. Eksp. i Teoret. Fiz., Vol. 39, No. 1, pp. 57-63, 1960. Bibliography: 16 titles. [Soviet Physics – JETP, Vol. 39(12), p. 41.]

G. M. Zverev and A. M. Prokhorov. Electronic Paramagnetic Resonance in the V^{+++} Ion in Corundum. Zhur. Eksp. i Teoret. Fiz., Vol. 40, No. 4, pp. 1016-1018, 1961. Bibliography: 6 titles. [Soviet Physics – JETP, Vol. 40(13), p. 714.]

QUANTUM ELECTRONICS

A. I. Barchukov and A. M. Prokhorov. Experimental Study of Millimeter Wave Disc Resonators. Radiotekh. i Elektron., Vol. 4, No. 12, pp. 2094-2095, 1959. Bibliography: 1 title.

A. I. Barchukov and A. M. Prokhorov. Use of Parallel-Plate Resonators in Masers, in coll. Quantum Electronics (Symposium, New York, 1959), 1960, pp. 45-46.

N. G. Basov and A. N. Oraevskii. Quantum Radio Technology. Izvest. Vysshikh. Ucheb. Zavedenii, Radiotekh., Vol. 2, No. 1, pp. 3-17, 1959. Bibliography: 52 titles.

N. G. Basov, B. M. Vul, and Yu. M. Popov. Quantum-Mechanical Semiconductor Generators and Amplifiers of Electromagnetic Oscillations. Zhur. Eksp. i Teoret. Fiz., Vol. 37, No. 2, pp. 587-588, 1959. Bibliography: 3 titles. [Soviet Physics – JETP, Vol. 37(10), p. 416.]

N. G. Basov, O. N. Krokhin, and Yu. M. Popov. Possibility of Using Indirect Transitions to Obtain a Negative Temperature in Semiconductors. Zhur. Eksp. i Teoret. Fiz., Vol. 39, No. 5, pp. 1486-1487, 1960. Bibliography: 10 titles. [Soviet Physics – JETP, Vol. 39(12), p. 1033.]

N. G. Basov, O. N. Krokhin, and Yu. M. Popov. Generation, Amplification, and Detection of Infrared and Optical Radiation by Quantum-Mechanical Systems. Uspekhi Fiz. Nauk, Vol. 72, No. 2, pp. 161-209, 1960. Bibliography: 80 titles. [Soviet Physics – Uspekhi, Vol. 3, p. 702.]

N. G. Basov and A. M. Prokhorov. Quantum Radio Physics (talk presented at the Annual Meeting of the Academy of Sciences of the USSR, February 1960). Vestnik. Akad. Nauk SSSR, No. 4, pp. 110-119, 1960.

N. G. Basov, O. N. Krokhin, and Yu. M. Popov. Semiconductor Amplifiers and Generators with Carriers Having Negative Effective Mass. Zhur. Eksp. i Teoret. Fiz., Vol. 38, No. 3, pp. 1001-1002, 1960. Bibliography: 6 titles. [Soviet Physics – JETP, Vol. 38(11), p. 720.]

N. G. Basov and N. V. Karlov. Wideband Radiometer with Quantum Spectral Transformer. Radiotekh. i Elektron. Vol. 5, No. 4, pp. 676-677, 1960.

N. G. Basov and O. N. Krokhin. Production of Negative Temperature States by Electron Excitation in a Gas Mixture. Zhur. Eksp. i Teoret. Fiz., Vol. 39, No. 6, pp. 1777-1780, 1960. Bibliography: 6 titles. [Soviet Physics – JETP, Vol. 29(12), p. 1240.]

N. G. Basov, O. N. Krokhin, and Yu. M. Popov. Oscillators and Light Amplifiers. Priroda, No. 12, pp. 16-25, 1961.

N. G. Basov, O. N. Krokhin, and Yu. M. Popov. Solid State Coherent Light Oscillators. Vestnik Akad. Nauk SSSR, No. 3, pp. 61-66, 1961.

N. G. Basov, O. N. Krokhin, and Yu. M. Popov. Use of Indirect Transitions in Semiconductors for the Determination of States with Negative Absorption Coefficients. Zhur. Eksp. i Teoret. Fiz., Vol. 40, No. 4, pp. 1203-1209, 1961. Bibliography: 15 titles. [Soviet Physics – JETP, Vol. 40(13), p. 845.]

N. G. Basov. Quantum Radioelectronics. Radio, No. 10, p. 8, 1961.

N. G. Basov. On the Road to Optical Radio. How Quantum Oscillators and Amplifiers Operate. Nauka i Zhizn, No. 7, pp. 34-36, 1961.

N. G. Basov, O. N. Krokhin, L. M. Lisitsyn, E. P. Markin, and B. D. Osipov. Negative Conductivity in Induced Transitions. Zhur. Eksp. i Teoret. Fiz., Vol. 41, No. 3, pp. 988-989, 1961. Bibliography: 3 titles. [Soviet Physics – JETP, Vol. 41 (14), p. 701.]

N. G. Basov, O. N. Krokhin, and Yu. M. Popov. Production of Negative-Temperature States in p-n Junctions of Degenerate Semiconductors. Zhur. Eksp. i Teoret. Fiz., Vol. 40, No. 6, pp. 1879-1880, 1961. Bibliography: 4 titles. [Soviet Physics – JETP, Vol. 40(13), p. 1320.]

N. G. Basov, B. D. Osipov, and A. N. Khvoshchev. Recombination Radiation from Indium Antimonide in Avalanche Breakdown. Zhur. Eksp. i Teoret. Fiz., Vol. 40, No. 6, p. 1882, 1961. Bibliography: 4 titles. [Soviet Physics – JETP, Vol. 40(13), p. 1323.]

F. V. Bunkin. Level Inversion During Zeeman Splitting. Radiotekh. i Elektron., Vol. 4, No. 5, pp. 886-890, 1959. Bibliography: 7 titles.

V. M. Kontorovich and A. M. Prokhorov. Nonlinear Effects of the Interaction of the Resonance Fields in the Molecular Generator and Amplifier. Zhur. Eksp. i Teoret. Fiz., Vol. 33, No. 6, pp. 1428-1430, 1957. Bibliography: 3 titles. [Soviet Physics – JETP, Vol. 33(6), p. 1100.]

A. M. Prokhorov. Molecular Amplifier and Generator for Submillimeter Waves. Zhur. Eksp. i Teoret. Fiz., Vol. 34, No. 6, pp. 1658-1659, 1958. [Soviet Physics – JETP, Vol. 34(7), p. 1140.]

A. M. Prokhorov. Quantum Counters. Zhur. Eksp. i Teoret. Fiz., Vol. 40, No. 5, pp. 1384-1396, 1961. Bibliography: 2 titles. [Soviet Physics – JETP, Vol. 40(13), p. 973.]

V. G. Veselago. Spin Oscillator. Radiotekh. i Elektron., Vol. 6, No. 5, pp. 849-851, 1961. Bibliography: 8 titles.

MOLECULAR OSCILLATORS AND AMPLIFIERS. FREQUENCY STANDARDS

N. G. Basov. Molecular Oscillators. Doctor's Thesis, Mathematical-Physical Sciences, Moscow, 1956. 144 pp. plus ill. (Academy of Sciences of the USSR, P. N. Lebedev Institute of Physics). Bibliography: pp. 141-144. Typewritten.

N. G. Basov and A. M. Prokhorov. Molecular Oscillators and Amplifiers. Uspekhi Fiz. Nauk, Vol. 57, No. 3, pp. 485-501, 1955. Bibliography: 12 titles.

N. G. Basov and A. M. Prokhorov. Possible Methods of Obtaining Active Molecules for a Molecular Oscillator. Zhur. Eksp. i Teoret. Fiz., Vol. 28, No. 2, pp. 249-250, 1955. [Soviet Physics – JETP, Vol. 28(1), p. 184.]

N. G. Basov and A. M. Prokhorov. Molecular Oscillator and Molecular Power Amplifier Theory. Doklady Akad. Nauk SSSR, Vol. 101, No. 1, pp. 47-49, 1955. Bibliography: 4 titles.

N. G. Basov and A. M. Prokhorov. Molecular Oscillator and Molecular Power Amplifier Theory. Trans. Faraday Soc., No. 19, pp. 96-99, 1955. Bibliography: 4 titles.

N. G. Basov. Molecular Oscillator. Radiotekh. i Elektron., Vol. 1, No. 6, pp. 752-757, 1956. Bibliography: 10 titles.

N. G. Basov, B. D. Osipov, and A. M. Prokhorov. A Molecular Oscillator Which Does Not Utilize a Molecular Beam (in connection with the article by N. G. Basov and A. M. Prokhorov, Molecular Oscillator and Amplifier, in Uspekhi Fiz. Nauk, Vol. 57, No. 3, 1955). Uspekhi Fiz. Nauk, Vol. 59, No. 2, p. 375, 1956.

N. G. Basov and A. M. Prokhorov. Theory of the Molecular Generator and Molecular Power Amplifier. Zhur. Eksp. i Teoret. Fiz., Vol. 30, No. 3, pp. 560-563, 1956. [Soviet Physics – JETP, Vol. 30(3), p. 426.]

N. G. Basov. A Study of Molecular Oscillator Operation. Pribory i Tekh. Eksp., No. 1, pp. 77-82, 1957. Bibliography: 9 titles.

N. G. Basov. Molecular Oscillator Using an Ammonia Molecular Beam. Pribory i Tekh. Eksp., No. 1, pp. 71-77, 1957. Bibliography: 15 titles.

N. G. Basov and A. M. Prokhorov. Molecular Oscillator and Amplifier. Priroda, No. 7, pp. 24-32, 1958. Printed in Czech: Pokroky mat. fys., astr.,Vol. 4, No. 4, pp. 439-48, 1959. In Roumanian: Analele Rom.-Sov. mat.-fiz., Vol. 13, No. 4, pp. 182-194, 1959.

N. G. Basov, I. D. Murin, A. P. Petrov, A. M. Prokhorov, and I. V. Shtranikh. Molecular Clock. Izvest. Vysshikh. Ucheb. Zavedenii, Radiofiz., Vol. 1, No. 3, pp. 50-53, 1958. Bibliography: 6 titles.

N. G. Basov and A. P. Petrov. Relative Frequency Stability of Molecular Oscillators. Radiotekh. i Elektron., Vol. 3, No. 2, pp. 298-299, 1958.

N. G. Basov. Self-Excitation Conditions for a Molecular Oscillator without Volume Resonance. Radiotekh. i Elektron., Vol. 3, No. 2, pp. 297-298, 1958. Bibliography: 8 titles.

N. G. Basov and A. N. Oraevskii, Possibility of Designing a Sealed Molecular Oscillator Using ND_3, NH_2D, and NHD_2 Molecules. Izvest. Vysshikh. Ucheb. Zavedenii, Radiofiz., Vol. 1, No. 4, pp. 63-68, 1958. Bibliography: 13 titles.

N. G. Basov and K. K.Svidzinskii. Design of a Molecular Oscillator Utilizing a ND_3 Beam. Izvest. Vysshikh. Ucheb. Zavedenii, Radiofiz., Vol. 1, No. 2, pp. 89-94, 1958. Bibliography: 10 titles.

N. G. Basov and A. N. Oraevskii. Absolute Stability of a Molecular Oscillator Utilizing an Ammonia Molecular Beam. Radiotekh. i Elektron., Vol. 4, No. 7, pp. 1185-1195, 1959. Bibliography: 19 titles.

N. G. Basov and A. N. Oraevskii. Use of Slow Molecules in a Maser. Zhur. Eksp. i Teoret. Fiz., Vol. 37, No. 4, pp. 1068-1071, 1959. Bibliography: 10 titles. [Soviet Physics – JETP, Vol. 37(10), p. 761.]

N. G. Basov, V. V. Nikitin, and A. N. Oraevskii. A Study of the Frequency Dependence of Molecular Oscillators upon Various Parameters. Part I. Theory of the Line J = 3, K = 2. Radiotekh. i Elektron., Vol. 6, No. 5, pp. 796-805, 1961. Bibliography: 6 titles.

N. G. Basov, G. M. Strakhovskii, and I. V. Cheremiskin. A Study of the Frequency Dependence of Molecular Oscillators upon Various Parameters. Part II. Theory of the Line J = 3, K = 3. Radiotekh. i Elektron., Vol. 6, No. 6, pp. 1020-1028, 1961. Bibliography: 8 titles.

N. G. Basov and V. S. Zuev. ND_3 Molecular Beam Oscillator. Pribory i Tekh. Eksp., No. 1, pp. 120-121, 1961. Bibliography: 7 titles. [Instr. and Exptl. Tech., No. 1, p. 122.]

N. G. Basov, O. N. Krokhin, A. N. Oraevskii, G. M. Strakhovskii,and B. M. Chikhachev. Possible Investigation of Relativistic Effects with the Aid of Molecular and Atomic Frequency Standards. Uspekhi Fiz. Nauk, Vol. 75, No. 1, pp. 3-59, 1961. Bibliography: 113 titles. [Soviet Physics – Uspekhi, Vol. 4, p. 641.]

A. V. Dudenkova.Sealed Off Molecular Generator. Pribory i Tekh. Eksp., No. 3, p. 180, 1961. [Instr. and Exptl. Tech., No. 3, p. 593.]

I. D. Murin. Electronic Circuits for the Molecular Clock at the Institute of Physics of the Academy of Sciences of the USSR. Izvest. Vysshikh. Ucheb. Zavedenii, Radiotekh., No. 5, pp. 555-564, 1958.

I. D. Murin. Using a Molecular Oscillator to Stabilize the Frequency of a Quartz Oscillator. Radiotekh. i Elektron. Vol. 4, No. 11, pp. 1941-1943, 1959. Bibliography: 5 titles.

A. N. Oraevskii. Molecular Oscillator Theory. Radiotekh. i Elektron., Vol. 4, No. 4, pp. 718-723, 1959. Bibliography: 12 titles.

A. M. Prokhorov. Influence of Resonator Q upon the Frequency of a Molecular Oscillator (in connection with the paper by N. G. Basov and A. M. Prokhorov, Molecular Oscillator and Amplifier, in Uspekhi Fiz. Nauk, Vol. 57, No. 3, 1955). Radiotekh. i Elektron., Vol. 2, No. 4, p. 510, 1957.

GAS RADIOSPECTROSCOPY

A. I. Barchukov. Modulation Method for Measuring the Coefficient of Absorption in Microwave Radiospectroscopy. Candidate's Thesis, Physical-Mathematical Sciences, Moscow, 1953, 128 pp. with ill. (Academy of Sciences, USSR, P. N. Lebedev Institute of Physics). Typewritten.

A. I. Barchukov, T. M. Minaeva, and A. M. Prokhorov. Microwave Spectrum of the C_2H_5Cl Molecule. Zhur. Eksp. i Teoret. Fiz., Vol. 29, No. 6, p. 892, 1955. [Soviet Physics – JETP, Vol. 29(2), p. 760.]

A. I. Barchukov and N. G. Basov. Measurement of the Frequency and Intensity of the Lines in the Hyperfine Structure of CH_3I (Transition $J = 0 \rightarrow 1$). Optika i Spektroskopiya, Vol. 4, No. 4, p. 532, 1958.

A. I. Barchukov and A. M. Prokhorov. Quadrupole Bond, Dipole Moment, and Internal Rotation Barrier in the CH_3GeH_3 Molecule and Its Rotational Spectrum. Optika i Spectroskopiya, Vol. 5, No. 5, pp. 530-534, 1958. Bibliography: 7 titles.

A. I. Barchukov and A. M. Prokhorov. Microwave Spectrum of the CH_3GeH_3 Molecule. Optika i Spektroskopiya, Vol. 4, No. 6, p. 799, 1958. Bibliography: 1 title.

A. I. Barchukov, T. M. Murin, and A. M. Prokhorov. Microwave Spectrum and Rotational Constant of the C_2H_5Cl Molecule. Optika i Spektroskopiya, Vol. 4, No. 4, pp. 521-523, 1958.

A. I. Barchukov and Yu. N. Petrov. Dipole Moment of the CH_3GeH_3 Molecule. Optika i Spektroskopiya, Vol. 9, No. 1, p. 129, 1961. Bibliography: 4 titles.

N. G. Basov. Radiospectroscopic Determination of Nuclear Moments. Candidate's Thesis, Mathematical-Physical Sciences, Moscow, 1953, 101 pp. with ill. (Academy of Sciences, USSR, P. N. Lebedev Institute of Physics). Typewritten.

N. G. Basov and A. M. Prokhorov. Molecular Beam Radiospectroscopic Investigation of Molecular Rotational Spectra. Zhur. Eksp. i Teoret. Fiz., Vol. 27, No. 4, pp. 431-438, 1954. Bibliography: 2 titles.

N. G. Basov, A. N. Oraevskii, and K. K. Svidzinskii. Theory of the Hyperfine Structure of Molecular Rotational Spectra Associated with the 2^4 Electric Nuclear Dipole Moment. Optika i Spektroskopiya, Vol. 1, No. 3, pp. 285-289, 1956. Bibliography: 6 titles.

N. G. Basov and B. D. Osipov. Line Associated with the $F = 5/2 \rightarrow 3/2$, $J = 1$, $K = 1$ Rotational Emission Spectrum of the CH_3I^{127} Molecule. Optika i Spektroskopiya, Vol. 4, No. 6, pp. 795-797, 1958. Bibliography: 4 titles.

N. A. Irisova. Frequency Stabilization of the Standard 3-cm Klystron Using the Spectral Line of a Gas. Candidate's Thesis, Mathematical-Physical Sciences, Moscow, 1950, 126 pp. with ill., 22 sections, table of fig. (Academy of Sciences, USSR, P. N. Lebedev Institute of Physics). Bibliography, p. 123. Typewritten.

N. A. Irisova. Radiospectroscopic Investigation of Molecules. Izvest. Akad. Nauk SSSR, Ser. Fiz., Vol. 18, No. 6, p. 663, 1954.

N. A. Irisova, M. E. Zhabotinskii, and V. G. Veselago. Using the Spectral Line to Stabilize the Frequency of the 3-cm Klystron. Radiotekhnika, Vol. 10, No. 4, pp. 26-35, 1955.

N. A. Irisova. Determination of the Rotational Constants of CH_3GeCl_3 from Its Ultrahigh-Frequency Absorption Spectrum (Materials of the Ninth All-Union Conference on the Theory of Spectroscopy, December, 1957). Izvest. Akad. Nauk, Ser. Fiz., Vol. 22, No. 11, p. 1307, 1958.

N. A. Irisova. Radiospectroscopic Investigation of CH_3GeCl_3. Optika i Spektroskopiya, Vol. 4, No. 4, pp. 543 to 546, 1958.

N. A. Irisova and E. M. Dianov. Ultrahigh-Frequency Absorption of CH_3GeF_3. Optika i Spektroskopiya, Vol. 9, No. 2, p. 261, 1960.

L. S. Mayants. The Theory of Characteristic Frequencies of Multiatomic Molecules. Doklady Akad. Nauk SSSR, Vol. 48, No. 6, pp. 416-419, 1945. Bibliography: 4 titles.

L. S. Mayants. The Theory of Characteristic Frequencies (paper presented at the All-Union Conference on Spectroscopy, Leningrad, 1946). Izvest. Akad. Nauk SSSR, Ser. Fiz., Vol. 11, No. 4, pp. 353-356, 1947. Bibliography: 8 titles.

L. S. Mayants. Simplified Method of Calculation of the Intensity and Polarization of Rotational Molecular Spectra. Zhur. Eksp. i Teoret. Fiz., Vol. 19, No. 7, pp. 627-632, 1948. Bibliography: 4 titles.

I. A. Mukhtarov. The Microwave Spectrum of 1,2-Fluorochloroethane. Doklady Akad. Nauk SSSR, Vol. 115, No. 3, pp. 486-487, 1957. [Soviet Physics – Doklady, Vol. 2, p. 357.]

T. M. Murina. Radiospectroscope to be Used in the Study of Rotational Molecular Spectra. Radiotekh. i Elektron., Vol. 2, No. 8, pp. 1271-1278, 1957. Bibliography: 4 titles.

T. M. Murina, A. M. Prokhorov, and É. A. Chayanova. Measurement of the Absolute Intensity of an Absorption Line. Radiotekh. i Elektron., Vol. 3, No. 11, pp. 1402-1404, 1958.

B. D. Osipov. Microwave Dispersion in Ammonia at a Pressure of 10^{-2} mm Hg. Zhur. Eksp. i Teoret. Fiz., Vol. 27, No. 1, p. 115, 1954. Bibliography: 4 titles.

B. D. Osipov and A. M. Prokhorov. Use of the (3,3) Absorption Line of Ammonia as a Standard When Measuring Frequencies in the 5-20 Megacycle Range to a Precision of 10^{-6}. Doklady Akad. Nauk SSSR, Vol. 102, No. 5, pp. 933-934, 1955. Bibliography: 5 titles.

B. D. Osipov. Some Phase Relationships in Radiospectroscopes. Author's Certificate, Candidate's Thesis, Mathematical-Physical Sciences, Moscow, 1955 (Academy of Sciences, USSR, P. N. Lebedev Institute of Physics). Bibliography: 10 titles.

B. D. Osipov. Some Phase Relationships in Radiospectroscopes. Candidate's Thesis, Mathematical-Physical Sciences, Moscow, 1955. 67 pp. (Academy of Sciences, USSR, P. N. Lebedev Institute of Physics). Typewritten.

B. D. Osipov. Hyperfine Structure of the Rotational Transition $J = 3 \rightarrow 4$ of the CH_3I^{127} Molecule. Optika i Spektroskopiya, Vol. 3, No. 1, p. 94, 1957. Bibliography: 3 titles.

B. D. Osipov, The 1-J Interaction in the CH_3I Molecule. Optika i Spektroskopiya, Vol. 8, No. 4, pp. 581-582, 1960.

A. M. Prokhorov and N. G. Basov. Determination of the O- and n/2 Nuclear Spins from the Microwave Molecular Spectra. Doklady Akad. Nauk SSSR, Vol. 90, No. 6, pp. 1003-1004, 1953.

A. M. Prokhorov and A. I. Barchukov. Method of Measuring Absorption Coefficients in Microwave Spectroscopy. Zhur. Eksp. i Teoret. Fiz., Vol. 26, No. 6, pp. 761-763, 1954.

A. M. Prokhorov. Radiospectroscopy and Molecular Structure. Priroda, No. 1, pp. 16-24, 1955.

A. M. Prokhorov and G. P. Shipulo. Radiospectroscopic Investigation of F_2BNH_3 and $F_3BN(CH_3)_3$ Molecules. Optika i Spektroskopiya, Vol. 8, No. 3, p. 419, 1960. Bibliography: 4 titles.

K. K. Svidzinskii. Theory of the Hyperfine Structure of Rotational Molecular Spectra of Gyrator-Type Molecules. Optika i Spektroskopiya, Vol. 6, No. 2, pp. 254-256, 1959. Bibliography: 7 titles.

K. K. Svidzinskii. Calculation of the Hyperfine Structure of Rotational Molecular Spectra. Optika i Spektroskopiya, Vol. 11, No. 6, pp. 713-723, 1961. Bibliography: 17 titles.

K. K. Svidzinskii. Theory of the Hyperfine Structure of Molecular Spectra. Author's Certificate, Candidate's Thesis, Mathematical-Physical Sciences, Moscow, 1961. 6 pp. (Ministry of Higher and Secondary Special Education of the RSFSR, Moscow Government University, Scientific Research Institute – Nuclear Physics).

V. G. Veselago and A. M. Prokhorov. The HDSe Microwave Spectrum. Zhur. Eksp. i Teoret. Fiz., Vol. 31, No. 4, p. 731, 1956. [Soviet Physics – JETP, Vol. 31(4), p. 750.]

V. G. Veselago. Dipole Moment of the HDSe Molecule. Zhur. Eksp. i Teoret. Fiz., Vol. 32, No. 3, p. 620, 1957. [Soviet Physics – JETP, Vol. 32(5), p. 513.]

V. G. Veselago and A. M. Prokhorov. Microwave Spectrum of the HDSe Molecule (paper presented at the Tenth All-Union Conference on Spectroscopy, July, 1956). Collected Papers, Physics, L'vov Institute, No. 3, p. 493, 1957.

V. G. Veselago and N. A. Irisova. Modulation Circuit for the Frequency Stabilization of a Reflex Klystron by Means of a Volume Wavemeter. Radiotekh. i Elektron., Vol. 2, No. 4, pp. 484-487, 1957. Bibliography: 4 titles.

V. G. Veselago. Radiospectroscopic Investigation of the HDSe Molecule. Author's Certificate, Candidate's Thesis, Mathematical-Physical Sciences, Moscow, 1958. 8 pp. (Academy of Sciences, USSR, P. N. Lebedev Institute of Physics).

V. G. Veselago. Radiospectroscopic Investigation of the HDSe Molecule. Candidate's Thesis, Mathematical-Physical Sciences, Moscow, 1958. 125 pp. with ill. (Academy of Sciences, USSR, P. N. Lebedev Institute of Physics). Bibliography, pp. 118-120 (52 titles). Typewritten.

V. G. Veselago. Determination of the Molecular Structure of HDSe from a Study of the Microwave Spectrum (paper presented at the Eleventh All-Union Conference on Spectroscopic Theory, December, 1957). Izvest. Akad. Nauk SSSR, Ser. Fiz., Vol. 22, No. 9, pp. 1150-1153, 1958. Bibliography: 8 titles.

V. G. Veselago. Determination of the Structure and Dipole Moment of the HDSe Molecule from a Study of the Microwave Spectrum. Optika i Spektroskopiya, Vol. 6, No. 4, pp. 450-456, 1959. Bibliography: 19 titles.

E. M. Zemskov and V. G. Veselago. The Stark Effect in the Rotational Spectrum of Asymmetric Gyrator-Type Molecules with a Quadrupole Bond (Case $\mu \varepsilon \approx eQq$). Optika i Spektroskopiya, Vol. 3, No. 2, pp. 183-186, 1957. Bibliography: 3 titles.

OSCILLATORY PHENOMENA

A. I. Barchukov, G. A. Vasil'ev, M. E. Zhabotinskii, and B. D. Osipov. Electromechanical Klystron Frequency Stabilizer. Radiotekhnika, Vol. 10, No. 3, pp. 29-32, 1955.

N. G. Basov, V. G. Veselago, and M. E. Zhabotinskii. Improvement of the Quality of a Cavity Resonator by Means of Regeneration. Zhur. Eksp. i Teoret. Fiz., Vol. 28, No. 2, p. 242, 1955. [Soviet Physics – JETP, Vol. 28(1), p. 177.]

N. G. Basov. The Transformation of Mechanical Energy into UHF-Oscillatory Energy and UHF-Motors. Radiotekh. i Elektron., Vol. 4, No. 7, pp. 1180-1184, 1959. Bibliography: 5 titles.

L. N. Borodovskaya and A. E. Salomonovich. Measuring the Amplitude of Oscillations by the Piezoelectric Interferometric Method. Zhur. Tekh. Fiz., Vol. 21, No. 2, pp. 221-224, 1951. Bibliography: 3 titles.

F. V. Bunkin. Parametric Excitation of Quantum Systems. Radiotekh. i Elektron., Vol. 5, No. 2, pp. 296-300, 1960. Bibliography: 2 titles.

M. A. Divil'kovskii and S. M. Rytov. Self-Excitation and Resonance in a System with Periodically Changing Self-Inductance. Zhur. Tekh. Fiz., Vol. 6, No. 3, pp. 474-482, 1936.

M. A. Divil'kovskii. Classical Theory of the Zeeman Effect in an Alternating Magnetic Field. Zhur. Eksp. i Teoret. Fiz., Vol. 7, No. 5, pp. 650-662, 1937.

M. A. Divil'kovskii. Problem of the Sphere in a Homogeneous Alternating Magnetic Field. Zhur. Tekh. Fiz., Vol. 9, No. 5, pp. 433-443, 1939. Bibliography: 13 titles. In French: J. Phys. USSR, Vol. 1, No. 5-6, pp. 471-478, 1939.

M. A. Divil'kovskii. Theory of Induction Heating. Zhur. Tekh. Fiz., Vol. 9, No. 14, pp. 1302-1314, 1939. Bibliography: 10 titles.

M. A. Divil'kovskii. High-Frequency Voltmeter. Zhur. Tekh. Fiz., Vol. 19, No. 7-8, pp. 450-455, 1943. Bibliography: 2 titles.

M. I. Filippov. Dielectric Losses in High-Frequency Fields and Debye Theory. Dissertation. Trudy Fiz. Inst. Akad. Nauk, Vol. 2, No. 2-3, pp. 123-155, 1942.

G. S. Gorelik. L. I. Mandelshtam and Other Scientists on Resonance. Izvest. Akad. Nauk SSSR, Ser. Fiz., Vol. 9, No. 1-2, pp. 67-76, 1945.

G. S. Gorelik. One Possible Method for Investigating the Rapid Energy Exchange Between the Degrees of Freedom of Gas Molecules. Doklady Akad. Nauk SSSR, Vol. 54, No. 9, pp. 783-785, 1946.

G. S. Gorelik. Possible Low-Inertial Photometry and Demodulational Analysis of Light. Doklady Akad. Nauk SSSR, Vol. 58, No. 1, pp. 45-47, 1947.

G. S. Gorelik. Interferometry, Diffraction, and Spectral Resolution in Optics and Radio. Uspekhi Fiz. Nauk, Vol. 36, No. 3, pp. 407-415, 1948.

G. S. Gorelik. Demodulational Analysis of Light. Uspekhi Fiz. Nauk, Vol. 34, No. 3, pp. 321-333, 1948.

B. N. Gorozhankin and Yu. A. Manteifel'. Symmetrical Circuits for Second-Order Resonance. Elektrosvyaz, Vol. 3, No. 3, pp. 33-44, 1941.

B. N. Gorozhankin. Mutual Resonance at a Distance. Izvest. Akad. Nauk SSSR, Ser. Fiz., Vol. 11, No. 2, pp. 147-153, 1947. Bibliography: 6 titles.

B. N. Gorozhankin. Model of a Radio-Frequency Electrometer. Zhur. Tekh. Fiz., Vol. 18, No. 10, pp. 1258 to 1264, 1948.

L. S. Gutkin and A. D. Kuz'min. Influence of Electron Inertia Upon Diode Detection and Frequency Transformation. Radiotekhnika, Vol. 10, No. 9, pp. 14-30, 1955.

N. A. Irisova. Frequency Standardization of a 3-cm Klystron by Means of a Spectral Gas Line. Candidate's Thesis, Mathematical-Physical Sciences, Moscow, 1950. 126 pp with ill. (Academy of Sciences, USSR, P. N. Lebedev Institute of Physics). Bibliography: p. 123. Typewritten.

N. V. Karlov and A. E. Salomonovich. A Study of the Dependence of the Crystal Detector Nonlinearity upon Small Shifts in Contact Potential. Zhur. Tekh. Fiz., Vol. 22, No. 12, pp. 1981-1984, 1952. Bibliography: 3 titles.

N. V. Karlov and A. E. Salomonovich. Automatic Null 3.2-cm Radiometer (Materials of the Scientific Conference on Radioelectronics, Gor'kii, January, 1956). Radiotekh. i Elektron., Vol. 1, No. 6, p. 886, 1956.

N. V. Karlov. Parasitic Modulation. Radiotekh. i Elektron., Vol. 1, No. 6, p. 852, 1956.

S. É. Khaikin. Half a Century of Theoretical Radio Engineering. Vestn. Svyazi, Élektrosvyaz', No. 5, pp. 34-36, 1945.

S. É. Khaikin. Oscillator. Radio, No. 4, pp. 45-47, 1948.

S. É. Khaikin. Oscillator. Radio, No. 6, pp. 54-56, 1948.

S. É. Khaikin, G. K. Demishev, and A. E. Salomonovich. Dynamic Method of Investigating Electrical Contact. Doklady Akad. Nauk SSSR, Vol. 70, No. 4, pp. 609-611, 1950. Bibliography: 3 titles.

S. É. Khaikin. Radio Transmission and Reception. Radio, No. 2, pp. 55-59, 1951.

S. É. Khaikin. Radio Tube Construction and Operation. Radio, No. 4, pp. 54-58, 1951.

S. É. Khaikin. Propagation of Electromagnetic Energy. Radio, No. 7, pp. 43-48; No. 8, pp. 58-62, 1951.

S. É. Khaikin. Capacitance and Inductance. Radio, No. 2, pp. 54-59, 1952.

S. É. Khaikin. Radio Waves. Radio, No. 1, pp. 56-58, 1952.

S. É. Khaikin. Radiotelephony. Radio, No. 8, pp. 60-63, 1952.

S. É. Khaikin. Electrical Oscillations. Radio, No. 3, pp. 52-57, 1952.

S. É. Khaikin. Detection. Radio, No. 1, pp. 54-57, 1953.

M. L. Kotlyarevskii and E. Ya. Pumper. Interferometric Investigation of the Oscillation of Piezoquartz Plates. Zhur. Tekh. Fiz., Vol. 11, No. 9, pp. 843-853, 1941. Bibliography: 10 titles.

A. D. Kuz'min. Modulation During Interferometric Radio Reception. Pribory i Tekh. Eksp., No. 5, pp. 67-68, 1957.

Ya. I. Likhter and S. É. Khaikin. The Influence of Rapid Temperature and Pressure Changes upon the Resistance of Electrolytes. Zhur. Eksp. i Teoret. Fiz., Vol. 18, No. 7, pp. 651-658, 1948. Bibliography: 5 titles.

L. A. Lyusternik and A. M. Prokhorov. Determination of the Eigenvalues and Eigenfunctions of Some Operators by Means of RC-Circuits. Doklady Akad. Nauk SSSR, Vol. 55, No. 7, pp. 579-582, 1947.

L. A. Lyusternik and A. M. Prokhorov. Experimental Determination of the Eigenvalues and Eigenfunctions of Some Operators by Means of RC-Circuits. Izvest. Akad. Nauk SSSR, Ser. Fiz., Vol. 11, No. 2, pp. 141-145, 1947.

L. I. Mandel'shtam and S. É. Khaikin. New Studies in Nonlinear Oscillations. Moscow, Radioizdat, 1936.

D. I. Mash and P. Enyshkov. The Magnetic Permeability of Iron in High-Frequency Fields. Zhur. Tekh. Fiz., Vol. 8, No. 22-23, pp. 1980-1991, 1938.

D. I. Mash. New Measurements of the Magnetic Permeability of Steel in High-Frequency Fields. Zhur. Tekh. Fiz., Vol. 9, No. 4, pp. 309-342, 1939.

D. I. Mash. The Dielectric Constant and Conductivity of Water in Water Solutions of Potassium Chloride in UHF Fields and Dispersion and Absorption of Electromagnetic Waves in Heavy Water. Candidate's Thesis, Mathematical-Physical Sciences. Kazan, 1941. 46 pp. Typewritten.

D. I. Mash. Losses and Dielectric Permittivity of Barium Titanate in UHF Fields. Zhur. Eksp. i Teoret. Fiz., Vol. 17, No. 6, pp. 537-539, 1947. Bibliography: 5 titles.

D. I. Mash, L. S. Mayants, and I. L. Fabelinskii. Measurement of the Temperature Dependence of the Dielectric Permittivity and Loss Angle in a Centimeter Wave Field. Zhur. Tekh. Fiz., Vol. 19, No. 10, pp. 1192-1198, 1949. Bibliography: 1 title.

D. I. Mash. Dielectric Permittivity and Loss Angle of Some Organic Substances in an UHF Field. Vestnik. Inform., 1951, No. 8 (39) (Byuro Novoi Tekhniki).

D. I. Mash. Rotation of the Vibration Plane of Millimeter Waves in a Magnetic Field (Faraday Effect). Zhur. Tekh. Fiz., Vol. 27, No. 2, pp. 360-363, 1957. [Soviet Physics – Technical Physics, Vol. 2, p. 324.]

D. I. Mash. The Temperature and Frequency Dependences of the Faraday Effect for Millimeter Wavelengths. Zhur. Tekh. Fiz., Vol. 28, No. 12, pp. 2713-2715, 1958. [Soviet Physics – Technical Physics, Vol. 3, p. 2483.]

D. I. Mash. Ferrite Valve for Millimeter Waves Utilizing the Field Displacement Effect in Waveguides. Radiotekh. i Elektron., Vol. 3, No. 7, pp. 958-959, 1958.

D. I. Mash and V. V. Nikol'skii. Measurement of the Complex Dielectric Susceptibility and the Magnetic Susceptibility Tensor of Ferrites in the Millimeter Region. Zhur. Tekh. Fiz., Vol. 29, No. 9, pp. 1070-1073, 1959. Bibliography: 5 titles. [Soviet Physics – Technical Physics, Vol. 4, p. 978.]

A. B. Melik'yan. A Rapid Method of Determining Variations in Gravitational Acceleration. Doklady Akad. Nauk SSSR, Vol. 21, No. 8, pp. 375-376, 1938.

A. B. Melik'yan. Magnetic Detection (Thesis). Trudy Fiz. Inst. Akad. Nauk, Vol. 2, No. 1, pp. 25-39, 1940.

V. V. Migulin. Autoparametric Excitation. Zhur. Tekh. Fiz., Vol. 6, No. 4, pp. 644-661, 1936.

V. V. Migulin. Combinational Resonance. Trudy Fiz. Inst. Akad. Nauk, Vol. 1, No. 3, pp. 71-101, 1936.

V. V. Migulin and Ya. L. Al'pert. About a Particular Case of Autoparametric Resonance. Zhur. Tekh. Fiz., Vol. 6, No. 5, pp. 813-818, 1936.

V. V. Migulin. Remarks on a Paper by Prof. M. A. Bonch-Bruevich Entitled "Remarks on the Parametric Excitation of Oscillations." Radiotekhnika, Vol. 5, No. 1, pp. 1-14, 1937.

V. V. Migulin. Resonance Phenomena in a System with Two Degrees of Freedom. Zhur. Tekh. Fiz., Vol. 7, No. 6, pp. 627-641, 1937.

A. M. Prokhorov. Frequency Stabilization in Small-Parameter Theory. Candidate's Thesis, Mathematical-Physical Sciences, Moscow, 1945, 83 pp. Typewritten.

A. M. Prokhorov. Frequency Capture of the External Force in a Self-Exciting Lecher System. Zhur. Tekh. Fiz., Vol. 16, No. 10, pp. 1145-1156, 1946. Bibliography: 9 titles.

E. Ya. Pumper. Methods Used in Measuring Ultrasonic Absorption. Doklady Akad. Nauk SSSR, Vol. 49, No. 8, pp. 581-583, 1945. Bibliography: 10 titles.

P. A. Ryazin. The Mechanism of Forced Synchronization. Zhur. Tekh. Fiz., Vol. 5, No. 10, pp. 1809-1833, 1935.

P. A. Ryazin. Processes Involved in Setting Up Oscillations in a Self-Oscillatory System During Total and Partial Frequency Increases. Zhur. Tekh. Fiz., Vol. 5, No. 1, pp. 38-52, 1935.

S. M. Rytov. n-th Order Resonance in a System with Two Degrees of Freedom and Strong Coupling. Zhur. Tekh. Fiz., Vol. 5, No. 1, pp. 3-37, 1935. Bibliography: 23 titles.

S. M. Rytov. Modulated Waves and Oscillations. Trudy Fiz. Inst. Akad. Nauk, Vol. 2, No. 1, pp. 25-142, 1940.

S. M. Rytov. Ultrasonics. Methods of Obtaining Ultrasonic Vibrations. Nauka i Zhizn, No. 3, pp. 24-29, 1943.

S. M. Rytov. Ultrasonics. Properties and Applications of Ultrasonics. Nauka i Zhizn, No. 4-5, pp. 18-27, 1943.

S. M. Rytov. Parametric Oscillations of a Steel Body in an Alternating Magnetic Field. Izvest. Akad. Nauk SSSR, Ser. Fiz., Vol. 8, No. 3, pp. 150-155, 1944.

S. M. Rytov. Parametric Oscillations of a Steel Body in an Alternating Magnetic Field. Zhur. Eksp. i Teoret. Fiz., Vol. 14, No. 9, pp. 370-378, 1944. Bibliography: 10 titles. In English: J. Phys. USSR, Vol. 8, No. 6, p. 383, 1944.

S. M. Rytov. L. I. Mandelshtam and Other Scientists on Modulation. Izvest. Akad. Nauk SSSR, Ser. Fiz., Vol. 9, No. 1-2, pp. 77-87, 1945.

S. M. Rytov, A. M. Prokhorov, and M. E. Zhabotinskii. Theory of Frequency Stabilization. I. Zhur. Eksp. i Teoret. Fiz., Vol. 15, No. 10, pp. 557-572, 1945. Bibliography: 4 titles.

S. M. Rytov, S. M. Prokhorov, and M. E. Zhabotinskii. Theory of Frequency Stabilization. II. Zhur. Eksp. i Teoret. Fiz., Vol. 15, No. 11, pp. 613-628, 1945.

S. M. Rytov. Extending the Range of Application of the Small-Parameter Method. Doklady Akad. Nauk SSSR, Vol. 47, No. 3, pp. 186-189, 1945. Bibliography: 6 titles.

S. M. Rytov. Some Paradoxes Associated with Spectral Resolution. Uspekhi Fiz. Nauk, Vol. 29, No. 1-2, pp. 147-160, 1946.

S. M. Rytov and M. E. Zhabotinskii. Application of the Method of Parametric Resolution to Systems with Properties Which Approach Those of Sturm-Liouville Systems. Izvest. Akad. Nauk SSSR, Ser. Fiz., Vol. 11, No. 2, pp. 135-140, 1947. Bibliography: 2 titles.

S. M. Rytov. Development of the Theory of Nonlinear Oscillations in the USSR. Radiotekhnika, Vol. 2, No. 8, pp. 58-62, 1947.

S. M. Rytov. Thermometric Method for Measuring the Electric Field Intensity of Centimeter Waves. Izvest. Akad. Nauk SSSR, Ser. Fiz., Vol. 11, No. 2, pp. 191-194, 1947. Bibliography: 4 titles.

S. M. Rytov, A. M. Prokhorov, and M. E. Zhabotinski. Stabilizing the Frequency of Vacuum-Tube Oscillators. I. Izvest. Akad. Nauk SSSR, Ser. Fiz., Vol. 12, No. 2, pp. 184-185, 1948. Bibliography: 5 titles.

S. M. Rytov, A. M. Prokhorov, and M. E. Zhabotinskii. Stabilizing the Frequency of Vacuum Tube Oscillators. II. Izvest. Akad. Nauk SSSR, Ser. Fiz., Vol. 18, No. 2, pp. 184-185, 1948.

S. M. Rytov. Development of Nonlinear Oscillation Theory In the USSR. Radiotekh. i Elektron., Vol. 2, No. 11, pp. 1435-1450, 1957. Bibliography: 198 titles.

S. M. Rytov. Diffraction of Light by Ultrasonic Waves (in German). Sow. Phys., Vol. 8, No. 6, pp. 626-643, 1935.

S. Rytov. Frequency Modulation (in French). Tech. Phys. USSR, Vol. 2, No. 2-3, pp. 215-231, 1935.

A. E. Salomonovich. Electrical and Mechanical Ferroresonance. Uspekhi Fiz. Nauk, Vol. 34, No. 3, pp. 415 to 439, 1948. Bibliography: 29 titles.

A. E. Salomonovich and L. Lisovskii. Force of Friction. Moscow, Goskul'tprosvetizdat, 1948, 72 pp.

A. E. Salomonovich and L. Lisovskii. Friction in Nature and Technology. Moscow, Goskul'tprosvetizdat, 1948.

A. E. Salomonovich. Dry Friction and Electrical Contact for Small Displacements. Candidate's Thesis, Mathematical-Physical Sciences, Moscow, 1949, 153 pp. (Academy of Sciences of the USSR, P. N. Lebedev Institute of Physics). Typewritten.

A. E. Salomonovich. Dry Friction and Electrical Contact for Small Displacements. Zhur. Eksp. i Teoret. Fiz., Vol. 20, No. 7, pp. 647-660, 1950. Bibliography: 21 titles.

A. E. Salomonovich. Self-Modulation during Ferroresonance. Zhur. Tekh. Fiz., Vol. 22, No. 2, pp. 245-258, 1952. Bibliography: 10 titles.

A. E. Salomonovich. Frequency Division Circuits with Nonlinear Capacitors. Zhur. Tekh. Fiz., Vol. 22, No. 7, pp. 1190-1194, 1952. Bibliography: 8 titles.

A. E. Salomonovich and T. A. Shmaonov. Choice of Modulation Frequency in a Modulated Radiometer, in Papers Presented at the Fifth Conference on Cosmogony. Izd-vo AN SSSR, 1956, pp. 127-129, questions and introductions, p. 130. Bibliography: 3 titles.

E. Ya. Shchegolev. Automatic Counter of the Cycle Changes during Changes in Phase. Izvest. Akad. Nauk SSSR, Ser. Fiz., Vol. 7, No. 3, pp. 52-63, 1943.

G. A. Vasil'ev and M. E. Zhabotinskii. Measuring the Power of Centimeter Waves by Thermoacoustic Methods. Zhur. Eksp. i Teoret. Fiz., Vol. 24, No. 5, pp. 570-574, 1953.

V. V. Vitkevich. Synchronization of Relaxation Oscillators. Zhur. Tekh. Fiz., Vol. 14, No. 1-2, pp. 70-91, 1944.

V. V. Vitkevich. Synchronization of Single-Cycle Relaxation Oscillators. II. Zhur. Tekh. Fiz., Vol. 15, No. 11, pp. 777-792, 1945. Bibliography: 4 titles.

V. V. Vitkevich. Geometrical Theory of Relaxation Oscillator Synchronization. III. Zhur. Tekh. Fiz., Vol. 15, No. 11, pp. 793-804, 1945. Bibliography: 6 titles.

V. V. Vitkevich. "Strict" Operation of Self-Excited Relaxation Oscillators (Abraham-Bloch Multivibrator). Zhur. Tekh. Fiz., Vol. 16, No. 3, pp. 308-314, 1946. Bibliography: 2 titles.

V. V. Vitkevich. The Frequency Spectrum of Aperiodic Functions. Zhur. Tekh. Fiz., Vol. 16, No. 3, pp. 317 to 320, 1946. Bibliography: 2 titles.

V. V. Vitkevich. Synchronization of Relaxation Self-Oscillations Using the Basic Frequency and Subharmonics of the External emf. Radiotekhnika, Vol. 4, No. 3, pp. 76-77, 1949.

V. V. Vitkevich. Synchronization and Subharmonics of the External emf. IV. Zhur. Tekh. Fiz., Vol. 20, No. 10, pp. 1245-1256, 1950. Bibliography: 8 titles.

M. E. Zhabotinskii. Special Case of a System with Two Degrees of Freedom. Zhur. Eksp. i Teoret. Fiz., Vol. 15, No. 10, pp. 573-586, 1945. Bibliography: 13 titles.

M. E. Zhabotinskii. Theory of Frequency Stabilization. Radiotekhnika, Vol. 1, No. 3-4, pp. 19-37, 1946. Bibliography: 8 titles.

M. E. Zhabotinskii and A. M. Prokhorov. Television. Nauka i Zhizn. No. 11-12, pp. 18-22, 1946.

M. E. Zhabotinskii. Periodic Solutions of Nonlinear Partial Differential Equations. Doklady Akad. Nauk SSSR, Vol. 56, No. 5, pp. 469-472, 1947. Bibliography: 4 titles.

M. E. Zhabotinskii. Self-Oscillatory Systems with Two Degrees of Freedom for the Case of Even Frequencies. Zhur. Eksp. i Teoret. Fiz., Vol. 20, No. 5, pp. 421-426, 1950. Bibliography: 5 titles.

M. E. Zhabotinskii. Klystron. Radio, No. 5, pp. 40-45, 1951.

M. E. Zhabotinskii and D. A. Lisichkin. Quartz Oscillators with Negative Feedback and Inertial Nonlinearity. Doklady Akad. Nauk SSSR, Vol. 95, No. 6, pp. 1197-1200, 1954.

M. E. Zhabotinskii, N. A. Irisova, and S. M. Rytov. The Effect of a Varying Frequency Signal upon a Linear Resonant Frequency. Trudy Fiz. Inst. Akad. Nauk, Vol. 8, pp. 5-12, 1956. Bibliography: 17 titles.

RADIO PROPAGATION

A. A. Ainberg. Distribution of Ionization with Height and the Coefficient of Recombination in the F-Layer of the Ionosphere. Zhur. Eksp. i Teoret. Fiz., Vol. 19, No. 6, pp. 515-520, 1949. Bibliography: 7 titles.

Ya. L. Al'pert, V. V. Migulin, and P. A. Ryazin. Dispersion of Electromagnetic Waves Over the Surface of the Earth. Doklady Akad. Nauk SSSR, Vol. 18, No. 9, pp. 635-638, 1938.

Ya. L. Al'pert, V. V. Migulin, and P. A. Ryazin. The Electromagnetic Field in the Vicinity of a Radiating Antenna. Zhur. Tekh. Fiz., Vol. 9, No. 9, pp. 824-830, 1939. In English: J. Phys. USSR, Vol. 1, No. 5-6, pp. 381-387, 1939.

Ya. L. Al'pert and V. V. Migulin. Phase Structure of an Electromagnetic Field Near the Earth's Surface. Izvest. Akad. Nauk SSSR, Ser. Fiz., Vol. 4, No. 3, pp. 458-467, 1940.

Ya. L. Al'pert and V. V. Migulin. Influence of the Earth's Surface upon the Phase Structure of the Electromagnetic Field of a Radiating Antenna. Doklady Akad. Nauk SSSR, Vol. 26, No. 9, pp. 878-881, 1940.

Ya. L. Al'pert, V. V. Migulin, and P. A. Ryazin. Investigation of the Phase Structure of the Electromagnetic Field and of Radio-Wave Velocity. Zhur. Tekh. Fiz., Vol. 11, No. 1-2, pp. 7-36, 1941. Bibliography: 22 titles.

Ya. L. Al'pert and B. N. Gorozhankin. Boundary Refraction. Zhur. Tekh. Fiz., Vol. 11, No. 13-14, pp. 1238 to 1244, 1941.

Ya. L. Al'pert. Radio Direction Finding. Nauka i Zhizn, No. 1-2, pp. 31-38, 1943.

Ya. L. Al'pert and V. L. Ginzburg. Ionospheric Radio-Wave Absorption. Izvest. Akad. Nauk SSSR, Ser. Fiz., Vol. 8, No. 2, pp. 42-67, 1944.

Ya. L. Al'pert. Radio and Solar Eclipses. Nauka i Zhizn, No. 10, pp. 16-20, 1944.

Ya. L. Al'pert and B. N. Gorozhankin. Solar Eclipses and Radio Investigations of the Ionosphere. Izvest. Akad. Nauk SSSR, Ser. Fiz., Vol. 8, No. 2, pp. 85-108, 1944.

Ya. L. Al'pert. The Effect of Solar Eruptions upon Radio. Nauka i Zhizn, No. 1, pp. 11-13, 1945.

Ya. L. Al'pert and B. N. Gorozhankin. Results of Radio Observations Made during the Solar Eclipse (Corpuscular and Ultraviolet), July 9, 1945. Doklady Akad. Nauk SSSR, Vol. 49, No. 4, pp. 260-264, 1945.

Ya. L. Al'pert. Investigation of the Phase Structure of the Electromagnetic Field in the Vicinity of a Transmitting Antenna. Trudy Fiz. Inst. Akad. Nauk, Vol. 3, No. 2, pp. 3-43, 1946.

Ya. L. Al'pert. On the Effect of Ionospheric Anisotropy. Doklady Akad. Nauk SSSR, Vol. 53, No. 8, pp. 763 to 765, 1946.

Ya. L. Al'pert. Some Physical Atmospheric Phenomena and Their Explanation (Sporadic F_2 Layer). Doklady Akad. Nauk SSSR, Vol. 53, No. 2, pp. 111-114, 1946.

Ya. L. Al'pert and B. N. Gorozhankin. Radio Observations during the Solar Eclipse of June 9, 1946. Izvest. Akad. Nauk SSSR, Ser. Fiz., Vol. 10, No. 3, pp. 245-251, 1946.

Ya. L. Al'pert and S. M. Rytov. The Electromagnetic Field Structure of Two Synchronously Operating Radiators. Zhur. Tekh. Fiz., Vol. 16, No. 4, pp. 469-484, 1946. Bibliography: 1 title.

Ya. L. Al'pert. A Study of the Effect of Ionospheric Anisotropy and Other Experimental Investigations of the F_2 Layer. Doctor's Thesis, Mathematical-Physical Sciences, Moscow, 1947, 138 pp. (Academy of Sciences, USSR, P. N. Lebedev Institute of Physics). Typewritten.

Ya. L. Al'pert and A. A. Ainberg. Concerning the Coefficient of Recombination of the Ionosphere and the Determination of Its Value during the Forthcoming Solar Eclipse on May 20, 1947 in Brazil. Izvest. Akad. Nauk SSSR, Ser. Geogr. i Geofiz., Vol. 11, No. 2, pp. 137-140, 1947.

Ya. L. Al'pert. The Sporadic F_2 Layer of the Ionosphere. Doklady Akad. Nauk SSSR, Vol. 55, No. 1, pp. 25-26, 1947.

Ya. L. Al'pert. Investigations Conducted on the F_2 Layer of the Ionosphere During the Total Solar Eclipse on May 20, 1947. Doklady Akad. Nauk SSSR, Vol. 58, No. 9, pp. 1919-1922, 1947.

Ya. L. Al'pert. Concerning the Atmospheric Structure and Processes Occurring in the F_2 Layer. Zhur. Eksp. i Teoret. Fiz., Vol. 18, No. 11, pp. 994-1011, 1948.

Ya. L. Al'pert. Concerning Investigations of the F_2 Layer of the Ionosphere during the Total Solar Eclipse of May 20, 1947 in Brazil. Izvest. Akad. Nauk SSSR, Ser. Fiz., Vol. 12, No. 1, pp. 44-48, 1948.

Ya. L. Al'pert. Trajectory of Rays in a Magnetoactive Ionized Medium – The Ionosphere. Izvest. Akad. Nauk SSSR, Ser. Fiz., Vol. 12, No. 3, pp. 242-266, 1948.

Ya. L. Al'pert. Contemporary State of the Question of Ionospheric Investigations. Uspekhi Fiz. Nauk, Vol. 34, No. 2, pp. 262-302; Vol. 36, No. 10, pp. 1-29, 1948.

Ya. L. Al'pert. Experimental Investigations of the So-Called Ionospheric Anisotropy Effect. Izvest. Akad. Nauk SSSR, Ser. Fiz., Vol. 12, No. 3, pp. 267-287, 1948.

Ya. L. Al'pert. Outline of Contemporary Data on the Investigation of Radio-Wave Distribution. Uspekhi Fiz. Nauk, Vol. 42, No. 4, pp. 505-506, 1950.

A. S. Berkman and D. I. Mash. Influence of Slotted Shielding upon Waveguide Field Structure. Izvest. Akad. Nauk SSSR, Otd. Tekh. Nauk, No. 10-11, pp. 1139-1144, 1945.

M. A. Divil'kovskii and M. I. Filippov. Measurement of the Field Intensity of an Ultrahigh-Frequency Field. Zhur. Eksp. i Teoret. Fiz., Vol. 5, No. 10, pp. 958-969, 1935.

M. A. Divil'kovskii and M. I. Filippov. Measurement of the Field Intensity of an Ultrahigh-Frequency Field. Doklady Akad. Nauk SSSR, Vol. 2, No. 8-9, pp. 521-527, 1935.

M. A. Divil'kovskii and M. I. Filippov. Determination of Dielectric Losses in Liquids at High Frequency. Zhur. Eksp. i Teoret. Fiz., Vol. 6, No. 1, pp. 93-98, 1936. Bibliography: 6 titles.

M. A. Divil'kovskii and M. I. Filippov. Setup for Measuring the Field Intensity of Ultrahigh-Frequency Fields. Zhur. Tekh. Fiz., Vol. 6, No. 11, pp. 1873-1884, 1936.

M. A. Divil'kovskii. Absolute Method for Measuring the Dielectric Constant and Losses in Fluids for Decimeter Waves. Doklady Akad. Nauk SSSR, Vol. 24, No. 5, pp. 433-436, 1939.

M. A. Divil'kovskii and D. I. Mash. Dispersion and Absorption of Electromagnetic Waves in Heavy Water. Doklady Akad. Nauk SSSR, Vol. 27, No. 8, pp. 802-814, 1940.

M. A. Divil'kovskii and D. I. Mash. Measurement of the Dispersion and Absorption of Electromagnetic Waves in Heavy Water. Zhur. Eksp. i Teoret. Fiz., Vol. 10, No. 8, pp. 903-907, 1940. Bibliography: 8 titles.

M. A. Divil'kovskii and D. I. Mash. Measurement of the Conductivity and Dielectric Constant of Water and of an Aqueous Solution of Potassium Chloride at Ultrahigh Frequencies. Zhur. Eksp. i Teoret. Fiz., Vol. 10, No. 5, pp. 520-541, 1940. French translation: J. Phys. USSR, Vol. 2, No. 5, pp. 385-407, 1940.

M. A. Divil'kovskii and D. I. Mash. Conductivity and Dielectric Constant of Water and of an Aqueous Solution of Potassium Chloride at Ultrahigh Frequencies. Zhur. Eksp. i Teoret. Fiz., Vol. 10, No. 11, pp. 1257-1262, 1940. French translation: J. Phys. USSR, Vol. 4, No. 1-2, pp. 59-65, 1941.

M. A. Divil'kovskii. Measurement of the Dielectric Losses of Solids at Ultrahigh Frequencies. Doklady Akad. Nauk SSSR, Vol. 32, No. 4, pp. 250-251, 1941. Bibliography: 4 titles.

M. A. Divil'kovskii. Some Data Regarding the Dispersion and Absorption of Electromagnetic Waves in H_2O, D_2O, and Aqueous Solutions of KCl at Ultrahigh Frequencies. Izvest. Akad. Nauk SSSR, Ser. Fiz., Vol. 5, No. 1, pp. 51-52, 1941.

A. A. Gorozhankina. Localization of Atmospheric Inhomogeneities. Doklady Akad. Nauk SSSR, Vol. 93, No. 3, pp. 459-461, 1953.

N. V. Karlov. Investigation of Time Delay Systems. Radiotekh. i Elektron., Vol. 6, No. 6, p. 1029, 1961. Bibliography: 4 titles.

V. Kobelev and A. Salomonovich. Waveguides. Radio, No. 2, pp. 19-23, 1952.

A. D. Kuz'min. Diaphragmatic Control of the Direction of Radiointerferometric Frequency Changes in a Heterodyne. Radiotekh. i Elektron., Vol. 3, No. 7, p. 943, 1957.

A. D. Kuz'min. Measuring the Intensity of Radioemission of Small-Angle Sources. Radiotekh. i Elektron., Vol. 3, No. 4, pp. 561-562, 1958.

A. D. Kuz'min. Joint Operation of Several Radiators of Varying Wavelength at the Focus of a Parabolic Reflector. Radiotekh. i Elektron., Vol. 3, No. 5, pp. 722-723, 1958. Bibliography: 1 title.

M. A. Leontovich and M. V. Fok. Solution of the Problem of Electromagnetic Wave Propagation Along the Surface of the Earth by Means of Parabolic Equations. Zhur. Eksp. i Teoret. Fiz., Vol. 16, No. 7, pp. 557 to 573, 1946. Bibliography: 3 titles.

M. A. Leontovich and M. L. Levin. The Excitation of Oscillations in Antenna Oscillators. Zhur. Tekh. Fiz., Vol. 14, No. 9, pp. 481-506, 1944.

M. A. Leontovich. A Solution of the Problem of Electromagnetic Wave Propagation Along the Surface of the Earth. Izvest. Akad. Nauk SSSR, Ser. Fiz., Vol. 8, No. 1, pp. 16-22, 1944. Bibliography: 1 title. English translation: J. Phys. USSR, Vol. 8, No. 6, p. 382, 1944.

M. A. Leontovich and M. L. Levin. Excitation of Antenna Oscillators. Izvest. Akad. Nauk SSSR, Ser. Fiz., Vol. 8, No. 3, pp. 156-163, 1944.

M. L. Levin and S. M. Rytov. The Transition to a Geometric Approximation in the Theory of Elasticity. Akust. Zhur., Vol. 2, No. 2, pp. 173-176, 1956. Bibliography: 4 titles [Soviet Physics – Acoustics, Vol. 2, p. 179.]

V. V. Migulin. Radio-Wave Interference. Dissertation, Moscow, 1944, 138 pp.

P. A. Ryazin. Electromagnetic Field of a Planar Vertical Half-Wave Antenna Located Above the Earth. Zhur. Tekh. Fiz., Vol, 7, No. 18-19, pp. 1871-1879, 1937.

P. A. Ryazin. Calculation of the Radiation from a Short-Range Rectilinear Antenna. Zhur. Tekh. Fiz., Vol. 7, No. 6, pp. 646-667, 1937.

P. A. Ryazin. Calculation of the Phase Structure of the Electromagnetic Field and the Velocity of Propagation of a Radio Wave in the Vicinity of the Earth's Surface. Izvest. Akad. Nauk SSSR, Ser. Fiz., Vol. 4, No. 3, pp. 434-453, 1940.

P. A. Ryazin. Radio-Wave Propagation in the Vicinity of the Earth's Surface. Doctoral Thesis, Mathematical-Physical Sciences, Moscow, 1941. 107 pp.

P. A. Ryazin and L. M. Brekhovskikh. Radio-Wave Field in the Space Between Two Semiconducting Media (paper presented at the Conference on the Propagation of Radio Waves and Oscillations, Moscow, 1945). Izvest. Akad. Nauk SSSR, Ser. Fiz., Vol. 10, No. 3, pp. 285-305, 1946. Bibliography: 16 titles.

P. A. Ryazin. Propagation of Radio Waves Along the Earth's Surface. Dissertation. Trudy Fiz. Inst. Akad. Nauk, Vol. 3, No. 2, pp. 47-120, 1946. Bibliography: 22 titles.

S. M. Rytov. Diffraction of Light by Ultrasound. Zhur. Eksp. i Teoret. Fiz., Vol. 5, No. 9, pp. 843-856, 1935.

S. M. Rytov. Diffraction of Light by Ultrasound. Doklady Akad. Nauk SSSR, Vol. 2, No. 6, pp. 223-226; Vol. 3, No. 4, pp. 151-156, 1936. Bibliography: 5 titles.

S. M. Rytov. Diffraction of Light by Ultrasound. Izvest. Akad. Nauk SSSR, Ser. Fiz., No. 2, pp. 223-259, 1937. Bibliography: 36 titles.

S. M. Rytov. Transition from Wave to Geometrical Optics. Doklady Akad. Nauk SSSR, Vol. 18, No. 4-5, pp. 263-266, 1938.

S. M. Rytov, V. V. Vladimirskii, and M. D. Galanin. Propagation of Sound in Dispersive Systems. Zhur. Eksp. i Teoret. Fiz., Vol. 8, No. 5, pp. 614-621, 1938. Bibliography: 4 titles.

S. M. Rytov. Calculation of Electromagnetic Wave Absorption in Waveguides. Zhur. Eksp. i Teoret. Fiz., Vol. 10, No. 2, pp. 176-179, 1940. Bibliography: 2 titles. English translation: J. Phys. USSR, Vol. 2, No. 2, pp. 187-190, 1940.

S. M. Rytov and F. S. Yudkevich. Reflection of Electromagnetic Waves from a Layer with a Negative Dielectric Constant. Zhur. Eksp. i Teoret. Fiz., Vol. 10, No. 8, pp. 887-902, 1940. Bibliography: 20 titles. German translation: J. Phys. USSR, Vol. 3, No. 2, pp. 111-124, 1940.

S. M. Rytov. Reflection of Electromagnetic Waves from a Layer with a Negative Dielectric Constant. Izvest. Akad. Nauk SSSR, Ser. Fiz., Vol. 4, No. 3, pp. 407-408, 1940.

S. M. Rytov. Calculation of the Skin Effect by the Excitation Method. Zhur. Eksp. i Teoret. Fiz., Vol. 10, No. 2, pp. 180-189, 1940. French translation: J. Phys. USSR, Vol. 2, No. 3, pp. 233-242, 1940.

S. M. Rytov. Field Excitation of a Spherical Resonator by a Dipole at the Center of the Resonator. Doklady Akad. Nauk SSSR, Vol. 51, No. 2, pp. 107-110, 1946.

S. M. Rytov. Theorems on the Group Velocity of Electromagnetic Waves. Zhur. Eksp. i Teoret. Fiz., Vol. 17, No. 10, pp. 930-936, 1947. Bibliography: 3 titles.

S. Rytov and M. Jabotinsky. Observation of Refracting Structures. J. Phys. USSR, Vol. 11, No. 1, p. 92, 1947.

S. M. Rytov. Some Relations for Group Velocity in a Magnetoactive Medium (paper presented at the Session of the All-Union Scientific Conference on Radiophysics and Radio Engineering at the OFMN of the Academy of Sciences of the USSR). Izvest. Akad. Nauk SSSR, Ser. Fiz., Vol. 12, No. 2, p. 189, 1948.

S. M. Rytov and I. L. Fabelinskii. New Appearance of a Phase Diffraction Lattice. Zhur. Eksp. i Teoret. Fiz., Vol. 20, No. 4, pp. 340-341, 1950. Bibliography: 4 titles.

S. M. Rytov. Phase Constant Method in Microscopy. Uspekhi Fiz. Nauk, Vol. 41, No. 4, pp. 425-451, 1950. Bibliography: 16 titles.

S. M. Rytov. Scattering by Means of a Charged Filament. Zhur. Eksp. i Teoret. Fiz., Vol. 22, No. 4, pp. 510 to 511, 1952.

S. M. Rytov. Magnetic Flux Created by a Dipole Located within a Ferromagnetic Round Wire. Zhur. Eksp. i Teoret. Fiz., Vol. 27, No. 3, pp. 307-312, 1954.

S. M. Rytov. Electromagnetic Properties of a Finely Stratified Medium. Zhur. Eksp. i Teoret. Fiz., Vol. 29, No. 5, pp. 605-616, 1955. Bibliography: 12 titles. [Soviet Physics – JETP, Vol. 29(2), p. 466.]

S. M. Rytov. Acoustical Properties of a Thinly Laminated Medium. Akust. Zhur., Vol. 2, No. 1, pp. 71-83, 1956. Bibliography: 7 titles. [Soviet Physics – Acoustics, Vol. 2, p. 67.]

A. Salomonovich. Waveguide in Ultrahigh-Frequency Technology. Radio, No. 3, pp. 16-19, 1952.

V. V. Vitkevich. Atmospheric Radio Interference Including a Study of the Factors Involved. Radiotekhnika, Vol. 1, No. 2, pp. 48-62, 1946. Bibliography: 26 titles.

V. V. Vitkevich. Experimental Study of Antennas by Means of Models. Radiotekhnika, Vol. 1, No. 3-4, pp. 70 to 78, 1946.

M. E. Zhabotinskii, M. L. Levin, and S. M. Rytov. The Fundamental Wave in Lossless Lines. Radiotekhnika, Vol. 4, No. 2, pp. 75-80, 1949.

M. E. Zhabotinskii, M. L. Levin, and S. M. Rytov. Telegrapher's Equation for a Low-Loss Generalized Line. Zhur. Tekh. Fiz., Vol. 20, No. 3, pp. 257-281, 1950. Bibliography: 6 titles.

M. E. Zhabotinskii. Coaxial Resonators with Capacitive Loads. Zhur. Tekh. Fiz., Vol. 21, No. 3, pp. 358-362, 1951. Bibliography: 1 title.

STATISTICAL RADIO PHYSICS

Ya. L. Al'pert and A. A. Ainberg. Statistical Nature of the Ionosphere. Zhur. Eksp. i Teoret. Fiz., Vol. 21, No. 3, pp. 389-400, 1951. Bibliography: 7 titles.

Ya. L. Al'pert and S. M. Rytov. Compensating for Radio Distrubances. Zhur. Tekh. Fiz., Vol. 14, No. 12, pp. 730-748, 1944.

V. I. Bunimovich and M. A. Leontovich. Distribution of Large Deviations during Electrical Fluctuations. Doklady Akad. Nauk SSSR, Vol. 53, No. 1, pp. 21-24, 1946. Bibliography: 1 title.

V. I. Bunimovich. Transformation of Fluctuations of a Nonlinear System. Zhur. Tekh. Fiz., Vol. 16, No. 6, pp. 635-650, 1946.

V. I. Bunimovich. Fluctuational Processes Considered as an Oscillation with Random Amplitude and Phase. Zhur. Tekh. Fiz., Vol. 19, No. 11, pp. 1231-1259, 1949. Bibliography: 23 titles.

F. V. Bunkin and N. V. Karlov. The Sensitivity of Radiometers. I. Zhur. Tekh. Fiz., Vol. 25, No. 3, pp. 430 to 435, 1955. Bibliography: 4 titles.

F. V. Bunkin and N. V. Karlov. The Sensitivity of Radiometers. II. Zhur. Tekh. Fiz., Vol. 25, No. 4, pp. 733 to 741, 1955. Bibliography: 7 titles.

F. V. Bunkin. Theory of Thermal Radiation of Anisotropic Media. Candidate's Thesis, Mathematical-Physical Sciences, Moscow, 1955, 108 pp. (Academy of Sciences, USSR, P. N. Lebedev Institute of Physics). Typewritten.

F. V. Bunkin. Statistical Nature of Remagnetization in Ferromagnets. Zhur. Tekh. Fiz., Vol. 26, No. 8, pp. 1782-1789, 1956. Bibliography: 6 titles. [Soviet Physics – Technical Physics, Vol. 1, p. 1727.]

F. V. Bunkin. The Theory of Thermal Radiation of an Ionized Gas in a Magnetic Field (paper presented at the Scientific Conference on Radioelectronics, Gor'kii, January, 1956). Radiotekh. i Elektron., Vol. 1, No. 6, p. 887, 1956.

F. V. Bunkin and N. V. Karlov. Sensitivity of a Modulational Radiometer in the Presence of Steady-State Fluctuations in the Coefficient of Amplification, in Papers Presented at the Fifth Conference on Cosmogony. Published by the Academy of Sciences of the USSR, 1956, pp. 81-87. Bibliography: 5 titles.

F. V. Bunkin. Periodic Remagnetization Noises in Ferromagnets. Zhur. Tekh. Fiz., Vol. 26, No. 8, pp. 1790 to 1798, 1956. Bibliography: 6 titles. [Soviet Physics – Technical Physics, Vol. 1, p. 1735.]

F. V. Bunkin. On Radiation in Anisotropic Media. Zhur. Eksp. i Teoret. Fiz., Vol. 32, No. 2, pp. 338-346, 1957. Bibliography: 7 titles. [Soviet Physics – JETP, Vol. 32(5), p. 277.]

F. V. Bunkin. Thermal Radiation from an Anisotropic Medium. Zhur. Eksp. i Teoret. Fiz., Vol. 32, No. 4, pp. 811-821, 1957. Bibliography: 20 titles. [Soviet Physics – JETP, Vol. 32(5), p. 665.]

F. V. Bunkin and L. I. Gudzenko. Uniform Amplitude and Phase Distributions in Steady-State Processes. Radiotekh. i Elektron., Vol. 3, No. 7, pp. 968-969, 1958. Bibliography: 6 titles.

F. V. Bunkin. Theory of the Barkhausen Effect in a Periodically Varying Field. Radiotekh. i Elektron., Vol. 4, No. 11, pp. 1913-1919, 1959. Bibliography: 13 titles.

F. V. Bunkin and A. N. Oraevskii. Spontaneous Molecular Irradiation within a Resonator. Izvest. Vysshikh. Ucheb. Zavedenii, Radiofiz., Vol. 2, No. 2, pp. 181-186, 1959. Bibliography: 6 titles.

F. V. Bunkin. Properties of the Modulation Envelope of a Steady Random Process. Radiotekh. i Elektron., Vol. 5, No. 9, pp. 1555-1556, 1960. Bibliography: 5 titles.

F. V. Bunkin. Theory of the Spontaneous Decay of Quantum Systems. Izvest. Vysshikh. Ucheb. Zavedenii, Radiotekh., Vol. 4, No. 5, pp. 892-902, 1961. Bibliography: 20 titles.

F. V. Bunkin. Contribution to the Theory of Electromagnetic Fluctuations in a Nonequilibrium Plasma. Zhur. Eksp. i Teoret. Fiz., Vol. 41, No. 1, pp. 288-293, 1961. Bibliography: 13 titles. [Soviet Physics – JETP, Vol. 41(14), p. 206.]

F. V. Bunkin. Contribution to the Theory of Electromagnetic Fluctuations in a Nonsteady-State Plasma. Zhur. Eksp. i Teoret. Fiz., Vol. 41, No. 6, pp. 1859-1867, 1961. Bibliography: 11 titles. [Soviet Physics – JETP, Vol. 41(14), p. 1322.]

F. V. Bunkin. The Concept of the Effective Temperature of Steady Nonequilibrium Systems. Izvest. Vysshikh. Ucheb. Zavedenii, Radiofiz., Vol. 4, No. 3, pp. 496-507, 1961. Bibliography: 16 titles.

F. V. Bunkin. Thermal Fluctuations in Nonlinear Systems. Radiotekh. i Elektron., Vol. 6, No. 1, pp. 3-8, 1961. Bibliography; 26 titles.

L. I. Gudzenko. Fluctuations in a Vacuum Tube Oscillator in the Presence of Grid Current. Radiotekh. i Elektron., Vol. 1, No. 9, pp. 1240-1254, 1956. Bibliography: 9 titles.

L. I. Gudzenko. Small Fluctuations in Essentially Nonlinear Autocorrelation System. Doklady Akad. Nauk SSSR, Vol. 125, No. 1, pp. 62-65, 1959. [Soviet Physics – Doklady, Vol. 4, p. 322.]

L. I. Gudzenko. Periodic Transients. Radiotekh. i Elektron., Vol. 4, No. 6, pp. 1062-1064, 1959. Bibliography: 2 titles.

L. I. Gudzenko. Velocity of Brownian Particles in the Vicinity of Stable Dynamic Equilibrium. Radiotekh. i Elektron., Vol. 4, No. 12, pp. 2061-2067, 1959. Bibliography: 6 titles.

L. I. Gudzenko. Amplitude Fluctuations of an Autonomous Vacuum Tube Oscillator. Radiotekh. i Elektron., Vol. 4, No. 1, pp. 97-108, 1959. Bibliography: 6 titles.

L. I. Gudzenko. Generalization of an Ergodic Theorem for Transient Behavior. Izvest. Vysshikh. Ucheb. Zavedenii, Radiofiz., Vol. 4, No. 2, pp. 267-274, 1961. Bibliography: 7 titles.

L. I. Gudzenko. Spectrum of Small Fluctuations Occurring during the Buildup of Oscillations in a Typical Oscillator. Izvest. Vysshikh. Zavedenii, Radiofiz.,Vol. 4, No. 4, pp. 671-679, 1961. Bibliography: 7 titles.

N. V. Karlov and A. E. Salomonovich. Automatic Centimeter Wave Null Radiometer for the Study of Weak Noise Signals. Radiotekh. i Elektron., Vol. 1, No. 1, pp. 121-122, 1956. Bibliography: 3 titles.

N. V. Karlov. Sensitivity of a Null Radiometer, in Papers Presented at the Fifth Conference on Cosmogony. Published by the Academy of Sciences of the USSR, 1956, pp. 88-93; appendices and questions, pp. 93-95. Bibliography: 2 titles.

N. V. Karlov. Sensitivity of a Radiometer with Automatic Gain Control. Radiotekh. i Elektron., Vol. 3, No. 1, pp. 74-79, 1958. Bibliography: 6 titles.

N. V. Karlov and B. M. Chikhachev. Sensitivity of a Radiometer with Low Noise Level (in the Quantum Level). Radiotekh. i Elektron., Vol. 4, No. 6, pp. 1047-1051, 1959. Bibliography: 4 titles.

N. V. Karlov and B. M. Chikhachev. Sensitivity of a Radio Telescope with Low Input Noise Level. Radiotekh. i Elektron., Vol. 4, No. 6, pp. 1052-1056, 1959. Bibliography: 3 titles.

A. D. Kuz'min and A. N. Khvoshchev. Wideband Noise Generator (Decimeter Range). Radiotekhnika, Vol. 13, No. 7, pp. 36-42, 1958. Bibliography: 6 titles.

M. A. Leontovich. Charge Density Fluctuations in an Electrolytic Solution. Doklady Akad. Nauk SSSR, Vol. 53, No. 2, pp. 115-118, 1946. Bibliography: 2 titles.

M. L. Levin and S. M. Rytov. Thermal Radiation of a Thin Rectilinear Antenna. Zhur. Tekh. Fiz., Vol. 25, No. 2, pp. 323-332, 1955.

E. Ya. Pumper. Method Developed for the Experimental Investigation of the Statistics of Electrical Fluctuations. Doklady Akad. Nauk SSSR, Vol. 53, No. 1, pp. 25-28, 1946. Bibliography: 4 titles.

E. Ya. Pumper. Some Peculiarities of Electrical Fluctuations. Doklady Akad. Nauk SSSR, Vol. 57, No. 8, pp. 775-778, 1947.

E. Ya. Pumper, Electrical Fluctuational Energy in Conductors. Zhur. Eksp. i Teoret. Fiz., Vol. 18, No. 12, pp. 1112-1129, 1948. Bibliography: 7 titles.

E. Ya. Pumper. Thermal Electrical Fluctuations in Electrolytes. Doklady Akad. Nauk SSSR, Vol. 59, No. 8, pp. 1415-1416, 1948.

E. Ya. Pumper, Electric Current Fluctuations in Electron Tubes with Tungsten Cathodes. Doklady Akad. Nauk SSSR, Vol. 59, No. 9, pp. 1559-1562, 1948.
E. Ya. Pumper. Measurement of Electrical Fluctuations in a Manner Similar to That Used in the Study of Metal Processes. Izvest. Akad. Nauk SSSR, Ser. Fiz., Vol. 13, No. 5, pp. 596-614, 1949.
E. Ya. Pumper. Level of Electrical Fluctuations in an Oscillatory Circuit (Oscillator). Doklady Akad. Nauk SSSR, Vol. 66, No. 1, pp. 41-44, 1949. Bibliography: 4 titles.
E. Ya. Pumper. Level of Electrical Fluctuations in Some Metals After Annealing. Doklady Akad. Nauk SSSR, Vol. 68, No. 2, pp. 277-279, 1949. Bibliography: 2 titles.
E. Ya. Pumper. Investigation of the Ordering Process in Cu_3Au Alloy by Measurement of the Electrical Fluctuations. Doklady Akad. Nauk SSSR, Vol. 72, No. 6, pp. 1033-1036, 1950. Bibliography: 5 titles.
S. M. Rytov. The Theory of Electrical Fluctuations and Thermal Radiation. Doklady Akad. Nauk SSSR, Vol. 87, No. 4, pp. 535-538, 1952. Bibliography: 5 titles.
S. M. Rytov. The Theory of Electrical Fluctuations and Thermal Radiation. Published by the Academy of Sciences, USSR, 1953. 232 pp.
S. M. Rytov. Thermal Radiation in Waveguides. Zhur. Eksp. i Teoret. Fiz., Vol. 27, No. 5, pp. 571-578, 1954.
S. M. Rytov. The Connection Between the Distribution of a Quasi-Monochromatic Stationary Process and the Distribution of Its Envelope. Zhur. Eksp. i Teoret. Fiz., Vol. 29, No. 5, pp. 702-703, 1955. [Soviet Physics—JETP, Vol. 29(2), p. 571.]
S. M. Rytov. Theory of Thermal Noise. Part I. Radiotekhnika, Vol. 10, No. 2, pp. 3-13, 1955; Part II. Radiotekhnika, Vol. 10, No. 3, pp. 3-13, 1955. Bibliography: 10 titles.
S. M. Rytov. Fluctuations in Oscillating Systems of the Thomson Type. I. Zhur. Eksp. i Teoret. Fiz., Vol. 29, No. 3, pp. 304-314, 1955. Bibliography: 7 titles. [Soviet Physics — JETP, Vol. 29(2), p. 217.]
S. M. Rytov. Fluctuations in Oscillating Systems of the Thomson Type. II. Zhur. Eksp. i Teoret. Fiz., Vol. 29, No. 3, pp. 315-328, 1955. Bibliography: 5 titles. [Soviet Physics — JETP, Vol. 29(2), p. 225.]
S. M. Rytov. Electrical Fluctuations and Thermal Radiation. Vestnik Akad. Nauk SSSR, No. 6, pp. 24-33, 1955.
S. M. Rytov. Electrical Fluctuations and Thermal Radiation (paper presented at the Conference to the Panel on Physical-Mathematical Sciences and the Panel on Technical Sciences of the Academy of Sciences of the USSR; Conference held in honor of the tenth anniversary of the death of L. I. Mandel'shtam, December 15, 1954). Uspekhi Fiz. Nauk, Vol. 55, No. 3, pp. 299-314, 1955. Bibliography: 13 titles.
S. M. Rytov. Phase Variation in Vacuum Tube Oscillators. Radiotekh. i Elektron., Vol. 1, No. 1, pp. 114-119, 1956. Bibliography: 3 titles.
S. M. Rytov. Thermal Agitation in Distributed Systems. Doklady Akad. Nauk SSSR, Vol. 110, No. 3, pp. 371 to 374, 1956. Bibliography: 11 titles. [Soviet Physics — Doklady, Vol. 1, p. 555.]
S. M. Rytov. Correlation Theory for Rayleigh Scattering of Light. I. Zhur. Eksp. i Teoret. Fiz., Vol. 33, No. 2, pp. 514-524, 1957. Bibliography: 14 titles. [Soviet Physics — JETP, Vol. 33(6), p. 401.]
S. M. Rytov. Correlation Theory for Rayleigh Scattering of Light. II. Zhur. Eksp. i Teoret. Fiz., Vol. 33, No. 3, pp. 669-682, 1957. Bibliography: 12 titles. [Soviet Physics — JETP, Vol. 33(6), p. 468.]
S. M. Rytov. Correlation Theory of Thermal Fluctuations in an Isotropic Medium. Zhur. Eksp. i Teoret. Fiz., Vol. 33, No. 1, pp. 166-178, 1957. Bibliography: 20 titles. [Soviet Physics — JETP, Vol. 33(6), p. 130.]
S. M. Rytov. Correlation Theory of Electrical Fluctuations in Thermal Radiation. Uspekhi Fiz. Nauk, Vol. 63, No. 4, pp. 657-672, 1957. Bibliography: 31 titles.
S. M. Rytov. Fluctuation Theory of Self-Oscillatory Systems with Piecewise-Linear Characteristics. Izvest. Vysshikh. Ucheb. Zavedenii, Radiofiz., Vol. 2, No. 1, pp. 50-62, 1959. Bibliography: 13 titles.
S. M. Rytov. Correlational Theory of Rayleigh Scattering of Light, in Investigations in Experimental and Theoretical Physics, issued in memory of Gregory Samuilovich Landsberg. Published by the Academy of Sciences of the USSR, 1959, pp. 175-191. Bibliography: 22 titles.
S. M. Rytov. Spectrum of a Quasi-Periodic Random Process. Izvest. Vysshikh. Ucheb. Zavedenii, Radiofiz., Vol. 2, No. 1, pp. 45-49, 1959. Bibliography: 1 title.
V. V. Vitkevich. Frequency Spectrum of a Disturbance Near and Far from the Noise Source. Zhur. Tekh. Fiz., Vol. 16, No. 3, pp. 315-316, 1956.
M. E. Zhabotinskii. Fluctuations in an Oscillator with Inertial Nonlinearities. Zhur. Eksp. i Teoret. Fiz., Vol. 26, No. 6, pp. 758-759, 1954. Bibliography: 6 titles.

ACCELERATORS

A. A. Andronov and G. S. Gorelik. Resonant Phenomena Occurring during the Motion of Relativistic Particles in a Cyclotron. Doklady Akad. Nauk SSSR, Vol. 49, No. 9, pp. 664-666, 1945. Bibliography: 2 titles.

A. Andronov and G. Gorelik. Nonlinear Resonance of a Relativistic Particle in a Cyclotron. Comptes rendus Acad. Sci., Vol. 221, No. 23, pp. 696-698, 1945.

T. M. Murina. Radio Telescope with a Disc Resonator. Radiotekh. i Elektron., Vol. 6, No. 9, pp. 1586-1588, 1961. Bibliography: 3 titles.

A. M. Prokhorov. Investigation of the Coherent Radiation Emitted by Accelerated Electrons in Synchrotron-Type Accelerators. Doctoral Thesis, Mathematical-Physical Sciences, Moscow, 1951, 142 pp. with drawings and ill. (Academy of Sciences, USSR, P. N. Lebedev Institute of Physics). Typewritten.

A. M. Prokhorov. Coherent Electron Radiation in the Centimeter Range for Synchrotron Electrons. Radiotekh. i Elektron., Vol. 1, No. 1, pp. 71-78, 1956. Bibliography: 17 titles.

P. A. Ryazin. Electron Capture during Betatron Acceleration, in Theses of the Papers Presented at the All-Union Conference on High-Energy Particle Physics, May 14-22, 1956. Published by the Academy of Sciences of the USSR, 1956, p. 161.

VARIOUS FIELDS

L. N. Borodovskaya and S. V. Lebedev. Dependence of the Electrical Conductivity and Electron Emission on the Energy of a Metal in the Process of Its Heating by a Current of High Density. Zhur. Eksp. i Teoret. Fiz., Vol. 28, No. 1, pp. 86-110, 1955. Bibliography: 12 titles. [Soviet Physics – JETP, Vol. 28(1), p. 71.]

G. S. Gorelik. Some Remarks Concerning the Style of N. D. Papaleksi's Scientific Creativity. Izvest. Akad. Nauk SSSR, Ser. Fiz., Vol. 12, No. 1, pp. 22-24, 1948.

S. V. Lebedev and S. É. Khaikin. Anomalous Electron Emission of Tungsten When Heated by a High-Density Pulsed Current. Zhur. Eksp. i Teoret. Fiz., Vol. 26, No. 6, pp. 723-735, 1954.

S. V. Lebedev. The Irregularity in the Boguslavski-Langmuir Law during the Heating of Tungsten Cathodes by High-Density Pulsed Currents. Zhur. Eksp. i Teoret. Fiz., Vol. 27, No. 4, pp. 487-500, 1954. Bibliography: 2 titles.

S. V. Lebedev and S. É. Khaikin. Some Anomalies in the Behavior of Metals When Heated by High-Density Pulsed Currents. Zhur. Eksp. i Teoret. Fiz., Vol. 26, No. 5, pp. 629-639, 1954. Bibliography: 8 titles.

S. V. Lebedev. Phenomena Occurring in Tungsten Wires Preceding Their Explosion during the Passage of Heavy Current. Zhur. Eksp. i Teoret. Fiz., Vol. 27, No. 5, pp. 605-614, 1954. Bibliography: 5 titles.

S. V. Lebedev. Explosion of a Metal by an Electric Current. Zhur. Eksp. i Teoret. Fiz., Vol. 32, No. 2, pp. 199 to 207, 1957. Bibliography: 19 titles. [Soviet Physics – JETP, Vol. 32(5), p. 243.]

S. V. Lebedev, S. L. Mandel'shtam, and G. M. Rodin. Shortwave Radiation from a Vacuum Spark. Zhur. Eksp. i Teoret. Fiz., Vol. 37, No. 2, pp. 349-354, 1959. Bibliography: 8 titles. [Soviet Physics – JETP, Vol. 32(5), p. 248.]

V. N. Lugovoi. Cyclotron Resonance in a Variable Magnetic Field. Zhur. Eksp. i Teoret. Fiz., Vol. 41, No. 5, pp. 1562-1565, 1961. Bibliography: 5 titles. [Soviet Physics – JETP, Vol. 41(14), p. 1113.]

I. D. Murin. Reversible Counter Using Decatrons OG-5. Pribory i Tekh. Eksp., No. 6, pp. 127-129, 1961. Bibliography: 6 titles. [Instr. and Exptl. Tech., No. 6, p. 1164.]

M. I. Podgoretskii, A. I. Barchukov, and D. F. Rakitin. Transitional Density Effects in Electronic-Nuclear Showers. Doklady Akad. Nauk SSSR, Vol. 73, No. 4, pp. 685-688, 1950. Bibliography: 3 titles.

V. A. Ponamarenko, Yu. P. Egorov, and G. Ya. Vzenkova. Preparation and Properties of Some Alkyldeuteriosilanes. Izvest. Akad. Nauk SSSR, Otd. Khim. Nauk, No. 1, pp. 54-58, 1958. Bibliography: 24 titles. [Bull. Acad. Sci. SSSR, Div. Chem. Sci., No. 1, p. 47, 1958.]

A. M. Prokhorov. Radio-Location Physics. Nauka i Zhizn, No. 8-9, pp. 2-5, 1946.

A. M. Prokhorov. Soviet Radio Physics. Nauka i Zhizn, No. 11, pp. 30-31, 1947.

A. M. Prokhorov. What is Radio-Location? Moscow. Goskul'tprosvetizdat, 1958. 32 pp. Bibliography: 8 titles.

A. E. Salomonovich. Radio Engineering at the Service of Contemporary Physics. Radio, No. 11, pp. 23-26, No. 12, pp. 20-23, 1950.

A. E. Voronkov, L. N. Korablev, I. D. Murin, and I. V. Shtranikh. Rapidly Acting, Multiple-Channel Amplitude Analyzer. Moscow, 1957, 64 pp. (All-Union Institute of Scientific-Technical Information). Bibliography: 7 titles.

M. E. Zhabotinskii. The Language of the Molecules (Combinational Light Scattering). Nauka i Zhizn, No. 11, pp. 10-12, 1948.

M. E. Zhabotinskii. Physics and the Flight Orientation of Birds. Nauka i Zhizn, No. 2, pp. 19-23, 1949.